Mathematical Statistics

Asymptotic Minimax Theory

Mathematical Statistics

Asymptotic Minimax Theory

Alexander Korostelev
Olga Korosteleva

Graduate Studies
in Mathematics
Volume 119

American Mathematical Society
Providence, Rhode Island

2010 *Mathematics Subject Classification.* Primary 62F12, 62G08;
Secondary 62F10, 62G05, 62G10, 62G20.

For additional information and updates on this book, visit
www.ams.org/bookpages/gsm-119

Library of Congress Cataloging-in-Publication Data

Korostelev, A. P. (Aleksandr Petrovich)
Mathematical statistics : asymptotic minimax theory / Alexander Korostelev, Olga Korosteleva.
p. cm. — (Graduate studies in mathematics ; v. 119)
Includes bibliographical references and index.
ISBN 978-0-8218-5283-5 (alk. paper)
1. Estimation theory. 2. Asymptotic efficiencies (Statistics) 3. Statistical hypothesis testing. I. Korostelev, Olga. II. Title.

QA276.8.K667 2011
519.5–dc22 2010037408

∞ The paper used in this book is acid-free and falls within the guidelines
established to ensure permanence and durability.
Visit the AMS home page at `http://www.ams.org/`

10 9 8 7 6 5 4 3 2 1 16 15 14 13 12 11

Contents

Preface

This book is based on the lecture notes written for the advanced Ph.D. level statistics courses delivered by the first author at the Wayne State University over the last decade. It has been easy to observe how the gap deepens between applied (computational) and theoretical statistics. It has become more difficult to direct and mentor graduate students in the field of mathematical statistics. The research monographs in this field are extremely difficult to use as textbooks. Even in the best published lecture notes the intensive material of original studies is typically included. On the other hand, the classical courses in statistics that cover the traditional parametric point and interval estimation methods and hypotheses testing are hardly sufficient for the teaching goals in modern mathematical statistics.

In this book, we tried to give a general overview of the key statistical topics, parametric and nonparametric, as a set of very special optimization problems. As a criterion for optimality of estimators we chose minimax risks, and we focused on asymptotically minimax rates of convergence for large samples. Definitely, the selection of models presented in this book follows our preferences. Many very important problems and examples are not included. The simplest models were deliberately selected for presentation, and we consciously concentrated on the detailed proofs of all propositions. We believe that mathematics students should be trained in proof-writing to be better prepared for applications in statistics.

This textbook can form a reasonable basis for a two-semester course in mathematical statistics. Every chapter is followed by a collection of exercises consisting partly of verification of technical results, and partly of important illustrative examples. In our opinion, the sufficient prerequisite is a standard course in advanced probability supported by undergraduate statistics and real analysis. We hope that students who successfully pass this course are prepared for reading original papers and monographs in the minimax estimation theory and can be easily introduced to research studies in this field.

This book is organized into three parts. Part 1 is comprised of Chapters 1-7 that contain fundamental topics of local asymptotic normality as well as irregular statistical models, change-point problem, and sequential estimation. For convenience of reference we also included a chapter on classical parametric linear regression with the concentration on the asymptotical properties of least-squares estimators. Part 2 (Chapters 8-12) focuses on estimation of nonparametric regression functions. We restrict the presentation to estimation at a point and in the quadratic and uniform norms, and consider deterministic as well as random designs. The last part of the book, Chapters 13-16, is devoted to special more modern topics such as influence of higher-dimension and structure in nonparametric regression models, problems of adaptive estimation, and testing of nonparametric hypotheses. We present the ideas through simple examples with the equidistant design.

Most chapters are weakly related to each other and may be covered in any order. Our suggestion for a two-semester course would be to cover the parametric part during the first semester and to cover the nonparametric part and selected topics in the second half of the course.

We are grateful to O. Lepskii for his advice and help with the presentation of Part 3.

The authors, October 2010

Part 1

Parametric Models

Chapter 1

The Fisher Efficiency

1.1. Statistical Experiment

A classical *statistical experiment* $\big(X_1, \dots, X_n;\, p(x, \theta);\, \theta \in \Theta\big)$ is composed of the following three elements: (i) a set of independent observations $X_1, \dots, X_n$ where n is the sample size, (ii) a family of probability densities $p(x, \theta)$ defined by a parameter θ, and (iii) a *parameter set* Θ of all possible values of θ.

Unless otherwise stated, we always assume that θ is one-dimensional, that is, $\Theta \subseteq \mathbb{R}$. For discrete distributions, $p\,(x, \theta)$ is the probability mass function. In this chapter we formulate results only for continuous distributions. Analogous results hold for discrete distributions if integration is replaced by summation. Some discrete distributions are used in examples.

Example 1.1. (a) If n independent observations $X_1, \dots, X_n$ have a normal distribution with an unknown mean θ and a known variance σ^2, that is, $X_i \sim \mathcal{N}(\theta, \sigma^2)$, then the density is

$$p\,(x, \theta) = (2\,\pi\,\sigma^2)^{-1/2}\,\exp\big\{-\,(x-\theta)^2/(2\sigma^2)\big\}\,,\ -\infty < x,\, \theta < \infty,$$

and the parameter set is the whole real line $\Theta = \mathbb{R}$.

(b) If n independent observations have a normal distribution with a known mean μ and an unknown variance θ, that is, $X_i \sim \mathcal{N}(\mu, \theta)$, then the density is

$$p(x, \theta) \;=\; (2\,\pi\,\theta)^{-1/2}\,\exp\big\{-\,(x-\mu)^2/(2\theta)\big\}\,,\ -\infty < x < \infty\,,\, \theta > 0,$$

and the parameter set is the positive half-axis $\Theta \;=\; \{\,\theta \in \mathbb{R} : \theta > 0\,\}$. $\square$

Example 1.2. Suppose n independent observations $X_1, \dots, X_n$ come from a distribution with density

$$p(x, \theta) = p_0(x - \theta), \; -\infty < x, \, \theta < \infty,$$

where p_0 is a fixed probability density function. Here θ determines the shift of the distribution, and therefore is termed the *location parameter*. The location parameter model can be written as $X_i = \theta + \varepsilon_i$, $i = 1, \dots, n$, where $\varepsilon_1, \dots, \varepsilon_n$ are independent random variables with a given density p_0, and $\theta \in \Theta = \mathbb{R}$. □

The independence of observations implies that the joint density of X_i's equals

$$p(x_1, \dots, x_n, \theta) = \prod_{i=1}^{n} p(x_i, \theta).$$

We denote the respective expectation by $\mathbb{E}_\theta[\,\cdot\,]$ and variance by $\mathbb{V}ar_\theta[\,\cdot\,]$.

In a statistical experiment, all observations are obtained under the same value of an unknown parameter θ. The goal of the parametric statistical estimation is to assess the true value of θ from the observations $X_1, \dots, X_n$. An arbitrary function of observations, denoted by $\hat{\theta} = \hat{\theta}_n = \hat{\theta}_n(X_1, \dots, X_n)$, is called an *estimator* (or a *point estimator*) of θ.

A random variable

$$l(X_i, \theta) = \ln p(X_i, \theta)$$

is referred to as a *log-likelihood function* related to the observation X_i. The *joint log-likelihood function of a sample of size* n (or, simply, the *log-likelihood function*) is the sum

$$L_n(\theta) = L_n(\theta \,|\, X_1, \dots, X_n) = \sum_{i=1}^{n} l(X_i, \theta) = \sum_{i=1}^{n} \ln p(X_i, \theta).$$

In the above notation, we emphasize the dependence of the log-likelihood function on the parameter θ, keeping in mind that it is actually a random function that depends on the entire set of observations $X_1, \dots, X_n$.

The parameter θ may be evaluated by the method of maximum likelihood estimation. An estimator θ_n^* is called the *maximum likelihood estimator* (MLE), if for any $\theta \in \Theta$ the following inequality holds:

$$L_n(\theta_n^*) \geq L_n(\theta).$$

If the log-likelihood function attains its unique maximum, then the MLE reduces to

$$\theta_n^* = \operatorname*{argmax}_{\theta \in \Theta} L_n(\theta).$$

If the function L is differentiable at its attainable maximum, then θ_n^* is a solution of the equation

$$\frac{\partial L_n(\theta)}{\partial \theta} = 0.$$

Note that if the maximum is not unique, this equation has multiple solutions.

The function

$$b_n(\theta) = b_n(\theta\,,\hat{\theta}_n) = \mathbb{E}_\theta\big[\,\hat{\theta}_n\,\big] - \theta = \mathbb{E}_\theta\big[\,\hat{\theta}_n(X_1,\dots,X_n)\big] - \theta$$

is called the *bias* of $\hat{\theta}_n$. An estimator $\hat{\theta}_n(X_1,\dots,X_n)$ is called an *unbiased* estimator of θ if its bias equals zero, or equivalently, $\mathbb{E}_\theta\big[\,\hat{\theta}_n\,\big] = \theta$ for all $\theta \in \Theta$.

Example 1.3. Assume that the underlying distribution of the random sample $X_1,\dots,X_n$ is Poisson with mean θ. The probability mass function is given by

$$p_n(x,\theta) = \frac{\theta^x}{x!}\,e^{-\theta}\,,\quad \theta > 0,\ x \in \{0,1,2,\dots\}.$$

Then the log-likelihood function has the form

$$L_n(\theta) = \sum_{i=1}^{n} X_i \ln\theta - n\theta - \sum_{i=1}^{n} \ln\,(X_i!).$$

Setting the derivative equal to zero yields the solution $\theta_n^* = \bar{X}_n$, where

$$\bar{X}_n = (X_1 + \cdots + X_n)/n$$

denotes the *sample mean*. In this example, the MLE is unbiased since $\mathbb{E}_\theta\big[\,\theta_n^*\big] = \mathbb{E}_\theta\big[\,\bar{X}_n\big] = \mathbb{E}_\theta\big[\,X_1\big] = \theta.$ □

Nonetheless, we should not take the unbiased MLE for granted. Even for common densities, its expected value may not exist. Consider the next example.

Example 1.4. For the exponential distribution with the density

$$p(x,\theta) = \theta\,\exp\big\{-\theta\,x\big\}\,,\quad x > 0,\ \ \theta > 0,$$

the MLE $\theta_n^* = 1/\bar{X}_n$ has the expected value $\mathbb{E}_\theta\big[\,\theta_n^*\big] = n\,\theta/(n-1)$ (see Exercise 1.6). In particular, for $n = 1$, the expectation does not exist since $\int_0^\infty x^{-1}\,\theta\,\exp\big\{-\theta\,x\big\}\,dx = \infty.$ □

In this example, however, an unbiased estimator may be found for $n > 1$. Indeed, the estimator $(n-1)\theta_n^*/n$ is unbiased. As the next example shows, an unbiased estimator may not exist at all.

Example 1.5. Let X be a Binomial$(n\,,\,\theta^2)$ observation, that is, a random number of successes in n independent Bernoulli trials with the probability of a success $p = \theta^2\,,\ 0 < \theta < 1$. An unbiased estimator of the parameter θ does not exist. In fact, if $\hat{\theta} = \hat{\theta}(X)$ were such an estimator, then its expectation would be an even polynomial of θ,

$$\mathbb{E}_\theta\big[\,\hat{\theta}(X)\,\big] \;=\; \sum_{k=0}^{n} \hat{\theta}(k)\binom{n}{k}\,\theta^{2k}\,(1-\theta^2)^{n-k},$$

which cannot be identically equal to θ. □

1.2. The Fisher Information

Introduce the *Fisher score function* as the derivative of the log-likelihood function with respect to θ,

$$l'(X_i\,,\theta) \;=\; \frac{\partial \ln p(X_i\,,\theta)}{\partial \theta} \;=\; \frac{\partial p(X_i\,,\theta)/\partial\theta}{p(X_i\,,\theta)}.$$

Note that the expected value of the Fisher score function is zero. Indeed,

$$\mathbb{E}_\theta\big[\,l'(X_i\,,\theta)\,\big] \;=\; \int_{\mathbb{R}} \frac{\partial p(x\,,\theta)}{\partial\theta}\,dx \;=\; \frac{\partial \int_{\mathbb{R}} p(x\,,\theta)\,dx}{\partial\theta} \;=\; 0.$$

The *total Fisher score function* for a sample $X_1\,,\ldots\,,X_n$ is defined as the sum of the score functions for each individual observation,

$$L_n'(\theta) \;=\; \sum_{i=1}^{n} l'(X_i\,,\theta).$$

The *Fisher information* of one observation X_i is the variance of the Fisher score function $l'(X_i\,,\theta)$,

$$I(\theta) \;=\; \mathbb{V}ar_\theta\big[\,l'(X_i\,,\theta)\,\big] \;=\; \mathbb{E}_\theta\Big[\,\big(l'(X_i\,,\theta)\big)^2\Big]$$

$$= \mathbb{E}_\theta\,\Big[\Big(\frac{\partial \ln\,p\,(X,\,\theta)}{\partial\,\theta}\Big)^2\Big] \;=\; \int_{\mathbb{R}}\Big(\frac{\partial \ln p(x\,,\theta)}{\partial\theta}\Big)^2\,p(x\,,\theta)\,dx$$

$$= \int_{\mathbb{R}} \frac{\big(\partial p(x\,,\theta)/\partial\theta\,\big)^2}{p(x\,,\theta)}\,dx.$$

Remark 1.6. In the above definition of the Fisher information, the density appears in the denominator. Thus, it is problematic to calculate the Fisher information for distributions with densities that may be equal to zero for some values of x; even more so, if the density turns into zero as a function of x on sets that vary depending on the value of θ. A more general approach to the concept of information that overcomes this difficulty will be suggested in Section 4.2. □

The *Fisher information for a statistical experiment of size* n is the variance of the total Fisher score function,

$$\begin{aligned} I_n(\theta) &= \mathbb{V}ar_\theta\big[\, L_n'(\theta)\,\big] = \mathbb{E}_\theta\big[\,\big(\, L_n'(\theta)\,\big)^2\,\big] \\ &= \mathbb{E}_\theta\Big[\Big(\frac{\partial \ln p\,(X_1,\dots,X_n,\,\theta)}{\partial\,\theta}\Big)^2\Big] \\ &= \int_{\mathbb{R}^n} \frac{(\partial\, p\,(x_1,\dots,x_n,\theta)/\partial\theta)^2}{p\,(x_1,\dots,x_n,\theta)}\, dx_1\,\dots\, dx_n\,. \end{aligned}$$

Lemma 1.7. *For independent observations, the Fisher information is additive. In particular, for any* $\theta \in \Theta$, *the equation holds* $I_n(\theta) = n\, I(\theta)$.

Proof. As the variance of the sum of n independent random variables,

$$\begin{aligned} I_n(\theta) &= \mathbb{V}ar_\theta\big[\, L_n'(\theta)\,\big] = \mathbb{V}ar_\theta\big[\, l'(X_1\,,\theta) + \dots + l'(X_n\,,\theta)\,\big] \\ &= n\,\mathbb{V}ar_\theta\big[\, l'(X_1\,,\theta)\,\big] = n\, I(\theta). \end{aligned}$$

□

In view of this lemma, we use the following definition of the Fisher information for a random sample of size n:

$$I_n(\theta) = n\,\mathbb{E}_\theta\Big[\Big(\frac{\partial \ln p\,(X,\,\theta)}{\partial\,\theta}\Big)^2\Big].$$

Another way of computing the Fisher information is presented in Exercise 1.1.

1.3. The Cramér-Rao Lower Bound

A statistical experiment is called *regular* if its Fisher information is continuous, strictly positive, and bounded for all $\theta \in \Theta$. Next we present an inequality for the variance of any estimator of θ in a regular experiment. This inequality is termed the *Cramér-Rao inequality*, and the lower bound is known as the *Cramér-Rao lower bound.*

Theorem 1.8. *Consider an estimator* $\hat\theta_n = \hat\theta_n(X_1,\dots,X_n)$ *of the parameter* θ *in a regular experiment. Suppose its bias* $b_n(\theta) = \mathbb{E}_\theta\big[\,\hat\theta_n\,\big] - \theta$ *is continuously differentiable. Let* $b_n'(\theta)$ *denote the derivative of the bias. Then the variance of* $\hat\theta_n$ *satisfies the inequality*

$$\mathbb{V}ar_\theta\big[\,\hat\theta_n\,\big] \geq \frac{\big(\,1 + b_n'(\theta)\,\big)^2}{I_n(\theta)}\,, \qquad \theta \in \Theta. \tag{1.1}$$

Proof. By the definition of the bias, we have that

$$\theta + b_n(\theta) = \mathbb{E}_\theta\big[\,\hat\theta_n\,\big] = \int_{\mathbb{R}^n} \hat\theta_n(x_1,\dots,x_n)\, p\,(x_1,\dots,x_n,\theta)\, dx_1 \dots dx_n.$$

In the regular case, the differentiation and integration are interchangeable, hence differentiating in θ, we get the equation,

$$1 + b_n'(\theta) = \int_{\mathbb{R}^n} \hat{\theta}_n(x_1, \dots, x_n) \left[\partial p(x_1, \dots, x_n, \theta)/\partial\theta\right] dx_1 \dots dx_n$$

$$= \int_{\mathbb{R}^n} \hat{\theta}_n(x_1, \dots, x_n) \left(\frac{\partial p(x_1, \dots, x_n, \theta)/\partial\theta}{p(x_1, \dots, x_n, \theta)} \right) p(x_1, \dots, x_n, \theta)\, dx_1 \dots dx_n$$

$$= \mathbb{E}_\theta\big[\, \hat{\theta}_n\, L_n'(\theta)\,\big] = \mathbb{C}ov_\theta\big[\,\hat{\theta}_n\,,\, L_n'(\theta)\,\big]$$

where we use the fact that $\mathbb{E}_\theta\big[\, L_n'(\theta)\,\big] = 0$. The correlation coefficient ρ_n of $\hat{\theta}_n$ and $L_n'(\theta)$ does not exceed 1 in its absolute value, so that

$$1 \geq \rho_n^2 = \frac{\big(\mathbb{C}ov_\theta\big[\,\hat{\theta}_n\,,\, L_n'(\theta)\,\big]\big)^2}{\mathbb{V}ar_\theta[\hat{\theta}_n]\;\mathbb{V}ar_\theta[L_n'(\theta)]} = \frac{(1 + b_n'(\theta))^2}{\mathbb{V}ar_\theta[\hat{\theta}_n]\; I_n(\theta)}. \qquad \square$$

1.4. Efficiency of Estimators

An immediate consequence of Theorem 1.8 is the formula for unbiased estimators.

Corollary 1.9. *For an unbiased estimator* $\hat{\theta}_n$, *the Cramér-Rao inequality* (1.1) *takes the form*

$$\mathbb{V}ar_\theta\big[\,\hat{\theta}_n\big] \geq \frac{1}{I_n(\theta)}, \quad \theta \in \Theta. \tag{1.2}$$

$\square$

An unbiased estimator $\theta_n^* = \theta_n^*(X_1, \dots, X_n)$ in a regular statistical experiment is called *Fisher efficient* (or, simply, *efficient*) if, for any $\theta \in \Theta$, the variance of θ_n^* reaches the Cramér-Rao lower bound, that is, the equality in (1.2) holds:

$$\mathbb{V}ar_\theta\big[\,\theta_n^*\big] = \frac{1}{I_n(\theta)}, \quad \theta \in \Theta.$$

Example 1.10. Suppose, as in Example 1.1(a), the observations $X_1, \dots, X_n$ are independent $\mathcal{N}(\theta,\, \sigma^2)$ where σ^2 is assumed known. We show that the sample mean $\bar{X}_n = (X_1 + \dots + X_n)/n$ is an efficient estimator of θ. Indeed, $\bar{X}_n$ is unbiased and $\mathbb{V}ar_\theta\big[\,\bar{X}_n\big] = \sigma^2/n$. On the other hand,

$$\ln p(X, \theta) = -\frac{1}{2}\ln(2\,\pi\,\sigma^2) - \frac{(X - \theta)^2}{2\sigma^2}$$

and

$$l'(X\,, \theta) = \frac{\partial \ln p(X, \theta)}{\partial\theta} = \frac{X - \theta}{\sigma^2}.$$

Thus, the Fisher information for the statistical experiment is

$$I_n(\theta) = n\,\mathbb{E}_\theta\big[\big(l'(X\,,\theta)\big)^2\big] = \frac{n}{\sigma^4}\,\mathbb{E}_\theta\big[(X - \theta)^2\big] = \frac{n\sigma^2}{\sigma^4} = \frac{n}{\sigma^2}.$$

Therefore, for any value of θ, the variance of $\bar{X}_n$ achieves the Cramér-Rao lower bound $1/I_n(\theta) = \sigma^2/n$. □

The concept of the Fisher efficiency seems to be nice and powerful. Indeed, besides being unbiased, an efficient estimator has the minimum possible variance uniformly in $\theta \in \Theta$. Another feature is that it applies to any sample size n. Unfortunately, this concept is extremely restrictive. It works only in a limited number of models. The main pitfalls of the Fisher efficiency are discussed in the next chapter.

Exercises

Exercise 1.1. Show that the Fisher information can be computed by the formula

$$I_n(\theta) = -\, n\, \mathbb{E}_\theta \Big[\frac{\partial^2 \ln p\,(X, \theta)}{\partial\, \theta^2} \Big].$$

Hint: Make use of the representation (show!)

$$\Big(\frac{\partial \ln p\,(x, \theta)}{\partial\, \theta} \Big)^2 p\,(x,\, \theta) \;=\; \frac{\partial^2\, p\,(x,\, \theta)}{\partial \theta^2} - \Big(\frac{\partial^2 \ln p\,(x,\, \theta)}{\partial \theta^2} \Big)\, p\,(x,\, \theta).$$

Exercise 1.2. Let $X_1, \dots, X_n$ be independent observations with the $\mathcal{N}(\mu,\, \theta)$ distribution, where μ has a known value (refer to Example 1.1(b)). Prove that

$$\theta_n^* \;=\; \frac{1}{n} \sum_{i=1}^{n} (X_i - \mu)^2$$

is an efficient estimator of θ. Hint: Use Exercise 1.1 to show that $I_n(\theta) = n/(2\,\theta^2)$. When computing the variance of θ_n^*, first notice that the variable $\sum_{i=1}^n (X_i - \mu)^2/\theta$ has a chi-squared distribution with n degrees of freedom, and, thus, its variance equals $2n$.

Exercise 1.3. Suppose that independent observations $X_1, \dots, X_n$ have a Bernoulli distribution with the probability mass function

$$p\,(x,\, \theta) = \theta^{\,x}\, (1 - \theta)^{1-x}, \quad x \in \{0, 1\}, \;\; 0 < \theta < 1.$$

Show that the Fisher information is of the form

$$I_n(\theta) \;=\; \frac{n}{\theta\,(1 - \theta)},$$

and verify that the estimator $\theta_n^* = \bar{X}_n$ is efficient.

Exercise 1.4. Assume that $X_1, \dots, X_n$ are independent observations from a Poisson distribution with the probability mass function

$$p\,(x,\,\theta) = \frac{\theta^{\,x}}{x!}\,e^{-\theta}, \quad x \in \{\,0, 1, \dots\}, \quad \theta > 0.$$

Prove that the Fisher information in this case is $I_n(\theta) = n/\theta$, and show that $\bar{X}_n$ is an efficient estimator of θ.

Exercise 1.5. Let $X_1, \dots, X_n$ be a random sample from an exponential distribution with the density

$$p\,(x,\,\theta) = \frac{1}{\theta}\,e^{-\,x/\theta}, \quad x > 0, \quad \theta > 0.$$

Verify that $I_n(\theta) = n/\theta^2$, and prove that $\bar{X}_n$ is efficient.

Exercise 1.6. Show that in the exponential model with the density $p(x\,,\theta) = \theta\,\exp\{-\theta\,x\}\,,\ x\,,\theta > 0$, the MLE $\theta_n^* = 1/\bar{X}_n$ has the expected value $\mathbb{E}_\theta[\,\theta_n^*\,] = n\,\theta/(n-1)$. What is the variance of this estimator?

Exercise 1.7. Show that for the location parameter model with the density $p(x\,,\theta) = p_0(x-\theta)$, introduced in Example 1.2, the Fisher information is a constant if it exists.

Exercise 1.8. In the Exercise 1.7, find the values of α for which the Fisher information exists if $p_0(x) = C\,\cos^{\alpha} x\,,\ \ -\pi/2 < x < \pi/2\,$, and $p_0(x) = 0$ otherwise, where $C = C(\alpha)$ is the normalizing constant. Note that p_0 is a probability density if $\alpha > -1\,$.

Chapter 2

The Bayes and Minimax Estimators

2.1. Pitfalls of the Fisher Efficiency

Fisher efficient estimators defined in the previous chapter possess two major unattractive properties, which prevent the Fisher efficiency from being widely used in statistical theory. First, the Fisher efficient estimators rarely exist, and second, they need to be unbiased. In effect, the Fisher efficiency does not provide an answer to how to compare biased estimators with different bias functions. A lesser issue is that the comparison of estimators is based on their variances alone.

Before we proceed to an illustrative example, we need several notions defined below. A function $w(u)$, $u \in \mathbb{R}$, is called a *loss function* if: (i) $w(0) = 0$, (ii) it is symmetric, $w(u) = w(-u)$, (iii) it is non-decreasing for $u > 0$, and (iv) it is not identically equal to zero. Besides, we require that w is bounded from above by a power function, that is, (v) $w(u) \leq k(1 + |u|^a)$ for all u with some constants $k > 0$ and $a > 0$.

The loss function $w(\hat{\theta}_n - \theta)$ measures the deviation of the estimator $\hat{\theta}_n = \hat{\theta}_n(X_1, \dots, X_n)$ from the true parameter θ. In this book, we do not go far beyond: (i) *quadratic loss function*, $w(u) = u^2$, (ii) *absolute loss function*, $w(u) = |u|$, or (iii) *bounded loss function*, $w(u) = \mathbb{I}(\,|u| > c\,)$ with a given positive c, where $\mathbb{I}(\cdot)$ denotes the indicator function.

The *normalized risk function* (or simply, the *normalized risk*) $R_n(\theta, \hat{\theta}_n, w)$ is the expected value of the loss function w evaluated at $\sqrt{I_n(\theta)}(\hat{\theta}_n - \theta)$, that is,

$$R_n(\theta, \hat{\theta}_n, w) = \mathbb{E}_\theta\Big[\, w\big(\sqrt{I_n(\theta)}(\hat{\theta}_n - \theta)\,\big)\Big]$$

$$= \int_{\mathbb{R}^n} w\Big(\sqrt{I_n(\theta)}(\hat{\theta}_n(x_1, \dots, x_n) - \theta)\Big)\, p(x_1, \dots, x_n\,, \theta)\, dx_1 \dots dx_n.$$

Example 2.1. For the quadratic loss function $w(u) = u^2$, the normalized risk (commonly termed the *normalized quadratic risk*) of an estimator $\hat{\theta}_n$ can be found as

$$R_n(\theta, \hat{\theta}_n, u^2) = \mathbb{E}_\theta\Big[\, I_n(\theta)\,(\hat{\theta}_n - \theta\,)^2\Big] = I_n(\theta)\,\mathbb{E}_\theta\Big[\,(\hat{\theta}_n - \mathbb{E}_\theta\,[\,\hat{\theta}_n\,] + \mathbb{E}_\theta\,[\,\hat{\theta}_n\,] - \theta)^2\Big]$$

$$= I_n(\theta)\,\Big[\,\mathbb{V}ar_\theta[\,\hat{\theta}_n\,] + b_n^2(\theta\,, \hat{\theta}_n)\,\Big] \tag{2.1}$$

where $b_n(\theta\,, \hat{\theta}_n) = \mathbb{E}_\theta\,[\,\hat{\theta}_n\,] - \theta$ is the bias of $\hat{\theta}_n$. □

By (2.1), for any unbiased estimator $\hat{\theta}_n$, the normalized quadratic risk function has the representation $R_n(\theta, \hat{\theta}_n, u^2) = I_n(\theta)\mathbb{V}ar_\theta[\,\hat{\theta}_n\,]$. The Cramér-Rao inequality (1.2) can thus be written as

$$R_n(\theta, \hat{\theta}_n, u^2) = \mathbb{E}_\theta\,\Big[\, I_n(\theta)\,(\hat{\theta}_n - \theta\,)^2\Big] \geq 1, \quad \theta \in \Theta, \tag{2.2}$$

with the equality attained for the Fisher efficient estimators θ_n^*,

$$R_n(\theta, \theta_n^*, u^2) = \mathbb{E}_\theta\,\Big[\, I_n(\theta)\,(\theta_n^* - \theta\,)^2\Big] = 1, \quad \theta \in \Theta. \tag{2.3}$$

Next, we present an example of a biased estimator that in a certain interval performs more efficiently than the Fisher efficient unbiased estimator, if we define a *more efficient estimator* as the one with a smaller normalized quadratic risk.

Example 2.2. Let $X_1, \dots, X_n$ be independent observations from the $\mathcal{N}(\theta,\, \sigma^2)$ distribution, where σ^2 is known. Consider two estimators: (i) $\theta_n^* = \bar{X}_n$, which is efficient by Example 1.10, and (ii) a constant-value estimator $\hat{\theta} = \theta_0$, where θ_0 is a fixed point. The normalized quadratic risk of θ_n^* equals the unity by (2.3), while that of $\hat{\theta}$ is

$$R_n(\theta, \hat{\theta}, u^2) = \mathbb{E}_\theta\big[\, I_n(\theta)(\hat{\theta} - \theta)^2\,\big] = \frac{n}{\sigma^2}\,(\theta_0 - \theta)^2.$$

Note that $\hat{\theta}$ is a biased estimator with the bias $b_n(\theta) = \theta_0 - \theta$.

It is impossible to determine which of the two normalized quadratic risks is smaller (refer to Figure 1). If θ is within $\theta_0 \pm \sigma/\sqrt{n}$, then $\hat{\theta}$ is more efficient, whereas for all other values of θ, θ^* is a more efficient estimator. □

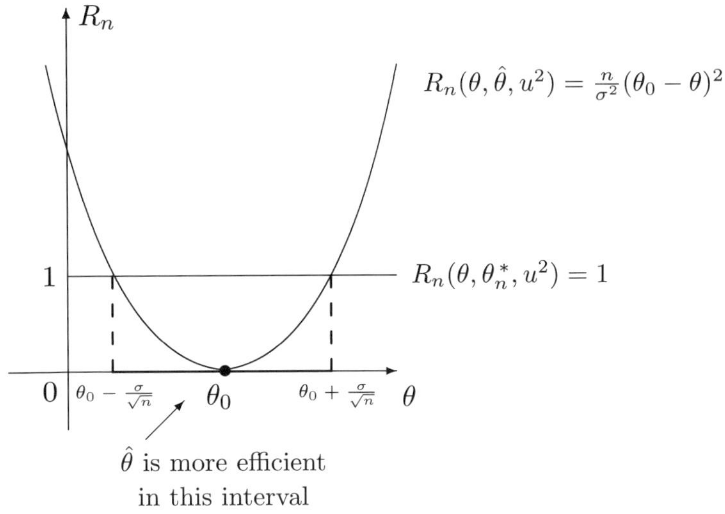

Figure 1. The normalized quadratic risk functions in Example 2.2.

This example illustrates the difficulty in comparing normalized risks of two estimators as functions of $\theta \in \Theta$. To overcome it, we could try to represent each risk function by a positive number. In statistics, there are two major ways to implement this idea. One approach is to integrate the normalized risk over the parameter set Θ, whereas the other one is to take the maximum value of the normalized risk function over Θ. These are called the Bayes and the minimax approaches, respectively. They are explored in the next three sections.

2.2. The Bayes Estimator

In what follows, we study only regular statistical models, which by definition have a strictly positive, continuous Fisher information.

Assume that there is a probability density $\pi(\theta)$ defined on the parameter set Θ. The density $\pi(\theta)$ is called a *prior density* of θ. It reflects the judgement of how likely values of θ are before the data are obtained. The *Bayes risk* of $\hat\theta_n$ is the integrated value of the normalized risk function,

$$\beta_n(\hat\theta_n,\, w,\, \pi) \;=\; \int_\Theta R_n(\theta,\hat\theta_n,w)\,\pi(\theta)\,d\theta\,. \tag{2.4}$$

An estimator $t_n = t_n(X_1,\dots,X_n)$ is called the *Bayes estimator* of θ, if for any other estimator $\hat\theta_n$, the following inequality holds:

$$\beta_n(t_n,\, w,\pi) \,\le\, \beta_n(\hat\theta_n,\, w,\pi).$$

In other words, the Bayes estimator minimizes the Bayes risk. Loosely speaking, we can understand the Bayes estimator as a solution of the minimization problem,

$$t_n = \operatorname{argmin}_{\hat{\theta}_n} \beta(\hat{\theta}_n, w, \pi),$$

though we should keep in mind that the minimum value may not exist or may be non-unique.

In the case of the quadratic loss $w(u) = u^2$, the Bayes estimator can be computed explicitly. Define the *posterior density* of θ as the conditional density, given the observations $X_1, \dots, X_n$; that is,

$$f(\theta \mid X_1, \dots, X_n) = C_n \, p(X_1, \dots, X_n \,, \theta) \, \pi(\theta), \quad \theta \in \Theta,$$

where $C_n = C_n\,(X_1, \dots, X_n)$ is the normalizing constant. Assuming that

$$\int_\Theta I_n(\theta) \, f(\theta \mid X_1 \,, \dots \,, X_n) \, d\theta < \infty,$$

we can introduce the *weighted posterior density* as

$$\tilde{f}(\theta \mid X_1 \,, \dots \,, X_n) = \tilde{C}_n \, I_n(\theta) \, f(\theta \mid X_1 \,, \dots \,, X_n), \quad \theta \in \Theta,$$

with the normalizing constant $\tilde{C}_n = \left[\int_\Theta I_n(\theta) \, f(\theta \mid X_1 \,, \dots \,, X_n) \, d\theta \right]^{-1}$, which is finite under our assumption.

Theorem 2.3. *If $w(u) = u^2$, then the Bayes estimator t_n is the weighted posterior mean*

$$t_n = t_n(X_1 \,, \dots \,, X_n) = \int_\Theta \theta \, \tilde{f}(\theta \mid X_1, \dots, X_n) \, d\theta.$$

In particular, if the Fisher information is a constant independent of θ, then the Bayes estimator is the non-weighted posterior mean,

$$t_n = t_n(X_1 \,, \dots \,, X_n) = \int_\Theta \theta \, f(\theta \mid X_1, \dots, X_n) \, d\theta.$$

Proof. The Bayes risk of an estimator $\hat{\theta}_n$ with respect to the quadratic loss can be written in the form

$$\beta_n(\hat{\theta}_n, \pi) = \int_\Theta \int_{\mathbb{R}^n} I_n(\theta) \, (\hat{\theta}_n - \theta)^2 \, p(x_1, \dots, x_n \,, \theta) \, \pi(\theta) \, dx_1 \dots dx_n \, d\theta$$

$$= \int_{\mathbb{R}^n} \left[\int_\Theta (\hat{\theta}_n - \theta)^2 \, \tilde{f}(\theta \mid x_1, \dots, x_n) \, d\theta \right] \tilde{C}_n^{-1}(x_1, \dots, x_n) \, dx_1 \dots dx_n.$$

Thus, the minimization problem of the Bayes risk is tantamount to minimization of the integral

$$\int_\Theta (\hat{\theta}_n - \theta)^2 \, \tilde{f}(\theta \mid x_1, \dots, x_n) \, d\theta$$

with respect to $\hat{\theta}_n$ for any fixed values $x_1, \ldots, x_n$. Equating to zero the derivative of this integral with respect to $\hat{\theta}_n$ produces a linear equation, satisfied by the Bayes estimator t_n,

$$\int_\Theta (t_n - \theta)\, \tilde{f}(\theta \,|\, x_1, \ldots, x_n)\, d\theta = 0.$$

Recalling that $\int_\Theta \tilde{f}(\theta \,|\, x_1, \ldots, x_n)\, d\theta = 1$, we obtain the result,

$$t_n = \int_\Theta \theta\, \tilde{f}(\theta \,|\, x_1, \ldots, x_n)\, d\theta. \qquad \square$$

In many examples, the weighted posterior mean t_n is easily computable if we choose a prior density $\pi(\theta)$ from a *conjugate family* of distributions. A *conjugate prior distribution* $\pi(\theta)$ is such that the posterior distribution belongs to the same family of distributions for any sample $X_1, \ldots, X_n$. If the posterior distribution allows a closed-form expression of expectations, then t_n can be found without integration. The following example illustrates the idea.

Example 2.4. Consider independent Bernoulli observations $X_1, \ldots, X_n$ with the probability mass function

$$p(x, \theta) = \theta^x (1-\theta)^{1-x}, \quad x \in \{0, 1\}, \quad 0 < \theta < 1,$$

where θ is assumed to be a random variable. The joint distribution function of the sample is

$$p(X_1, \ldots, X_n, \theta) = \theta^{\sum X_i} (1-\theta)^{n - \sum X_i}.$$

As a function of θ, it has an algebraic form or a beta distribution. Thus, we select a beta density as a prior density,

$$\pi(\theta) = C(\alpha, \beta)\, \theta^{\alpha-1} (1-\theta)^{\beta-1}, \quad 0 < \theta < 1,$$

where α and β are positive parameters, and $C(\alpha, \beta)$ is the normalizing constant. The posterior density is then also a beta density,

$$f(\theta \,|\, X_1, \ldots, X_n) = C(\alpha, \beta)\, \theta^{\sum X_i + \alpha - 1} (1-\theta)^{n - \sum X_i + \beta - 1}, \quad 0 < \theta < 1.$$

By Exercise 1.3, the Fisher information is equal to $I_n(\theta) = n/[\theta(1-\theta)]$. Thus, the weighted posterior density is a beta density as well,

$$\tilde{f}(\theta \,|\, X_1, \ldots, X_n) = \tilde{C}_n \theta^{\sum X_i + \alpha - 2} (1-\theta)^{n - \sum X_i + \beta - 2}, \quad 0 < \theta < 1,$$

where $\alpha > 1$ and $\beta > 1$. The weighted posterior mean therefore is equal to

$$t_n = \frac{\sum X_i + \alpha - 1}{\sum X_i + \alpha - 1 + n - \sum X_i + \beta - 1} = \frac{\sum X_i + \alpha - 1}{n + \alpha + \beta - 2}.$$

More examples of the conjugate families are in the exercises.

2.3. Minimax Estimator. Connection Between Estimators

Define a *maximum normalized risk* of an estimator $\hat{\theta}_n = \hat{\theta}_n(X_1, \dots, X_n)$ with respect to a loss function w by

$$r_n(\hat{\theta}_n,\, w) \,=\, \sup_{\theta\in\Theta} R_n(\theta, \hat{\theta}_n\,, w) \,=\, \sup_{\theta\in\Theta} \mathbb{E}_\theta\Big[w\big(\sqrt{I_n(\theta)}\,(\hat{\theta}_n - \theta)\big)\Big].$$

An estimator $\theta_n^* = \theta_n^*(X_1, \dots, X_n)$ is called *minimax* if its maximum normalized risk does not exceed that of any other estimator $\hat{\theta}_n$. That is, for any estimator $\hat{\theta}_n$,

$$r_n(\theta_n^*,\, w) \le r_n(\hat{\theta}_n,\, w).$$

The maximum normalized risk of a minimax estimator, $r_n(\theta_n^*,\, w)$, is called the *minimax risk*.

In contrast with the Bayes estimator, the minimax estimator represents a different concept of the statistical optimality. The Bayes estimator is optimal in the averaged (integrated) sense, whereas the minimax one takes into account the "worst-case scenario".

It follows from the above definition that a minimax estimator θ_n^* solves the optimization problem

$$\sup_{\theta\in\Theta} \mathbb{E}_\theta\Big[w\big(\sqrt{I_n(\theta)}\,(\hat{\theta}_n - \theta)\big)\Big] \to \inf_{\hat{\theta}_n}.$$

Finding the infimum over all possible estimators $\hat{\theta}_n = \hat{\theta}_n(X_1, \dots, X_n)$, that is, over all functions of observations $X_1, \dots, X_n$, is not an easily tackled task. Even for the most common distributions, such as normal or binomial, the direct minimization is a hopeless endeavor. This calls for an alternative route in finding minimax estimators.

In this section we establish a connection between the Bayes and minimax estimators that will lead to some advances in computing the latter. The following theorem shows that if the Bayes estimator has a constant risk, then it is also minimax.

Theorem 2.5. *Let $t_n = t_n(X_1, \dots, X_n)$ be a Bayes estimator with respect to a loss function w. Suppose that the normalized risk function of the Bayes estimator is a constant for any $\theta \in \Theta$, that is,*

$$R_n(\theta, t_n\,, w) \,=\, \mathbb{E}_\theta\Big[w\big(\sqrt{I_n(\theta)}\,(t_n - \theta)\big)\Big] \,=\, c$$

for some $c > 0$. Then t_n is also a minimax estimator.

Proof. Notice that since the risk function of t_n is a constant, the Bayes and maximum normalized risks of t_n are the same constants. Indeed, letting

$\pi(\theta)$ denote the corresponding prior density, we write

$$\beta_n(t_n\,,\, w, \pi) = \int_{\Theta} R_n(\theta, t_n\,, w)\,\pi(\theta)\,d\theta = c\int_{\Theta}\pi(\theta)\,d\theta = c$$

and

$$r_n(t_n,\, w) = \sup_{\theta\in\Theta} R_n(\theta, t_n\,, w) \;=\; \sup_{\theta\in\Theta}\; c = c.$$

Further, for any estimator $\hat{\theta}_n$,

$$r_n(\hat{\theta}_n, w) = \sup_{\theta\in\Theta} R_n(\theta, \hat{\theta}_n\,, w) \geq \int_{\Theta} R_n(\theta, \hat{\theta}_n\,, w)\,\pi(\theta)\,d\theta$$

$$= \beta_n(\hat{\theta}_n\,,\, w, \pi) \geq \beta_n(t_n\,,\, w, \pi) = c = r_n(t_n\,, w).$$ □

Unfortunately, Theorem 2.5 does not provide a recipe for choosing a prior density for which the normalized risk function is a constant on Θ. Moreover, constant-risk priors rarely exist. Below we give two examples where we try to explain why it happens.

Example 2.6. Consider independent Bernoulli observations $X_1, \dots, X_n$ with parameter θ. As shown in Example 2.4, the weighted posterior mean of θ is

$$t_n \;=\; \frac{\sum X_i + \alpha - 1}{n + \alpha + \beta - 2}.$$

If we now select $\alpha = \beta = 1$, then t_n becomes the sample mean $\bar{X}_n$. From Exercise 1.3 we know that $\bar{X}_n$ is an efficient estimator of θ, and therefore its weighted quadratic risk is equal to 1, a constant. However, $\alpha = \beta = 1$ is not a legitimate choice in this instance, because the weighted posterior density

$$\tilde{f}(\theta\,|\,X_1, \dots, X_n) \;=\; \tilde{C}_n \theta^{\sum X_i - 1}\,(1-\theta)^{\,n - \sum X_i - 1}$$

does not exist for $\sum X_i \;=\; 0$. Indeed, $\theta^{-1}\,(1-\theta)^{\,n-1}$ is not integrable at zero, and therefore the normalizing constant $\tilde{C}_n$ does not exist. □

Example 2.7. Let $X_1, \dots, X_n$ be independent observations from the $\mathcal{N}(\theta,\, 1)$ distribution. If we choose the prior density of θ to be $\mathcal{N}(0,\, b^2)$ for some positive real b, then, by Exercise 2.10, the weighted posterior distribution is also normal,

$$\mathcal{N}\Big(\frac{n\,b^2\,\bar{X}_n}{n\,b^2 + 1},\; \frac{b^2}{n\,b^2 + 1}\Big).$$

Here the weighted posterior mean $t_n \;=\; n\,b^2\,\bar{X}_n/(n\,b^2 + 1)$ is the Bayes estimator with respect to the quadratic loss function. If we let $b \to \infty$, then t_n equals $\bar{X}_n$, which is Fisher efficient (see Example 1.10) and thus has a constant normalized quadratic risk. The flaw in this argument is that no normal prior density exists with infinite b. □

2.4. Limit of the Bayes Estimator and Minimaxity

Assume that we can find a family of prior distributions with the densities $\pi_b(\theta)$ indexed by a positive real number b. If the Bayes risks of the respective Bayes estimators have a limit as b goes to infinity, then this limit guarantees a minimax lower bound. A rigorous statement is presented in the following theorem.

Theorem 2.8. *Let $\pi_b(\theta)$ be a family of prior densities on Θ that depend on a positive real parameter b, and let $t_n(b) = t_n(X_1, \dots, X_n, b)$ be the respective Bayes estimators for a loss function w. Suppose that the Bayes risk $\beta_n\big(t_n(b), w, \pi_b\big)$ has a limit,*

$$\lim_{b\to\infty} \beta_n\big(t_n(b), w, \pi_b\big) = c > 0.$$

Then the minimax lower bound holds for any n,

$$\inf_{\hat\theta_n} r_n(\hat\theta_n\,,\, w) = \inf_{\hat\theta_n}\sup_{\theta\in\Theta} \mathbb{E}_\theta\Big[\, w\big(\sqrt{I_n(\theta)}\,(\hat\theta_n - \theta)\big)\Big] \geq c.$$

Proof. As in the proof of Theorem 2.5, for any estimator $\hat\theta_n$, we can write

$$r_n(\hat\theta_n\,, w) = \sup_{\theta\in\Theta} R_n(\theta, \hat\theta_n\,, w) \geq \int_\Theta R_n(\theta, \hat\theta_n\,, w)\,\pi_b(\theta)\,d\theta$$

$$= \beta_n(\hat\theta_n\,, w, \pi_b) \geq \beta_n\big(t_n(b), w, \pi_b\big).$$

Now take the limit as $b\to\infty$. Since the left-hand side is independent of b, the theorem follows. $\square$

Example 2.9. Let $X_1, \dots, X_n$ be independent $\mathcal{N}(\theta,\, 1)$ observations. We will show that conditions of Theorem 2.8 are satisfied under the quadratic loss function $w(u) = u^2$, and therefore the lower bound for the corresponding minimax risk holds:

$$\inf_{\hat\theta_n} r_n(\hat\theta_n\,,\, w\,) = \inf_{\hat\theta_n}\sup_{\theta\in\mathbb{R}} E_\theta\Big[\big(\sqrt{n}\,(\hat\theta_n - \theta)\big)^2\Big] \geq 1.$$

As shown in Example 2.7, for a $\mathcal{N}(0,\, b^2)$ prior density, the weighted posterior mean $t_n(b) = n\,b^2\,\bar X_n/(n\,b^2+1)$ is the Bayes estimator with respect to the quadratic loss function. Now we will compute its Bayes risk. This estimator has the variance

$$\mathbb{V}ar_\theta\big[t_n(b)\big] = \frac{n^2\,b^4\,\mathbb{V}ar_\theta\big[\bar X_n\big]}{(n\,b^2+1)^2} = \frac{n\,b^4}{(n\,b^2+1)^2}$$

and the bias

$$b_n\big(\theta, t_n(b)\big) = \mathbb{E}_\theta\big[t_n(b)\big] - \theta = \frac{n^2\,b^2\,\theta}{n\,b^2+1} - \theta = -\frac{\theta}{n\,b^2+1}.$$

Therefore, its normalized quadratic risk is expressed as

$$R_n(\theta\,, t_n(b), w) = E_\theta\Big[\big(\sqrt{n}\,(t_n(b)\;-\;\theta)\big)^2\Big] = n\Big[\mathbb{V}ar_\theta[\,t_n(b)\,] + b_n^2(\theta\,,\,t_n(b))\Big]$$
$$= \frac{n^2\,b^4}{(n\,b^2+1)^2} + \frac{n\,\theta^2}{(n\,b^2+1)^2}.$$

With the remark that $\int_{\mathbb{R}} \theta^2\,\pi_b\,(\theta)\,d\theta = b^2$, the Bayes risk of $t_n(b)$ equals

$$\beta_n\big(t_n(b), w, \pi_b\big) \;=\; \int_{\mathbb{R}} \Big[\,\frac{n^2\,b^4}{(n\,b^2+1)^2} + \frac{n\,\theta^2}{(n\,b^2+1)^2}\,\Big]\,\pi_b\,(\theta)\,d\theta$$
$$= \frac{n^2\,b^4}{(n\,b^2+1)^2} + \frac{n\,b^2}{(n\,b^2+1)^2} \to 1 \text{ as } b\to\infty.$$

Applying Theorem 2.8, we obtain the result with $c = 1$. Taking a step further, note that the minimax lower bound is attained for the estimator $\bar{X}_n$, which is thus minimax. Indeed, $\mathbb{E}_\theta\Big[\big(\sqrt{n}\,(\bar{X}_n - \theta)\big)^2\Big] = 1$. □

In subsequent chapters we present additional useful applications of Theorem 2.8.

Exercises

Exercise 2.9. Suppose the random observations $X_1, \dots, X_n$ come from a Poisson distribution with the probability mass function

$$p(x\,,\theta) = \frac{\theta^{\,x}\,e^{-\theta}}{x!}\,,\quad x\in\{\,0\,,1\,,\dots\},$$

where θ is a random variable. Show that the conjugate prior density of θ is a gamma density, $\pi(\theta) = C(\alpha,\,\beta)\,\theta^{\,\alpha-1}\,e^{-\,\beta\,\theta}$, $\theta > 0$, for some positive parameters α and β, and the normalizing constant $C(\alpha,\,\beta)$. Find the weighted posterior mean of θ.

Exercise 2.10. Consider a set of independent observations $X_1, \dots, X_n \sim \mathcal{N}(\theta, \sigma^2)$, where θ is assumed random with the prior density $\mathcal{N}(\mu, \sigma_\theta^2)$. Show that the weighted posterior distribution of θ is also normal with the mean $\big(n\,\sigma_\theta^2\,\bar{X}_n + \mu\sigma^2\big)/\big(n\,\sigma_\theta^2 + \sigma^2\big)$ and variance $\sigma^2\,\sigma_\theta^2/(n\sigma_\theta^2+\sigma^2)$. Note that the family of normal distributions is self-conjugate.

Exercise 2.11. Find a conjugate distribution and the corresponding Bayes estimator for the parameter θ in the exponential model with $p(x,\theta) = \theta\exp\{-\,\theta\,x\}$, $\;x\,,\theta > 0$.

Exercise 2.12. Consider n independent Bernoulli observations $X_1, \dots, X_n$ with $p(x, \theta) = \theta^x (1-\theta)^{1-x}$, $x \in \{0, 1\}$, and $\Theta = (0,1)$. Define the estimator

$$\theta_n^* = \frac{\sum X_i + \sqrt{n}/2}{n + \sqrt{n}}.$$

(i) Verify that θ_n^* is the non-weighted posterior mean with respect to the conjugate prior density $\pi(\theta) = C\left[\theta(1-\theta)\right]^{\sqrt{n}/2 - 1}$, $0 < \theta < 1$.

(ii) Show that the non-normalized quadratic risk of θ_n^* (with the factor $\sqrt{I_n(\theta)}$ omitted) is equal to

$$\mathbb{E}_\theta\left[(\theta_n^* - \theta)^2\right] = \frac{1}{4(1+\sqrt{n})^2}.$$

(iii) Verify that Theorem 2.5 is valid for a non-normalized risk function, and argue that θ_n^* is minimax in the appropriate sense.

Exercise 2.13. Refer to the Bernoulli model in Example 2.4. Show that the prior beta distribution with $\alpha = \beta = 1 + b^{-1}$ defines the weighted posterior mean $t_n(b)$ which is minimax for $b = \infty$.

Chapter 3

Asymptotic Minimaxity

In this chapter we study the asymptotic minimaxity of estimators as the sample size n increases.

3.1. The Hodges Example

An estimator $\hat{\theta}_n$ is called *asymptotically unbiased* if it satisfies the limiting condition

$$\lim_{n\to\infty} \mathbb{E}_\theta\big[\,\hat{\theta}_n\big] = \theta, \qquad \theta \in \Theta.$$

In many cases when an unbiased estimator of θ does not exist, an asymptotically unbiased estimator is easy to construct.

Example 3.1. In Example 1.4, the MLE $\theta_n^* = 1/\bar{X}_n$, though biased for any $n > 1$, is asymptotically unbiased. Indeed,

$$\lim_{n\to\infty} \mathbb{E}_\theta\big[\,\theta_n^*\big] = \lim_{n\to\infty} \frac{n\,\theta}{n-1} = \theta. \qquad \square$$

Example 3.2. In Example 1.5, there is no unbiased estimator. The estimator $\hat{\theta}_n = \sqrt{X/n}$, however, is asymptotically unbiased (see Exercise 3.14.) $\square$

In the previous chapter, we explained why the Fisher approach fails as a criterion for finding the most efficient estimators. Now we are planning to undertake another desperate, though futile, task of rescuing the concept of Fisher efficiency at least in an asymptotic form. The question is: Can we define a sequence of *asymptotically Fisher efficient* estimators $\theta_n^* = \theta_n^*(X_1, \dots, X_n)$ by requiring that they: (i) are asymptotically unbiased and (ii) satisfy the equation (compare to (2.3)):

$$\lim_{n\to\infty} \mathbb{E}_\theta \left[I_n(\theta) \big(\hat\theta_n - \theta \big)^2 \right] = 1, \quad \theta \in \Theta \,? \tag{3.1}$$

The answer to this question would be positive, if for any sequence of asymptotically unbiased estimators $\hat\theta_n$, the following analogue of the Cramér-Rao lower bound (2.2) were true,

$$\lim_{n\to\infty} \mathbb{E}_\theta \left[I_n(\theta) \big(\hat\theta_n - \theta \big)^2 \right] \geq 1, \quad \theta \in \Theta \,. \tag{3.2}$$

Indeed, if (3.2) held, then the estimator that satisfies (3.1) would be asymptotically the most efficient one. However, it turns out that this inequality is not valid even for $\mathcal{N}(\theta,\, 1)$ observations. A famous Hodges example is presented below.

Example 3.3. Consider independent observations $X_1, \dots, X_n$ from $\mathcal{N}(\theta,\, 1)$ distribution, $\theta \in \mathbb{R}$. Define the sequence of estimators

$$\hat\theta_n = \begin{cases} \bar X_n & \text{if } |\, \bar X_n \,| \geq n^{-1/4}, \\ 0 & \text{otherwise.} \end{cases} \tag{3.3}$$

Note that in this example, $I_n(\theta) = n$. It can be shown (see Exercise 3.15) that this sequence is asymptotically unbiased, and that the following equalities hold:

$$\begin{cases} \lim_{n\to\infty} \mathbb{E}_\theta \left[n \,\big(\hat\theta_n - \theta \big)^2 \right] = 1 & \text{if } \theta \neq 0, \\ \lim_{n\to\infty} \mathbb{E}_\theta \left[n\, \hat\theta_n^2 \right] = 0 & \text{if } \theta = 0. \end{cases} \tag{3.4}$$

Thus, the sequence $\hat\theta_n$ is asymptotically more efficient than any asymptotically Fisher efficient estimator defined by (3.1). In particular, it is better than the sample mean $\bar X_n$. Sometimes the Hodges estimator is called *superefficient*, and the point at which the Cramér-Rao lower bound is violated, $\theta = 0$, is termed the *superefficient point.* □

The above example explains why the asymptotic theory of parameter estimation should be based on methods other than the pointwise asymptotic Fisher efficiency. We start introducing these methods in the next section.

3.2. Asymptotic Minimax Lower Bound

Recall from Section 2.3 that a minimax estimator corresponding to the quadratic loss function solves the minimization problem

$$\sup_{\theta\in\Theta} \left[I_n(\theta) \int_{\mathbb{R}^n} \Big(\hat\theta_n(x_1, \dots, x_n) - \theta \Big)^2 p(x_1, \dots, x_n, \theta)\, dx_1 \dots dx_n \right] \to \inf_{\hat\theta_n}.$$

The minimization is carried over all arbitrary functions $\hat{\theta}_n = \hat{\theta}(x_1, \dots, x_n)$. As discussed earlier, this problem is impenetrable from the point of view of standard analytic methods of calculus. In this section we will learn a bypassing approach based on the asymptotically minimax lower bound. Consider the maximum normalized risk of an estimator $\hat{\theta}_n$ with respect to the quadratic loss function

$$r_n(\hat{\theta}_n, u^2) = \sup_{\theta \in \Theta} R_n(\theta, \hat{\theta}_n, u^2)$$

$$= \sup_{\theta \in \Theta} \mathbb{E}_\theta \Big[I_n(\theta) (\hat{\theta}_n - \theta)^2 \Big] = n \sup_{\theta \in \Theta} I(\theta) \mathbb{E}_\theta \Big[(\hat{\theta}_n - \theta)^2 \Big].$$

Suppose we can show that for any estimator $\hat{\theta}_n$ the inequality

$$\liminf_{n \to \infty} r_n(\hat{\theta}_n, u^2) \geq r_* \tag{3.5}$$

holds with a positive constant r_* independent of n. This inequality implies that for any estimator $\hat{\theta}_n$ and for all large enough n, the maximum of the quadratic risk is bounded from below,

$$\sup_{\theta \in \Theta} I(\theta) \mathbb{E}_\theta \Big[\big(\hat{\theta}_n - \theta \big)^2 \Big] \geq \frac{r_* - \varepsilon}{n}$$

with arbitrarily small $\varepsilon > 0$. We call the inequality (3.5) the *asymptotically minimax lower bound.* If, in addition, we can find an estimator θ_n^*, which for all large n satisfies the upper bound

$$\sup_{\theta \in \Theta} I(\theta) \mathbb{E}_\theta \Big[\big(\theta_n^* - \theta \big)^2 \Big] \leq \frac{r^*}{n}$$

with a positive constant r^*, then for all large enough n, the minimax risk is sandwiched between two positive constants,

$$r_* \leq \inf_{\hat{\theta}_n} \sup_{\theta \in \Theta} \mathbb{E}_\theta \Big[\big(\sqrt{n I(\theta)} \, (\hat{\theta}_n - \theta) \big)^2 \Big] \leq r^*. \tag{3.6}$$

In this special case of the quadratic loss function $w(u) = u^2$, we define the *asymptotically minimax rate of convergence* as $1/\sqrt{n}$ (or, equivalently, $O(1/\sqrt{n})$ as $n \to \infty$). This is the fastest possible decrease rate of $\hat{\theta}_n - \theta$ in the mean-squared sense as $n \to \infty$. This rate is not improvable by any estimator.

More generally, we call a deterministic sequence ψ_n the *asymptotically minimax rate of convergence*, if for some positive constants r_* and r^*, and for all sufficiently large n, the following inequalities hold:

$$r_* \leq \inf_{\hat{\theta}_n} \sup_{\theta \in \Theta} \mathbb{E}_\theta \Big[w\Big(\frac{\hat{\theta}_n - \theta}{\psi_n} \Big) \Big] \leq r^* < \infty. \tag{3.7}$$

If $r_* = r^*$, these bounds are called *asymptotically sharp.*

In the following lemma we explain the idea of how the asymptotically minimax lower bound (3.5) may be proved. We consider only normally distributed observations, and leave some technical details out of the proof.

Lemma 3.4. *Take independent observations* $X_1, \dots, X_n \sim \mathcal{N}(\theta, \sigma^2)$ *where* σ^2 *is known. Let* $\theta \in \Theta$ *where* Θ *is an open interval containing the origin* $\theta = 0$. *Then for any estimator* $\hat{\theta}_n$, *the following inequality holds:*

$$\liminf_{n\to\infty} r_n(\hat{\theta}_n, u^2) = \liminf_{n\to\infty} \frac{n}{\sigma^2} \sup_{\theta\in\Theta} \mathbb{E}_\theta\Big[\big(\hat{\theta}_n - \theta\big)^2\Big] \ge r_* = 0.077.$$

Remark 3.5. Under the assumptions of Lemma 3.4, the maximum normalized risk $r_n(\hat{\theta}_n, u^2)$ admits the asymptotic upper bound $r^* = 1$, guaranteed by the sample mean estimator $\bar{X}_n$. □

Proof of Lemma 3.4. Without loss of generality, we can assume that $\sigma^2 = 1$ (hence $I(\theta) = 1$), and that Θ contains points $\theta_0 = 0$ and $\theta_1 = 1/\sqrt{n}$. Introduce the *log-likelihood ratio* associated with these values of the parameter θ,

$$\Delta L_n = \Delta L_n(\theta_0, \theta_1) = L_n(\theta_1) - L_n(\theta_0)$$

$$= \ln \frac{p(X_1, \dots, X_n, \theta_1)}{p(X_1, \dots, X_n, \theta_0)} = \sum_{i=1}^n \ln \frac{p(X_i, 1/\sqrt{n})}{p(X_i, 0)}$$

$$= \sum_{i=1}^n \Big[-\frac{1}{2}\big(X_i - \frac{1}{\sqrt{n}}\big)^2 + \frac{1}{2} X_i^2 \Big] = \frac{1}{\sqrt{n}} \sum_{i=1}^n X_i - \frac{1}{2} = Z - \frac{1}{2}$$

where Z is a $\mathcal{N}(0,1)$ random variable with respect to the distribution $\mathbb{P}_{\theta_0}$.

Further, by definition, for any random function $f(X_1, \dots, X_n)$, and for any values θ_0 and θ_1, the basic likelihood ratio identity relating the two expectations holds:

$$\mathbb{E}_{\theta_1}\big[f(X_1, \dots, X_n)\big] = \mathbb{E}_{\theta_0}\Big[f(X_1, \dots, X_n)\frac{p(X_1, \dots, X_n, \theta_1)}{p(X_1, \dots, X_n, \theta_0)}\Big]$$

$$= \mathbb{E}_{\theta_0}\Big[f(X_1, \dots, X_n)\exp\big\{\Delta L_n(\theta_0, \theta_1)\big\}\Big]. \tag{3.8}$$

Next, for any fixed estimator $\hat{\theta}_n$, the supremum over $\mathbb{R}$ of the normalized risk function is not less than the average of the normalized risk over the two points θ_0 and θ_1. Thus, we obtain the inequality

$$n \sup_{\theta\in\mathbb{R}} \mathbb{E}_\theta\Big[(\hat{\theta}_n - \theta)^2\Big] \ge n \max_{\theta\in\{\theta_0, \theta_1\}} \mathbb{E}_\theta\Big[(\hat{\theta}_n - \theta)^2\Big]$$

$$\ge \frac{n}{2}\Big\{\mathbb{E}_{\theta_0}\Big[(\hat{\theta}_n - \theta_0)^2\Big] + \mathbb{E}_{\theta_1}\Big[(\hat{\theta}_n - \theta_1)^2\Big]\Big\}$$

$$= \frac{n}{2}\mathbb{E}_{\theta_0}\Big[(\hat{\theta}_n - \theta_0)^2 + (\hat{\theta}_n - \theta_1)^2\exp\big\{\Delta L_n(\theta_0, \theta_1)\big\}\Big] \qquad \text{(by (3.8))}$$

$$\geq \frac{n}{2}\,\mathbb{E}_{\theta_0}\Big[\Big((\hat{\theta}_n-\theta_0)^2+(\hat{\theta}_n-\theta_1)^2\Big)\,\mathbb{I}\Big(\Delta L_n(\theta_0\,,\theta_1)\geq 0\Big)\Big]$$

$$\geq \frac{n}{2}\,\frac{(\theta_1-\theta_0)^2}{2}\,\mathbb{P}_{\theta_0}\Big(\Delta L_n(\theta_0\,,\theta_1)\geq 0\Big)$$

$$=\frac{n}{4}\Big(\frac{1}{\sqrt{n}}\Big)^2\,\mathbb{P}_{\theta_0}\big(Z-1/2\geq 0\big)=\frac{1}{4}\,\mathbb{P}_{\theta_0}\big(Z\geq 1/2\big).$$

In the above, if the log-likelihood ratio $\Delta L_n(\theta_0\,,\theta_1)$ is non-negative, then its exponent is at least 1. At the last stage we used the elementary inequality

$$(x-\theta_0)^2+(x-\theta_1)^2\geq \frac{1}{2}\,(\theta_1-\theta_0)^2,\quad x\in\mathbb{R}.$$

As shown previously, Z is a standard normal random variable with respect to the distribution $\mathbb{P}_{\theta_0}$, therefore, $\mathbb{P}_{\theta_0}\big(Z\geq 1/2\big)=0.3085$. Finally, the maximum normalized risk is bounded from below by $0.3085/4>0.077$. □

Remark 3.6. Note that computing the mean value of the normalized risk over two points is equivalent to finding the Bayes risk with respect to the prior distribution that is equally likely concentrated at these points. Thus, in the above proof, we could have taken a Bayes prior concentrated not at two but at three or more points, then the lower bound constant r_* would be different from 0.077. □

The normal distribution of the observations in Lemma 3.4 is used only in the explicit formula for the log-likelihood ratio $\Delta L_n(\theta_0\,,\theta_1)$. A generalization of this lemma to the case of a statistical experiment with an arbitrary distribution is stated in the theorem below. The proof of the theorem is analogous to that of the lemma, and therefore is left as an exercise (see Exercise 3.16).

Theorem 3.7. *Assume that an experiment* $(X_1,\dots,X_n\,;\,p(x\,,\theta)\,;\,\Theta)$ *is such that for some points* θ_0 *and* $\theta_1=\theta_0+1/\sqrt{n}$ *in* Θ, *the log-likelihood ratio*

$$\Delta L_n(\theta_0\,,\theta_1)=\sum_{i=1}^{n}\ln\frac{p(X_i\,,\theta_1)}{p(X_i\,,\theta_0)}$$

satisfies the condition

$$\mathbb{P}_{\theta_0}\Big(\Delta L_n(\theta_0\,,\theta_1)\geq z_0\Big)\geq p_0$$

with the constants z_0 *and* p_0 *independent of* n. *Assume that* $z_0\leq 0$. *Then for any estimator* $\hat{\theta}_n$, *the lower bound of the minimax risk holds:*

$$\liminf_{n\to\infty}\sup_{\theta\in\mathbb{R}}\mathbb{E}_\theta\Big[I_n(\theta)\,\big(\hat{\theta}_n-\theta\big)^2\Big]\geq\frac{1}{4}\,I_*\,p_0\,\exp\{z_0\}.$$

where $I_*=\min\big[I(\theta_0)\,,\,I(\theta_1)\big]>0$.

3.3. Sharp Lower Bound. Normal Observations

Lemma 3.4 leaves a significant gap between the lower and upper constants in (3.6). Indeed, $r_* = 0.077$, while $r^* = 1$ by Remark 3.5. It should not come as a surprise that in such a regular case as normal observations it can be shown that $r_* = r^*$. In this section, we prove the sharp lower bound with $r_* = r^* = 1$ for the normal observations. To do this, we have to overcome the same technical difficulties and we will need the same ideas as in the case of more general observations discussed in the next section.

Theorem 3.8. *Under the assumptions of Lemma 3.4, for any estimator $\hat{\theta}_n$, the following lower bound holds:*

$$\liminf_{n\to\infty} \frac{n}{\sigma^2} \sup_{\theta\in\Theta} \mathbb{E}_\theta\Big[\big(\hat{\theta}_n - \theta\big)^2\Big] \geq 1.$$

Proof. As in the proof of Lemma 3.4, we can take $\sigma^2 = 1$. The idea of the proof is based on the substitution of the maximum normalized risk by the Bayes risk with the uniform prior distribution in an interval $\big[-b/\sqrt{n}\,, b/\sqrt{n}\big]$ where b will be chosen later. Under the assumption on Θ, it contains this interval for all sufficiently large n. Proceeding as in the proof of Lemma 3.4, we obtain the inequalities

$$\sup_{\theta\in\mathbb{R}} \mathbb{E}_\theta\Big[n\big(\hat{\theta}_n - \theta\big)^2\Big] \geq \frac{\sqrt{n}}{2b}\int_{-b/\sqrt{n}}^{b/\sqrt{n}} \mathbb{E}_\theta\Big[n\big(\hat{\theta}_n - \theta\big)^2\Big]\,d\theta$$

$$= \frac{1}{2b}\int_{-b}^{b} \mathbb{E}_{t/\sqrt{n}}\Big[\big(\sqrt{n}\,\hat{\theta}_n - t\big)^2\Big]\,dt \quad \text{(by substitution } t = \sqrt{n}\,\theta)$$

$$= \frac{1}{2b}\int_{-b}^{b} \mathbb{E}_0\Big[\big(\sqrt{n}\,\hat{\theta}_n - t\big)^2 \exp\big\{\Delta L_n\big(0, \tfrac{t}{\sqrt{n}}\big)\big\}\Big]\,dt. \tag{3.9}$$

Here the same trick is used as in the proof of Lemma 3.4 with the change of the distribution by means of the log-likelihood ratio, which in this case is equal to

$$\Delta L_n\big(0, \frac{t}{\sqrt{n}}\big) = L_n\big(\frac{t}{\sqrt{n}}\big) - L_n(0) = \sum_{i=1}^{n}\Big[-\frac{1}{2}\big(X_i - \frac{t}{\sqrt{n}}\big)^2 + \frac{1}{2}X_i^2\Big]$$

$$= \frac{t}{\sqrt{n}}\sum_{i=1}^{n} X_i - \frac{t^2}{2} = t\,Z - \frac{t^2}{2} = \frac{Z^2}{2} - \frac{(t-Z)^2}{2}$$

where $Z \sim \mathcal{N}(0,1)$ under $\mathbb{P}_0$. Thus, the latter expression can be written as

$$\mathbb{E}_0\Big[e^{Z^2/2}\,\frac{1}{2b}\int_{-b}^{b}\big(\sqrt{n}\,\hat{\theta}_n - t\big)^2 e^{-(t-Z)^2/2}\,dt\Big]$$

$$\geq \mathbb{E}_0\Big[e^{Z^2/2}\,\mathbb{I}(\,|Z| \leq a\,)\,\frac{1}{2b}\int_{-b}^{b}\big(\sqrt{n}\,\hat{\theta}_n - t\big)^2 e^{-(t-Z)^2/2}\,dt\Big] \tag{3.10}$$

where a is a positive constant, $a < b$. The next step is to change the variable of integration to $u = t - Z$. The new limits of integration are $[-b - Z\,, b - Z\,]$. For any Z that satisfies $|Z| \leq a$, this interval includes the interval $[-(b-a)\,, b-a\,]$, so that the integral over $[-b\,, b\,]$ with respect to t can be estimated from below by the integral in u over $[-(b-a)\,, b-a\,]$. Hence, for $|Z| \leq a$,

$$\int_{-b}^{b}\big(\sqrt{n}\,\hat{\theta}_n - t\big)^2 e^{(t-Z)^2/2}\,dt \geq \int_{-(b-a)}^{b-a}\big(\sqrt{n}\,\hat{\theta}_n - Z - u\big)^2 e^{-u^2/2}\,du$$

$$= \int_{-(b-a)}^{b-a}\big[\big(\sqrt{n}\,\hat{\theta}_n - Z\big)^2 + u^2\big]\,e^{-u^2/2}\,du \geq \int_{-(b-a)}^{b-a} u^2\,e^{-u^2/2}\,du. \tag{3.11}$$

Here the cross term disappears because $\int_{-(b-a)}^{b-a} u\,\exp\{-u^2/2\}\,du = 0$.

Further, we compute the expected value

$$\mathbb{E}_0\Big[e^{Z^2/2}\,\mathbb{I}(\,|Z| \leq a\,)\Big] = \int_{-a}^{a} e^{z^2/2}\frac{1}{\sqrt{2\pi}}\,e^{-z^2/2}\,dz = \frac{2a}{\sqrt{2\pi}}. \tag{3.12}$$

Putting together (3.11) and (3.12), and continuing from (3.10), we arrive at the lower bound

$$\sup_{\theta\in\mathbb{R}} \mathbb{E}_\theta\Big[n\,\big(\hat{\theta}_n - \theta\big)^2\Big] \geq \frac{2a}{\sqrt{2\pi}}\,\frac{1}{2b}\int_{-(b-a)}^{b-a} u^2\,e^{-u^2/2}\,du$$

$$= \frac{a}{b}\,\mathbb{E}\Big[\,Z_0^2\,\mathbb{I}(\,|Z_0| \leq b-a\,)\Big] \tag{3.13}$$

where Z_0 is a standard normal random variable. Choose a and b such that $a/b \to 1$ and $b - a \to \infty$, for example, put $a = b - \sqrt{b}$ and let $b \to \infty$. Then the expression in (3.13) can be made however close to $\mathbb{E}\big[\,Z_0^2\,\big] = 1$. □

The quadratic loss function is not critical in Theorem 3.8. The next theorem generalizes the result to any loss function.

Theorem 3.9. *Under the assumptions of Theorem 3.8, for any loss function* w *and any estimator* $\hat{\theta}_n$*, the following lower bound holds:*

$$\liminf_{n\to\infty}\,\sup_{\theta\in\Theta} \mathbb{E}_\theta\Big[w\Big(\sqrt{\frac{n}{\sigma^2}}\,(\hat{\theta}_n - \theta)\Big)\Big] \geq \int_{-\infty}^{\infty}\frac{w(u)}{\sqrt{2\pi}}\,e^{-u^2/2}\,du.$$

Proof. In the proof of Theorem 3.8, the quadratic loss function was used only to demonstrate that for any $\sqrt{n}\,\hat{\theta}_n - Z$, the following inequality holds:

$$\int_{-(b-a)}^{b-a} \big(\sqrt{n}\,\hat{\theta}_n \,-\, Z \,-\, u\big)^2 e^{-u^2/2}\,du \,\geq\, \int_{-(b-a)}^{b-a} u^2\, e^{-u^2/2}\,du.$$

We can generalize this inequality to any loss function as follows (see Exercise 3.18). The minimum value of the integral $\int_{-(b-a)}^{b-a} w(\,c-u\,)\,e^{-u^2/2}\,du$ over $c \in \mathbb{R}$ is attained at $c = 0$, that is,

$$\int_{-(b-a)}^{b-a} w(\,c-u\,)\,e^{-u^2/2}\,du \,\geq\, \int_{-(b-a)}^{b-a} w(u)\,e^{-u^2/2}\,du. \tag{3.14}$$ □

Remark 3.10. Note that in the proof of Theorem 3.8 (respectively, Theorem 3.9), we considered the values of θ not in the whole parameter set Θ, but only in the interval $\big[-b/\sqrt{n}, b/\sqrt{n}\big]$ of however small the length. Therefore, it is possible to formulate a local version of Theorem 3.9 with the proof remaining the same. For any loss function w, the inequality holds

$$\lim_{\delta\to 0}\,\liminf_{n\to\infty}\,\sup_{|\theta-\theta_0|<\delta}\,\mathbb{E}_\theta\Big[\,w\Big(\sqrt{\frac{n}{\sigma^2}}\,(\hat{\theta}_n-\theta)\Big)\Big] \,\geq\, \int_{-\infty}^{\infty}\frac{w(u)}{\sqrt{2\pi}}\,e^{-u^2/2}\,du.$$ □

3.4. Local Asymptotic Normality (LAN)

The sharp lower bounds of the previous section were proved under the restrictive assumption of normal observations. How far can these results be extended? What is to be required from a statistical experiment to ensure an asymptotically sharp lower bound of the minimax risk similar to the one in Theorem 3.9? The answers to these and related questions comprise an essential part of modern mathematical statistics. Below we present some ideas that stay within the scope of this book.

Assume that a statistical experiment is regular in the sense that the Fisher information $I(\theta)$ is positive, continuous and bounded in Θ. Let θ and $\theta + t/\sqrt{n}$ belong to Θ for all t in some compact set. If the Taylor expansion below is legitimate, we obtain that

$$\Delta L_n\big(\theta,\theta+t/\sqrt{I_n(\theta)}\big) \,=\, L_n\big(\theta+t/\sqrt{I_n(\theta)}\,\big) \,-\, L_n(\theta)$$

$$= \sum_{i=1}^{n}\Big[\,l\big(X_i,\theta+t/\sqrt{nI(\theta)}\,\big) \,-\, l(X_i,\theta)\Big]$$

$$= \frac{t}{\sqrt{nI(\theta)}}\sum_{i=1}^{n} l'(X_i,\theta) \,+\, \frac{t^2}{2nI(\theta)}\sum_{i=1}^{n} l''(X_i,\theta)\,\big(1+o_n(1)\big) \tag{3.15}$$

where $o_n(1) \to 0$ as $n \to \infty$.

By the computational formula for the Fisher information (see Exercise 1.1),

$$\mathbb{E}_\theta\Big[\sum_{i=1}^{n} l''(X_i\,,\,\theta)\Big] \;=\; -\,nI(\theta),$$

and therefore, the Law of Large Numbers ensures the convergence of the second term in (3.15),

$$\frac{t^2}{2nI(\theta)}\sum_{i=1}^{n} l''(X_i,\theta)\big(1\,+\,o_n(1)\big) \to -\frac{t^2}{2} \quad \text{as } n\to\infty.$$

Thus, at the heuristic level of understanding, we can expect that for any $t \in \mathbb{R}$, the log-likelihood ratio satisfies

$$\Delta L_n\big(\theta, \theta + t/\sqrt{I_n(\theta)}\,\big) \;=\; z_n(\theta)\,t \;-\; t^2/2 \,+\, \varepsilon_n(\theta, t) \tag{3.16}$$

where, by the Central Limit Theorem, the random variable

$$z_n(\theta) \;=\; \frac{1}{\sqrt{nI(\theta)}}\sum_{i=1}^{n} l'(X_i,\theta)$$

converges in $\mathbb{P}_\theta$-distribution to a standard normal random variable $z(\theta)$, and for any $\delta > 0$,

$$\lim_{n\to\infty} \mathbb{P}_\theta\big(\,|\,\varepsilon_n(\theta,t)\,| \geq \delta\,\big) \;=\; 0. \tag{3.17}$$

A family of distributions for which the log-likelihood ratio has the representation (3.16) under constraint (3.17) is said to satisfy the *local asymptotic normality (LAN) condition*. It can actually be derived under less restrictive assumptions. In particular, we do not need to require the existence of the second derivative l''.

To generalize Theorem 3.9 to the distributions satisfying the LAN condition, we need to justify that the remainder term $\varepsilon_n(\theta, t)$ may be ignored in the expression for the likelihood ratio,

$$\exp\big\{\,\Delta L_n\big(\theta, \theta + t/\sqrt{I_n(\theta)}\,\big)\,\big\} \;\approx\; \exp\big\{\,z_n(\theta)\,t \;-\; t^2/2\,\big\}.$$

To do this, we have to guarantee that the following approximation holds, as $n \to \infty$,

$$\mathbb{E}_\theta\Big[\Big|\,\exp\big\{\,z_n(\theta)\,t - t^2/2 + \varepsilon_n(\theta,t)\,\big\} - \exp\big\{\,z_n(\theta)\,t - t^2/2\,\big\}\Big|\Big] \to 0. \tag{3.18}$$

Unfortunately, the condition (3.17) that warrants that $\varepsilon_n(\theta, t)$ vanishes in probability does not imply (3.18). The remedy comes in the form of LeCam's theorem stated below.

Theorem 3.11. *Under the LAN conditions* (3.16) *and* (3.17), *there exists a sequence of random variables* $\tilde{z}_n(\theta)$ *such that* $|\, z_n(\theta) - \tilde{z}_n(\theta)\,| \to 0$ *in* $\mathbb{P}_\theta$*-probability, as* $n \to \infty$, *and for any* $c > 0$,

$$\lim_{n\to\infty} \sup_{-c\le t\le c} \mathbb{E}_\theta\Big[\Big|\exp\big\{z_n(\theta)t - t^2/2 + \varepsilon_n(\theta,t)\big\} - \exp\big\{\tilde{z}_n(\theta)t - t^2/2\big\}\Big|\Big] = 0.$$

To ease the proof, we split it into lemmas proved as the technical results below.

Lemma 3.12. *Under the LAN condition* (3.16), *there exists a truncation of* $z_n(\theta)$ *defined by*

$$\tilde{z}_n(\theta) = z_n(\theta)\,\mathbb{I}(z_n(\theta) \le c_n),$$

with the properly chosen sequence of constants c_n, *such that the following equations hold:*

$$\tilde{z}_n(\theta) - z_n(\theta) \to 0 \text{ as } n \to \infty \tag{3.19}$$

and

$$\lim_{n\to\infty} \sup_{-c\le t\le c} \mathbb{E}_\theta\Big[\Big|\exp\big\{\tilde{z}_n(\theta)\,t - t^2/2\big\} - 1\Big|\Big] = 0. \tag{3.20}$$

Introduce the notations

$$\xi_n(t) = \exp\big\{z_n(\theta)\,t - t^2/2 + \varepsilon_n(\theta,t)\big\},$$

$$\tilde{\xi}_n(t) = \exp\big\{\tilde{z}_n(\theta)\,t - t^2/2\big\}, \text{ and } \xi(t) = \exp\big\{z(\theta)\,t - t^2/2\big\}$$

where $\tilde{z}_n(\theta)$ is as defined in Lemma 3.12, and $z(\theta)$ is a standard normal random variable.

Lemma 3.13. *Under the LAN condition* (3.16), *the tails of* $\xi_n(t)$ *and* $\tilde{\xi}_n(t)$ *are small, uniformly in* n *and* $t \in [-c, c]$, *in the sense that*

$$\lim_{A\to\infty} \sup_{n\ge 1} \sup_{-c\le t\le c} \mathbb{E}_\theta\Big[\xi_n(t)\,\mathbb{I}(\xi_n(t) > A)\Big]$$

$$= \lim_{A\to\infty} \sup_{n\ge 1} \sup_{-c\le t\le c} \mathbb{E}_\theta\Big[\tilde{\xi}_n(t)\,\mathbb{I}(\tilde{\xi}_n(t) > A)\Big] = 0. \tag{3.21}$$

Now we are in the position to prove the LeCam theorem.

Proof of Theorem 3.11. We have to show that for any $t \in [-c, c]$, the convergence takes place:

$$\lim_{n\to\infty} \mathbb{E}_\theta\Big[\,|\,\xi_n(t) - \tilde{\xi}_n(t)\,|\,\Big] = 0. \tag{3.22}$$

From the triangle inequality, we obtain that

$$\mathbb{E}_\theta\Big[\,|\,\xi_n(t) - \tilde{\xi}_n(t)\,|\,\Big] \le \mathbb{E}_\theta\Big[\Big|\xi_n(t)\,\mathbb{I}\big(\xi_n(t) \le A\big) - \xi(t)\,\mathbb{I}\big(\xi(t) \le A\big)\Big|\Big]$$
$$+\,\mathbb{E}_\theta\Big[\Big|\tilde{\xi}_n(t)\,\mathbb{I}\big(\tilde{\xi}_n(t) \le A\big) - \xi(t)\,\mathbb{I}\big(\xi(t) \le A\big)\Big|\Big]$$

$$(3.23) \qquad + \mathbb{E}_\theta\Big[\, \xi_n(t)\, \mathbb{I}\big(\, \xi_n(t) > A\,\big)\Big] \,+\, \mathbb{E}_\theta\Big[\, \tilde{\xi}_n(t)\, \mathbb{I}\big(\, \tilde{\xi}_n(t) > A\,\big)\Big].$$

Due to Lemma 3.13, we can choose A so large that the last two terms do not exceed a however small positive δ. From Lemma 3.12, $\tilde{\xi}_n(t) - \xi(t) \to 0$ in $\mathbb{P}_\theta$-distribution, and by the LAN condition, $\xi_n(t) - \xi(t) \to 0$, therefore, for a fixed A, the first two terms on the right-hand side of (3.23) are vanishing uniformly over $t \in [-c, c]$ as $n \to \infty$. □

Finally, we formulate the result analogous to Theorem 3.9 (for the proof see Exercise 3.20).

Theorem 3.14. *If a statistical model satisfies the LAN condition* (3.16), *then for any loss function w, the asymptotic lower bound of the minimax risk holds:*

$$\liminf_{n\to\infty} \inf_{\hat{\theta}_n} \sup_{\theta\in\Theta} \mathbb{E}_\theta\Big[\, w\big(\, \sqrt{I_n(\theta)}\, (\hat{\theta}_n \,-\, \theta)\,\big)\Big] \,\geq\, \int_{-\infty}^{\infty} \frac{w(u)}{\sqrt{2\pi}}\, e^{-u^2/2}\, du.$$

3.5. The Hellinger Distance

Though this section may seem rather technical, it answers an important statistical question. Suppose that the statistical experiment with the family of densities $p(x, \theta)$ is such that $p(x, \theta_0) = p(x, \theta_1)$ for some $\theta_0 \neq \theta_1$, where $\theta_0, \theta_1 \in \Theta$, and all $x \in \mathbb{R}$. Clearly, no statistical observations can distinguish between θ_0 and θ_1 in this case.

Thus, we have to require that the family of probability densities $\{p(x, \theta)\}$ is such that for $\theta_0 \neq \theta_1$, the densities $p(\,\cdot\,, \theta_0)$ and $p(\,\cdot\,, \theta_1)$ are essentially different in some sense. How can the difference between these two densities be measured?

For any family of densities $\{p(x, \theta)\}$, the set $\{\sqrt{p(\,\cdot\,, \theta)},\ \theta \in \Theta\}$ presents a parametric curve on the surface of a unit sphere in L_2-space, the space of square integrable functions in x variable. Indeed, for any θ, the square of the L_2-norm is

$$\|\, \sqrt{p\,(\,\cdot\,, \theta)}\, \|_2^2 \,=\, \int_{\mathbb{R}} \Big(\, \sqrt{p(x, \theta)}\,\Big)^2 dx \,=\, \int_{\mathbb{R}} p(x, \theta)\, dx \,=\, 1.$$

The *Hellinger distance* $H(\theta_0, \theta_1)$ between $p(\,\cdot\,, \theta_0)$ and $p(\,\cdot\,, \theta_1)$ is defined as

$$(3.24) \qquad H(\theta_0, \theta_1) \,=\, \|\, \sqrt{p(\,\cdot\,, \theta_0)} \,-\, \sqrt{p(\,\cdot\,, \theta_1)}\, \|_2^2, \quad \theta_0,\, \theta_1 \in \Theta.$$

Lemma 3.15. *For the Hellinger distance* (3.24), *the following identities hold:*

$$(i)\ \ H(\theta_0, \theta_1) \,=\, 2\Big(1 \,-\, \int_{\mathbb{R}} \sqrt{p(x, \theta_0)\, p(x, \theta_1)}\, dx\Big)$$

and

$$(ii)\ \ \mathbb{E}_{\theta_0}\Big[\, \sqrt{Z_1(\theta_0, \theta_1)}\,\Big] \,=\, 1 \,-\, \frac{1}{2}\, H(\theta_0, \theta_1)$$

where $Z_1(\theta_0,\theta_1) = p(X,\theta_1)/p(X,\theta_0)$ *denotes the likelihood ratio for a single observation.*

Proof. (i) We write by definition

$$H(\theta_0,\theta_1) = \int_{\mathbb{R}} \Big(\sqrt{p(x,\theta_0)} - \sqrt{p(x,\theta_1)} \Big)^2 dx$$

$$= \int_{\mathbb{R}} p(x,\theta_0)\,dx + \int_{\mathbb{R}} p(x,\theta_1)\,dx - 2\int_{\mathbb{R}} \sqrt{p(x,\theta_0)\,p(x,\theta_1)}\,dx$$

$$= 2\Big(1 - \int_{\mathbb{R}} \sqrt{p(x,\theta_0)\,p(x,\theta_1)}\,dx\Big).$$

(ii) By definition of $Z_1(\theta_0,\theta_1)$, we have

$$\mathbb{E}_{\theta_0}\Big[\sqrt{Z_1(\theta_0,\theta_1)}\Big] = \int_{\mathbb{R}} \sqrt{\frac{p(x,\theta_1)}{p(x,\theta_0)}}\, p(x,\theta_0)\,dx$$

$$= \int_{\mathbb{R}} \sqrt{p(x,\theta_0)\,p(x,\theta_1)}\,dx = 1 - \frac{1}{2}\,H(\theta_0,\theta_1)$$

where the result of part (i) is applied. □

Lemma 3.16. *Let* $Z_n(\theta_0,\theta_1)$ *be the likelihood ratio for a sample of size* n,

$$Z_n(\theta_0,\theta_1) = \prod_{i=1}^{n} \frac{p(X_i,\theta_1)}{p(X_i,\theta_0)}, \quad \theta_0,\ \theta_1 \in \Theta.$$

Then the following identity is true:

$$\mathbb{E}_{\theta_0}\Big[\sqrt{Z_n(\theta_0,\theta_1)}\Big] = \Big(1 - \frac{1}{2}\,H(\theta_0,\theta_1)\Big)^n.$$

Proof. In view of independence of observations and Lemma 3.15 (ii), we have

$$\mathbb{E}_{\theta_0}\Big[\sqrt{Z_n(\theta_0,\theta_1)}\Big] = \prod_{i=1}^{n} \mathbb{E}_{\theta_0}\Big[\sqrt{\frac{p(X_i,\theta_1)}{p(X_i,\theta_0)}}\Big]$$

$$= \Big(1 - \frac{1}{2}\,H(\theta_0,\theta_1)\Big)^n.$$ □

Assumption 3.17. There exists a constant $a > 0$ such that for any $\theta_0,\ \theta_1 \in \Theta \subseteq \mathbb{R}$, the inequality $H(\theta_0,\ \theta_1) \geq a\,(\theta_0 - \theta_1)^2$ holds. □

Example 3.18. If X_i's are independent $\mathcal{N}(0,\sigma^2)$ random variables, then by Lemma 3.15 (i),

$$H(\theta_0,\theta_1) = 2\Big(1 - \frac{1}{\sqrt{2\pi\sigma^2}} \int_{-\infty}^{\infty} \exp\Big\{ -\frac{(x-\theta_0)^2}{4\sigma^2} - \frac{(x-\theta_1)^2}{4\sigma^2} \Big\}\,dx\Big)$$

$$= 2\Big(1 - \exp\Big\{ -\frac{(\theta_0-\theta_1)^2}{8\sigma^2} \Big\} \frac{1}{\sqrt{2\pi\sigma^2}} \int_{-\infty}^{\infty} \exp\Big\{ -\frac{(x-\bar{\theta})^2}{2\sigma^2} \Big\}\,dx\Big)$$

where $\bar{\theta} = (\theta_0 + \theta_1)/2$. As the integral of the probability density, the latter one equals 1. Therefore,

$$H(\theta_0, \theta_1) = 2\Big(1 - \exp\Big\{-\frac{(\theta_0 - \theta_1)^2}{8\sigma^2}\Big\}\Big).$$

If Θ is a bounded interval, then $(\theta_0 - \theta_1)^2/(8\sigma^2) \le C$ with some constant $C > 0$. In this case,

$$H(\theta_0, \theta_1) \ge a\,(\theta_0 - \theta_1)^2, \quad a = (1 - e^{-C})/(4\,C\,\sigma^2),$$

where we used the inequality $(1 - e^{-x}) \ge (1 - e^{-C})\,x/C$ if $0 \le x \le C$. □

3.6. Maximum Likelihood Estimator

In this section we study regular statistical experiments, which have continuous, bounded, and strictly positive Fisher information $I(\theta)$.

We call θ_n^* an *asymptotically minimax estimator*, if for any loss function w, and all sufficiently large n, the following inequality holds:

$$\sup_{\theta \in \Theta} \mathbb{E}_\theta\Big[\,w\Big(\frac{\theta_n^* - \theta}{\psi_n}\Big)\Big] \le r^* < \infty$$

where ψ_n and r^* are as in (3.7).

Recall from Section 1.1 that an estimator θ_n^* is called the maximum likelihood estimator (MLE), if for any $\theta \in \Theta$,

$$L_n(\theta_n^*) \ge L_n(\theta).$$

It turns out that Assumption 3.17 guarantees the asymptotic minimax property of the MLE with $\psi_n = 1/\sqrt{nI(\theta)}$. This result is proved in Theorem 3.20. We start with a lemma, the proof of which is postponed until the next section.

Lemma 3.19. *Under Assumption 3.17, for any $\theta \in \Theta$ and any $c > 0$, the MLE θ_n^* satisfies the inequality*

$$\mathbb{P}_\theta\big(\sqrt{n}\,|\theta_n^* - \theta| \ge c\big) \le C\exp\big\{-a\,c^2/4\big\}$$

where the constant $C = 2 + 3\sqrt{\pi I^*/a}$ with $I^* = \sup_{\theta \in \Theta} I(\theta) < \infty$.

At this point we are ready to prove the asymptotic minimaxity of the MLE.

Theorem 3.20. *Under Assumption 3.17, the MLE is asymptotically minimax. That is, for any loss function w and for any $\theta \in \Theta$, the normalized risk function of the MLE is finite,*

$$\limsup_{n \to \infty} \mathbb{E}_\theta\Big[\,w\big(\sqrt{nI(\theta)}\,(\theta_n^* - \theta)\big)\Big] = r^* < \infty.$$

Proof. Since $w(u)$ is an increasing function for $u \geq 0$, we have

$$\mathbb{E}_\theta\Big[\, w\big(\sqrt{nI(\theta)}\,(\theta_n^* - \theta)\,\big)\,\Big]$$

$$\leq \sum_{m=0}^{\infty} w(m+1)\,\mathbb{P}_\theta\Big(m \leq \sqrt{nI(\theta)}\,|\theta_n^* - \theta| \leq m+1\Big)$$

$$\leq \sum_{m=0}^{\infty} w(m+1)\,\mathbb{P}_\theta\Big(\sqrt{n}\,|\theta_n^* - \theta| \geq m/\sqrt{I(\theta)}\,\Big).$$

By definition, the loss w is bounded from above by a power function, while the probabilities decrease exponentially fast by Lemma 3.19. Therefore, the latter sum is finite. □

To find the sharp upper bound for the MLE, we make an additional assumption that allows us to prove a relatively simple result. As shown in the next theorem, the normalized deviation of the MLE from the true value of the parameter, $\sqrt{nI(\theta)}\,(\theta_n^* - \theta\,)$, converges in distribution to a standard normal random variable. Note that this result is sufficient to claim the asymptotically sharp minimax property for all bounded loss functions.

Theorem 3.21. *Let Assumption 3.17 and the LAN condition* (3.16) *hold. Moreover, suppose that for any $\delta > 0$ and any $c > 0$, the remainder term in* (3.16) *satisfies the equality:*

$$\lim_{n\to\infty} \sup_{\theta\in\Theta} \mathbb{P}_\theta\Big(\sup_{-c\leq t\leq c} |\,\varepsilon_n(\theta,t)\,| \geq \delta\,\Big) = 0. \tag{3.25}$$

Then for any $x \in \mathbb{R}$, uniformly in $\theta \in \Theta$, the MLE satisfies the limiting equation:

$$\lim_{n\to\infty} \mathbb{P}_\theta\big(\sqrt{n\,I(\theta)}\,(\theta_n^* - \theta\,) \leq x\,\big) = \Phi(x)$$

where Φ denotes the standard normal cumulative distribution function.

Proof. Fix a large c such that $c > |\,x\,|$, and a small $\delta > 0$. Put

$$t_n^* = \sqrt{n\,I(\theta)}\,(\theta_n^* - \theta\,).$$

Define two random events

$$\mathcal{A}_n = \mathcal{A}_n(c,\delta) = \Big\{\sup_{-2c\leq t\leq 2c} |\,\varepsilon_n(\theta,t)\,| \geq \delta\,\Big\}$$

and

$$\mathcal{B}_n = \mathcal{B}_n(c) = \big\{\,|\,t_n^*\,| \geq c\,\big\}.$$

Note that under the condition (3.25), we have that $\mathbb{P}_\theta(\mathcal{A}_n) \to 0$ as $n \to \infty$. Besides, as follows from the Markov inequality and Theorem 3.20 with $w(u) = |u|$,

$$\mathbb{P}_\theta(\mathcal{B}_n) \leq \mathbb{E}_\theta\big[\,|\,t_n^*\,|\,\big]/c \leq r^*/c.$$

Let $\overline{\mathcal{A}}_n$ and $\overline{\mathcal{B}}_n$ denote the complements of the events $\mathcal{A}$ and $\mathcal{B}$, respectively. We will use the following inclusion (for the proof see Exercise 3.21)

$$\overline{\mathcal{A}}_n \cap \overline{\mathcal{B}}_n \subseteq \left\{ \, | \, t_n^* - z_n(\theta) \, | \leq 2\sqrt{\delta} \, \right\} \tag{3.26}$$

or, equivalently,

$$\left\{ \, | \, t_n^* - z_n(\theta) \, | > 2\sqrt{\delta} \, \right\} \subseteq \mathcal{A}_n \cup \mathcal{B}_n \tag{3.27}$$

where $z_n(\theta)$ is defined in (3.16). Elementary inequalities and (3.26) imply that

$$\begin{aligned} \mathbb{P}_\theta\big(t_n^* \leq x \big) &\leq \mathbb{P}_\theta\big(\{t_n^* \leq x\} \cap \overline{\mathcal{A}}_n \cap \overline{\mathcal{B}}_n \big) + \mathbb{P}_\theta\big(\mathcal{A}_n \big) + \mathbb{P}_\theta\big(\mathcal{B}_n \big) \\ &\leq \mathbb{P}_\theta\big(z_n(\theta) \leq x + 2\sqrt{\delta} \big) + \mathbb{P}_\theta\big(\mathcal{A}_n \big) + \mathbb{P}_\theta\big(\mathcal{B}_n \big). \end{aligned}$$

Taking the limit as $n \to \infty$, we obtain that

$$\limsup_{n\to\infty} \mathbb{P}_\theta\big(t_n^* \leq x \big) \leq \Phi(x + 2\sqrt{\delta}) + r^*/c \tag{3.28}$$

where we use the fact that $z_n(\theta)$ is asymptotically standard normal. Next,

$$\begin{aligned} \mathbb{P}_\theta\big(t_n^* \leq x \big) &\geq \mathbb{P}_\theta\big(\{ t_n^* \leq x \} \cap \{ \, | \, t_n^* - z_n(\theta) \, | \leq 2\sqrt{\delta} \, \} \big) \\ &\geq \mathbb{P}_\theta\big(\{ z_n(\theta) \leq x - 2\sqrt{\delta} \} \cap \{ \, | \, t_n^* - z_n(\theta) \, | \leq 2\sqrt{\delta} \, \} \big) \\ &\geq \mathbb{P}_\theta\big(z_n(\theta) \leq x - 2\sqrt{\delta} \big) - \mathbb{P}_\theta\big(\, | \, t_n^* - z_n(\theta) \, | > 2\sqrt{\delta} \big) \\ &\geq \mathbb{P}_\theta\big(z_n(\theta) \leq x - 2\sqrt{\delta} \big) - \mathbb{P}_\theta\big(\mathcal{A}_n \big) - \mathbb{P}_\theta\big(\mathcal{B}_n \big) \end{aligned}$$

where at the last stage we have applied (3.27). Again, taking $n \to \infty$, we have that

$$\liminf_{n\to\infty} \mathbb{P}_\theta\big(t_n^* \leq x \big) \geq \Phi(x - 2\sqrt{\delta}) - r^*/c. \tag{3.29}$$

Now we combine (3.28) and (3.29) and take into account that c is however large and δ is arbitrarily small. Thus, the theorem follows. □

3.7. Proofs of Technical Lemmas

Proof of Lemma 3.12. Define $\tilde{z}_n(\theta, A) = z_n(\theta) \, \mathbb{I}(z_n(\theta) \leq A)$ where A is a large positive constant. Note that $z_n(\theta, A)$ converges in distribution as n increases to $z(\theta, A) = z(\theta) \, \mathbb{I}(z(\theta) \leq A)$ with a standard normal $z(\theta)$. Thus, for any k, $k = 1, 2, \dots$, we can find a constant A_k and an integer n_k so large that for all $n \geq n_k$,

$$\sup_{-c \leq t \leq c} \mathbb{E}_\theta\Big[\Big| \exp\big\{ \tilde{z}_n(\theta, A_k) \, t - t^2/2 \big\} - 1 \Big| \Big] \leq 1/k.$$

Without loss of generality, we can assume that n_k is an increasing sequence, $n_k \to \infty$ as $k \to \infty$. Finally, put $c_n = A_k$ if $n_k \leq n < n_{k+1}$. From this definition, (3.19) and (3.20) follow. □

Proof of Lemma 3.13. First we will prove (3.21) for $\tilde{\xi}_n(t)$. Note that $\tilde{\xi}_n(t)$, $n = 1, \dots,$ are positive random variables. By Lemma 3.12, for any $t \in [-c, c]$, the convergence takes place

$$\tilde{\xi}_n(t) \to \xi(t) \text{ as } n \to \infty. \tag{3.30}$$

Hence, the expected value of $\tilde{\xi}_n(t)$ converges as $n \to \infty$,

$$\sup_{-c \le t \le c} \left| \mathbb{E}_\theta\big[\,\tilde{\xi}_n(t)\,\big] - \mathbb{E}_\theta\big[\,\xi(t)\,\big] \right| = \sup_{-c \le t \le c} \left| \mathbb{E}_\theta\big[\,\tilde{\xi}_n(t)\,\big] - 1 \right| \to 0. \tag{3.31}$$

Choose an arbitrarily small $\delta > 0$. There exists $A(\delta)$ such that uniformly over $t \in [-c, c]$, the following inequality holds:

$$\mathbb{E}_\theta\Big[\,\xi(t)\,\mathbb{I}\big(\,\xi(t) > A(\delta)\,\big)\,\Big] \le \delta. \tag{3.32}$$

Next, we can choose $n = n(\delta)$ so large that for any $n \ge n(\delta)$ and all $t \in [-c, c]$, the following inequalities are satisfied:

$$\left| \mathbb{E}_\theta\big[\,\tilde{\xi}_n(t)\,\big] - \mathbb{E}_\theta\big[\,\xi(t)\,\big] \right| \le \delta \tag{3.33}$$

and

$$\Big| \mathbb{E}_\theta\big[\,\tilde{\xi}_n(t)\,\mathbb{I}\big(\,\tilde{\xi}_n(t) \le A(\delta)\,\big)\,\big] - \mathbb{E}_\theta\big[\,\xi(t)\,\mathbb{I}\big(\,\xi(t) \le A(\delta)\,\big)\,\big] \Big| \le \delta. \tag{3.34}$$

To see that the latter inequality holds, use the fact that $A(\delta)$ is fixed and $\tilde{\xi}_n(t) \to \xi(t)$ as $n \to \infty$.

The triangle inequality and the inequalities (3.32)-(3.34) imply that for any $A \ge A(\delta)$,

$$\begin{aligned}
\mathbb{E}_\theta\Big[\,\tilde{\xi}_n(t)\,\mathbb{I}\big(\,\tilde{\xi}_n(t) > A\,\big)\Big] &\le \mathbb{E}_\theta\Big[\,\tilde{\xi}_n(t)\,\mathbb{I}\big(\,\tilde{\xi}_n(t) > A(\delta)\,\big)\Big] \\
&= \mathbb{E}_\theta\Big[\,\tilde{\xi}_n(t) - \tilde{\xi}_n(t)\,\mathbb{I}\big(\,\tilde{\xi}_n(t) \le A(\delta)\,\big)\Big] \\
- \mathbb{E}_\theta\Big[\,\xi(t) - \xi(t)\,\mathbb{I}\big(\,\xi(t) \le A(\delta)\,\big) &- \xi(t)\,\mathbb{I}\big(\,\xi(t) > A(\delta)\,\big)\Big] \\
&\le \left| \mathbb{E}_\theta\big[\,\tilde{\xi}_n(t)\,\big] - \mathbb{E}_\theta\big[\,\xi(t)\,\big] \right| \\
+ \Big| \mathbb{E}_\theta\Big[\,\tilde{\xi}_n(t)\,\mathbb{I}\big(\,\xi_n(t) \le A(\delta)\,\big)\Big] &- \mathbb{E}_\theta\Big[\,\xi(t)\,\mathbb{I}\big(\,\xi(t) \le A(\delta)\,\big)\Big] \Big| \\
+ \mathbb{E}_\theta\Big[\,\xi(t)\,\mathbb{I}\big(\,\xi(t) > A(\delta)\,\big)\Big] &\le 3\delta.
\end{aligned} \tag{3.35}$$

There are finitely many n such that $n \le n(\delta)$. For each $n \le n(\delta)$, we can find A_n so large that for all $A \ge A_n$, the following expected value is bounded: $\mathbb{E}_\theta\Big[\,\tilde{\xi}_n(t)\,\mathbb{I}\big(\,\tilde{\xi}_n(t) > A\,\big)\Big] \le 3\delta$. Put $A_0 = \max\big(\,A_1, \dots, A_{n(\delta)}, A(\delta)\,\big)$. By definition, for any $A \ge A_0$, and all $t \in [-c, c]$, we have that

$$\sup_{n \ge 1} \mathbb{E}_\theta\Big[\,\tilde{\xi}_n(t)\,\mathbb{I}\big(\,\tilde{\xi}_n(t) > A\,\big)\Big] \le 3\delta. \tag{3.36}$$

Thus, the lemma follows for $\tilde{\xi}_n(t)$.

The proof for $\xi_n(t)$ is simpler. Similarly to $\tilde{\xi}_n(t)$, the random variables $\xi_n(t)$, $n = 1, \dots,$ are positive, and the convergence analogous to (3.30) is valid from the LAN condition, $\xi_n(t) \to \xi(t)$ as $n \to \infty$. But since $\mathbb{E}_\theta\big[\,\xi_n(t)\,\big] = 1$, the convergence (3.31) of the expected values is replaced by exact equality, $\big|\,\mathbb{E}_\theta\big[\,\tilde{\xi}_n(t)\,\big] - \mathbb{E}_\theta\big[\,\xi(t)\,\big]\,\big| = 0$. Therefore, (3.35) and (3.36) hold for $\xi_n(t)$ with the upper bound replaced by 2δ, and the result of the lemma follows. □

Proof of Lemma 3.19. The proof of this lemma (and Theorem 3.20) is due to A.I. Sakhanenko (cf. Borovkov [**Bor99**], Chapter 2, §23). The proof is based on two results which we state and prove below.

Introduce the likelihood ratio

$$Z_n(\theta, \theta + t) = \prod_{i=1}^{n} \frac{p(X_i, \theta + t)}{p(X_i, \theta)}.$$

Result 1. *Put* $\mathfrak{z}_n(t) = \big[Z_n(\theta, \theta + t)\big]^{3/4}$. *Under Assumption 3.17, for any* $\theta,\ \theta + t \in \Theta$, *the following inequalities hold:*

$$\mathbb{E}_\theta\Big[\sqrt{Z_n(\theta, \theta + t)}\,\Big] \le \exp\big\{ -a\,n\,t^2/2\,\big\}, \tag{3.37}$$

$$\mathbb{E}_\theta\big[\,\mathfrak{z}_n(t)\,\big] \le \exp\big\{ -a\,n\,t^2/4\,\big\}, \tag{3.38}$$

and

$$\mathbb{E}_\theta\big[\,\mathfrak{z}_n'(t)\,\big] \le \frac{3}{4}\sqrt{I_n(\theta + t)}\,\exp\big\{ -a\,n\,t^2/4\,\big\} \tag{3.39}$$

where $\mathfrak{z}_n'(t) = d\mathfrak{z}_n(t)/dt$.

Proof. From Lemma 3.16 and Assumption 3.17, we obtain (3.37),

$$\mathbb{E}_\theta\Big[\sqrt{Z_n(\theta, \theta + t)}\,\Big] = \Big(1 - \frac{1}{2}\,H(\theta_0, \theta_1)\Big)^n$$
$$\le \exp\Big\{ -\frac{n}{2}\,H(\theta_0, \theta_1)\Big\} \le \exp\big\{ -a\,n\,t^2/2\,\big\}.$$

To prove (3.38), we use the Cauchy-Schwarz inequality and (3.37),

$$\mathbb{E}_\theta\Big[\,\mathfrak{z}_n(t)\Big] = \mathbb{E}_\theta\Big[\,\big(\,Z_n(\theta,\,\theta + t)\,\big)^{3/4}\Big]$$
$$= \mathbb{E}_\theta\Big[\,\big(Z_n(\theta,\,\theta + t)\big)^{1/2}\,\big(Z_n(\theta,\,\theta + t)\big)^{1/4}\Big]$$
$$\le \Big(\mathbb{E}_\theta\Big[\,Z_n(\theta, \theta + t)\,\Big]\Big)^{1/2}\,\Big(\mathbb{E}_\theta\Big[\,\big(\,Z_n(\theta, \theta + t)\,\big)^{1/2}\Big]\Big)^{1/2}$$
$$= \Big(\mathbb{E}_\theta\Big[\,\big(\,Z_n(\theta, \theta + t)\,\big)^{1/2}\Big]\Big)^{1/2} \le \exp\big\{ -a\,n\,t^2/4\,\big\}.$$

Here we used the identity (show!) $\mathbb{E}_\theta\big[\,Z_n(\theta, \theta + t)\,\big] = 1$.

The proof of (3.39) requires more calculations. We write

$$\mathbb{E}_\theta\Big[\mathfrak{z}_n'(t)\Big] = \mathbb{E}_\theta\Big[\frac{d}{dt}\exp\Big\{\frac{3}{4}\big(L_n(\theta+t) - L_n(\theta)\big)\Big\}\Big]$$

$$= \mathbb{E}_\theta\Big[\frac{3}{4}L_n'(\theta+t)\,\big(Z_n(\theta,\theta+t)\big)^{3/4}\Big]$$

$$= \frac{3}{4}\mathbb{E}_\theta\Big[\Big(L_n'(\theta+t)\sqrt{Z_n(\theta,\theta+t)}\Big)\big(Z_n(\theta,\theta+t)\big)^{1/4}\Big]$$

$$\le \frac{3}{4}\Big(\mathbb{E}_\theta\Big[\big(L_n'(\theta+t)\big)^2 Z_n(\theta,\theta+t)\Big]\Big)^{1/2}\Big(\mathbb{E}_\theta\Big[\big(Z_n(\theta,\theta+t)\big)^{1/2}\Big]\Big)^{1/2}$$

$$= \frac{3}{4}\Big(\mathbb{E}_{\theta+t}\Big[\big(L_n'(\theta+t)\big)^2\Big]\Big)^{1/2}\Big(\mathbb{E}_\theta\Big[\big(Z_n(\theta,\theta+t)\big)^{1/2}\Big]\Big)^{1/2} \qquad \text{(by (3.8))}$$

$$\le \frac{3}{4}\sqrt{I_n(\theta+t)}\,\exp\big\{-a\,n\,t^2/4\big\}.$$

The last inequality sign is justified by the definition of the Fisher information, and (3.37). □

Result 2. *Let Assumption 3.17 be true. Then for any positive constants γ and c, the following inequality holds:*

$$\mathbb{P}_\theta\Big(\sup_{|t|\ge c/\sqrt{n}} Z_n(\theta,\theta+t) \ge e^\gamma\Big) \le C\,e^{-3\gamma/4}\exp\{-a\,c^2/4\}$$

where $C = 2 + 3\sqrt{\pi I^*/a}$ *with* $I^* = \sup_{\theta\in\Theta} I(\theta) < \infty$.

Proof. Consider the case $t > 0$. Note that

$$\sup_{t\ge c/\sqrt{n}} \mathfrak{z}_n(t) = \mathfrak{z}_n(c/\sqrt{n}) + \sup_{t> c/\sqrt{n}}\int_{c/\sqrt{n}}^t \mathfrak{z}_n'(u)\,du$$

$$\le \mathfrak{z}_n(c/\sqrt{n}) + \sup_{t> c/\sqrt{n}}\int_{c/\sqrt{n}}^t |\mathfrak{z}_n'(u)|\,du$$

$$\le \mathfrak{z}_n(c/\sqrt{n}) + \int_{c/\sqrt{n}}^\infty |\mathfrak{z}_n'(u)|\,du.$$

Applying Result 1, we find that

$$\mathbb{E}_\theta\Big[\sup_{t\ge c/\sqrt{n}}\mathfrak{z}_n(t)\Big] \le \exp\{-a\,c^2/4\} + \frac{3}{4}\sqrt{n\,I^*}\int_{c/\sqrt{n}}^\infty \exp\{-a\,n\,u^2/4\}\,du$$

$$= \exp\{-a\,c^2/4\} + \frac{3}{4}\sqrt{n\,I^*}\sqrt{\frac{4\pi}{a\,n}}\int_{c\sqrt{a/2}}^\infty \frac{1}{\sqrt{2\pi}}e^{-u^2/2}\,du$$

$$= \exp\{-a\,c^2/4\} + \frac{3}{2}\sqrt{\frac{\pi I^*}{a}}\Big[1 - \Phi(c\sqrt{a/2})\Big]$$

where Φ denotes the cumulative distribution function of the standard normal random variable. The inequality $1 - \Phi(x) \leq e^{-x^2/2}$ yields

$$\exp\big\{ - a\, c^2/4 \big\} + \frac{3}{2}\sqrt{\frac{\pi I^*}{a}}\,\Big[1 - \Phi(c\sqrt{a/2}) \Big]$$

$$\leq \Big(1 + \frac{3}{2}\sqrt{\frac{\pi I^*}{a}} \Big) \exp\big\{ - a\, c^2/4 \big\} = \frac{1}{2} C \exp\big\{ - a\, c^2/4 \big\}.$$

The same inequality is true for $t < 0$ (show!),

$$\mathbb{E}_\theta\Big[\sup_{t \leq -c/\sqrt{n}} \mathfrak{z}_n(t) \Big] \leq \frac{1}{2} C \exp\big\{ - a\, c^2/4 \big\}.$$

Further,

$$\mathbb{P}_\theta\Big(\sup_{|t| \geq c/\sqrt{n}} Z_n(\theta, \theta + t) \geq e^\gamma \Big)$$

$$\leq \mathbb{P}_\theta\Big(\sup_{t \geq c/\sqrt{n}} Z_n(\theta, \theta + t) \geq e^\gamma \Big) + \mathbb{P}_\theta\Big(\sup_{t \leq -c/\sqrt{n}} Z_n(\theta, \theta + t) \geq e^\gamma \Big)$$

$$= \mathbb{P}_\theta\Big(\sup_{t \geq c/\sqrt{n}} \mathfrak{z}_n(t) \geq e^{3\gamma/4} \Big) + \mathbb{P}_\theta\Big(\sup_{t \leq -c/\sqrt{n}} \mathfrak{z}_n(t) \geq e^{3\gamma/4} \Big),$$

and the Markov inequality $\mathbb{P}(X > x) \leq \mathbb{E}[\,X\,]\,/\,x$ completes the proof,

$$\leq \frac{1}{2} C\, e^{-3\gamma/4} \exp\big\{ - a\, c^2/4 \big\} + \frac{1}{2} C\, e^{-3\gamma/4} \exp\big\{ - a\, c^2/4 \big\}$$

$$= C\, e^{-3\gamma/4} \exp\big\{ - a\, c^2/4 \big\}. \qquad \square$$

Now we are in the position to prove Lemma 3.19. Applying the inclusion

$$\Big\{ \sqrt{n}\, |\theta_n^* - \theta| \geq c \Big\} = \Big\{ \sup_{|t| \geq c/\sqrt{n}} Z_n(\theta, \theta + t) \geq \sup_{|t| < c/\sqrt{n}} Z_n(\theta, \theta + t) \Big\}$$

$$\subseteq \Big\{ \sup_{|t| \geq c/\sqrt{n}} Z_n(\theta, \theta + t) \geq Z_n(\theta, \theta) = 1 \Big\},$$

and using the Result 2 with $\gamma = 0$, we obtain

$$\mathbb{P}_\theta\Big(\sqrt{n}\, |\theta_n^* - \theta| \geq c \Big) \leq \mathbb{P}_\theta\Big(\sup_{|t| \geq c/\sqrt{n}} Z_n(\theta, \theta + t) \geq 1 \Big)$$

$$\leq C \exp\big\{ - a\, c^2/4 \big\}. \qquad \square$$

Exercises

Exercise 3.14. Verify that in Example 1.5, the estimator $\hat{\theta}_n = \sqrt{X/n}$ is an asymptotically unbiased estimator of θ. Hint: Note that $|\sqrt{X/n} - \theta| = |X/n - \theta^2| \,/\, |\sqrt{X/n} + \theta|$, and thus, $\mathbb{E}_\theta\big[\,|\sqrt{X/n} - \theta|\,\big] \le \theta^{-1}\mathbb{E}_\theta\big[\,|X/n - \theta^2|\,\big]$. Now use the Cauchy-Schwarz inequality to finish off the proof.

Exercise 3.15. Show that the Hodges estimator defined by (3.3) is asymptotically unbiased and satisfies the identities (3.4).

Exercise 3.16. Prove Theorem 3.7.

Exercise 3.17. Suppose the conditions of Theorem 3.7 hold, and a loss function w is such that $w(1/2) > 0$. Show that for any estimator $\hat{\theta}_n$ the following lower bound holds:

$$\sup_{\theta \in \Theta} \mathbb{E}_\theta\Big[\, w\big(\sqrt{n}\,(\hat{\theta}_n - \theta)\big)\,\Big] \ge \frac{1}{2}\, w(1/2)\, p_0\, \exp\{z_0\}.$$

Hint: Use Theorem 3.7 and the inequality (show!)

$$w\left(\sqrt{n}\,(\hat{\theta}_n - \theta)\right) + w\left(\sqrt{n}\,(\hat{\theta}_n - \theta) - 1\right) \ge w(1/2), \quad \text{for any } \theta \in \Theta.$$

Exercise 3.18. Prove (3.14). Hint: First show this result for bounded loss functions.

Exercise 3.19. Prove the local asymptotic normality (LAN) for

(i) exponential model with the density

$$p(x\,,\theta) \;=\; \theta\,\exp\{-\,\theta\, x\}\,, \quad x\,,\theta \,>\, 0;$$

(ii) Poisson model with the probability mass function

$$p(x\,,\theta) \;=\; \frac{\theta^{\,x}}{x!}\,\exp\{-\theta\}\,, \quad \theta > 0\,,\; x \in \{0\,,1\,,\dots\}\,.$$

Exercise 3.20. Prove Theorem 3.14. Hint: Start with a truncated loss function $w_C(u) = \min(w(u), C)$ for some $C > 0$. Applying Theorem 3.11, obtain an analogue of (3.9) of the form

$$\sup_{\theta \in \mathbb{R}} \mathbb{E}_\theta\Big[\, w_C\big(\sqrt{nI(\theta)}\,(\hat{\theta}_n - \theta)\big)\,\Big]$$

$$\ge \frac{1}{2b}\int_{-b}^{b} \mathbb{E}_0\Big[\, w_C\big(\,\sqrt{a_n}\,\hat{\theta}_n - t\,\big)\,\exp\big\{\,\tilde{z}_n(0)\,t - t^2/2\,\big\}\Big]\,dt \,+\, o_n(1)$$

where $a_n = nI\big(t/\sqrt{nI(0)}\big)$, $\tilde{z}_n(0)$ is an asymptotically standard normal random variable, and $o_n(1) \to 0$ as $n \to \infty$. Then follow the lines of Theorems 3.8 and 3.9, and, finally, let $C \to \infty$.

Exercise 3.21. Consider a distorted parabola $zt - t^2/2 + \varepsilon(t)$ where z has a fixed value and $-2c \le t \le 2c$. Assume that the maximum of this function is attained at a point t^* that lies within the interval $[-c, c]$. Suppose that the remainder term satisfies $\sup_{-2c \le t \le 2c} |\,\varepsilon(t)\,| \le \delta$. Show that $|\,t^* - z\,| \le 2\sqrt{\delta}$.

Chapter 4

Some Irregular Statistical Experiments

4.1. Irregular Models: Two Examples

As shown in the previous chapters, in regular models, for any estimator $\hat{\theta}_n$, the normalized deviation $\sqrt{nI(\theta)}\,(\hat{\theta}_n - \theta)$ either grows or stays bounded in the minimax sense, as n increases. In particular, we have shown that the quadratic risk $\mathbb{E}_\theta\big[\,(\hat{\theta}_n - \theta)^2\big]$ decreases not faster than at the rate $O(n^{-1})$ as $n \to \infty$. This result has been obtained under some regularity conditions. The easiest way to understand their importance is to look at some *irregular experiments* commonly used in statistics, for which the regularity conditions are violated and the quadratic risk converges faster than $O(n^{-1})$. We present two examples below.

Example 4.1. Suppose the observations $X_1, \dots, X_n$ come from the uniform distribution on $[\,0\,,\,\theta\,]$. The family of probability densities can be defined as $p(x\,,\theta) \;=\; \theta^{\,-1}\,\mathbb{I}\big(\,0\;\le\; x\;\le\;\theta\,\big)$. In this case, the MLE of θ is the maximum of all observations (see Exercise 4.22), that is, $\hat{\theta}_n \;=\; X_{(n)} \;=\; \max\,\big(X_1, \dots, X_n\big)$. The estimator

$$\theta_n^* \;=\; \frac{n+1}{n}\,X_{(n)}$$

is an unbiased estimator of θ with the variance

$$\mathbb{V}ar_\theta\,\big[\,\theta_n^*\,\big] \;=\; \frac{\theta^2}{n(n+2)} \;=\; O\big(n^{-2}\big)\;\text{ as }\; n\to\infty. \qquad \Box$$

Example 4.2. Consider a model with observations $X_1, \dots, X_n$ which have a shifted exponential distribution with the density

$$p(x\,,\theta) \;=\; e^{-\,(x-\theta)}\;\mathbb{I}(x \geq \theta),\;\; \theta \in \mathbb{R}.$$

It can be shown (see Exercise 4.23) that the MLE of θ is $\hat{\theta}_n \;=\; X_{(1)} \;=\; \min(X_1, \dots, X_n)$, and that $\theta_n^* \;=\; X_{(1)} \;-\; n^{-1}$ is an unbiased estimator of θ with the variance $\mathbb{V}ar_\theta\left[\,\theta_n^*\,\right] \;=\; n^{-2}$. □

The unbiased estimators in the above examples violate the Cramér-Rao lower bound (1.2) since their variances decrease faster than $O\big(n^{-1}\big)$. Why does it happen? In the next section we explain that in these examples the Fisher information does not exist, and therefore, the Cramér-Rao inequality is not applicable.

4.2. Criterion for Existence of the Fisher Information

For any probability density $p(x\,,\theta)$, consider the set $\left\{\,\sqrt{p(\,\cdot\,,\,\theta)},\;\theta \in \Theta\,\right\}$. It has been shown in Section 3.5 that for any fixed θ, $\sqrt{p(\,\cdot\,,\,\theta)}$ has a unit L_2-norm, that is,

$$\|\sqrt{p(\,\cdot\,,\,\theta)}\,\|_2^2 \;=\; \int_{\mathbb{R}}\Big(\,\sqrt{p(\,x\,,\theta)}\,\Big)^2\,dx \;=\; 1.$$

The existence of the Fisher information is equivalent to the smoothness of this curve as a function of θ. We show that the Fisher information exists if this curve is differentiable with respect to θ in the L_2-space.

Theorem 4.3. *The Fisher information is finite if and only if the L_2-norm of the derivative $\|\,\partial\sqrt{p(\,\cdot\,,\,\theta)}\,/\,\partial\theta\,\|_2$ is finite. The Fisher information is computed according to the formula*

$$I(\theta) \;=\; 4\,\|\,\frac{\partial}{\partial\theta}\,\sqrt{p(\,\cdot\,,\,\theta)}\,\|_2^2.$$

Proof. The proof is straightforward:

$$\|\,\frac{\partial}{\partial\theta}\,\sqrt{p(\,\cdot\,,\,\theta)}\,\|_2^2 \;=\; \int_{\mathbb{R}}\Big(\frac{\partial}{\partial\theta}\,\sqrt{p(\,x,\theta)}\Big)^2\,dx$$
$$=\;\int_{\mathbb{R}}\Big(\frac{\partial p(x,\theta)/\partial\theta}{2\,\sqrt{p(x,\theta)}}\Big)^2\,dx \;=\; \frac{1}{4}\int_{\mathbb{R}}\Big(\frac{\partial p(x,\theta)/\partial\theta}{p(x,\theta)}\Big)^2\,p(x,\theta)\,dx$$
$$=\;\frac{1}{4}\int_{\mathbb{R}}\Big(\frac{\partial\,\ln p(x,\theta)}{\partial\theta}\Big)^2\,p(x,\theta)\,dx \;=\; \frac{1}{4}\,I(\theta).$$
□

Example 4.4. The family of the uniform densities in Example 4.1 is not differentiable in the sense of Theorem 4.3. By definition,

$$\|\,\frac{\partial}{\partial\theta}\,\sqrt{p(\,\cdot\,,\theta)}\,\|_2^2 \;=\; \lim_{\Delta\theta\to 0}\,(\Delta\theta)^{-2}\,\|\,\sqrt{p(\,\cdot\,,\theta+\Delta\theta)}\;-\;\sqrt{p(\,\cdot\,,\theta)}\,\|_2^2$$

$$= \lim_{\Delta\theta \to 0} (\Delta\theta)^{-2} \, \Big\| \, \frac{1}{\sqrt{\theta + \Delta\theta}} \, \mathbb{I}_{[0,\,\theta+\Delta\theta]}(\,\cdot\,) \, - \, \frac{1}{\sqrt{\theta}} \, \mathbb{I}_{[0,\,\theta]}(\,\cdot\,) \, \Big\|_2^2.$$

A finite limit exists if and only if

$$\Big\| \, \frac{1}{\sqrt{\theta + \Delta\theta}} \, \mathbb{I}_{[0,\,\theta+\Delta\theta]}(\,\cdot\,) \, - \, \frac{1}{\sqrt{\theta}} \, \mathbb{I}_{[0,\,\theta]}(\,\cdot\,) \, \Big\|_2^2 = O\big((\Delta\theta)^2 \big) \text{ as } \Delta\theta \to 0.$$

However, the L_2-norm decreases at a lower rate. To see this, assume $\Delta\theta$ is positive and write

$$\Big\| \, \frac{1}{\sqrt{\theta + \Delta\theta}} \, \mathbb{I}_{[0,\,\theta+\Delta\theta]}(\,\cdot\,) \, - \, \frac{1}{\sqrt{\theta}} \, \mathbb{I}_{[0,\,\theta]}(\,\cdot\,) \, \Big\|_2^2$$

$$= \int_{\mathbb{R}} \Big[\frac{1}{\sqrt{\theta + \Delta\theta}} \, \mathbb{I}\big(0 \le x \le \theta + \Delta\theta\big) \, - \, \frac{1}{\sqrt{\theta}} \, \mathbb{I}\big(0 \le x \le \theta\big) \Big]^2 dx$$

$$= \int_0^{\theta} \Big(\frac{1}{\sqrt{\theta + \Delta\theta}} - \frac{1}{\sqrt{\theta}} \Big)^2 dx + \int_{\theta}^{\theta+\Delta\theta} \Big(\frac{1}{\sqrt{\theta + \Delta\theta}} \Big)^2 dx$$

$$= \frac{\big(\sqrt{\theta} - \sqrt{\theta + \Delta\theta} \big)^2}{\theta + \Delta\theta} + \frac{\Delta\theta}{\theta + \Delta\theta}$$

$$= 2 \Big(1 - \big(1 + \Delta\theta/\theta \big)^{-1/2} \Big) = \Delta\theta/\theta + o(\Delta\theta/\theta) \gg O\big((\Delta\theta)^2 \big) \text{ as } \Delta\theta \to 0.$$

Hence, in this example, $\sqrt{p(\,\cdot\,,\theta)}$ is not differentiable as a function of θ, and the finite Fisher information does not exist. □

A similar result is true for the shifted exponential model introduced in Example 4.2 (see Exercise 4.24).

If we formally write the Fisher information as $I(\theta) = \infty$, then the right-hand side of the Cramér-Rao inequality (1.2) becomes zero, and there is no contradiction with the faster rate of convergence.

4.3. Asymptotically Exponential Statistical Experiment

What do the two irregular models considered in the previous sections have in common? First of all, $\sqrt{p(\,\cdot\,,\,\theta)}$ is not differentiable in the sense of Theorem 4.3, and $\| \sqrt{p(\,\cdot\,,\,\theta + \Delta\theta)} - \sqrt{p(\,\cdot\,,\,\theta)} \|_2^2 = O(\Delta\theta)$ as $\Delta\theta \to 0$. For the uniform model, this fact is verified in Example 4.4, while for the shifted exponential distribution, it is assigned as Exercise 4.24.

Another feature that these models share is the limiting structure of the likelihood ratio

$$Z_n(\theta_0, \theta_1) = \exp\big\{ L_n(\theta_1) - L_n(\theta_0) \big\}$$

$$= \prod_{i=1}^{n} \frac{p(X_i, \theta_1)}{p(X_i, \theta_0)}, \quad \theta_0, \, \theta_1 \in \Theta.$$

A statistical experiment is called *asymptotically exponential* if for any $\theta \in \Theta$, there exists an asymptotically exponential random variable T_n such that

$$\lim_{n\to\infty} \mathbb{P}(T_n \geq \tau) = \exp\{-\lambda(\theta)\tau\}, \quad \tau > 0,$$

and either

$$\text{(i)}\ Z_n(\theta, \theta + t/n) = \exp\{-\lambda(\theta)\, t\}\, \mathbb{I}(t \geq -T_n) + o_n(1)$$

or

$$\text{(ii)}\ Z_n(\theta, \theta + t/n) = \exp\{\lambda(\theta)\, t\}\, \mathbb{I}(t \leq T_n) + o_n(1)$$

where $\lambda(\theta)$ is a continuous positive function of θ, $\theta \in \Theta$, and $o_n(1) \to 0$ in $\mathbb{P}_\theta$-probability as $n \to \infty$.

Both uniform and shifted exponential models are special cases of the asymptotically exponential statistical experiment, as stated in Propositions 4.5 and 4.6 below.

Proposition 4.5. *The uniform statistical experiment defined in Example 4.1 is asymptotically exponential with* $\lambda(\theta) = 1/\theta$.

Proof. The likelihood ratio for the uniform distribution is

$$Z_n(\theta, \theta + t/n) = \Big(\frac{\theta}{\theta + t/n}\Big)^n \prod_{i=1}^n \frac{\mathbb{I}(X_i \leq \theta + t/n)}{\mathbb{I}(X_i \leq \theta)}$$

$$= \Big(\frac{\theta}{\theta + t/n}\Big)^n \frac{\mathbb{I}(X_{(n)} \leq \theta + t/n)}{\mathbb{I}(X_{(n)} \leq \theta)}.$$

Note that the event $\{X_{(n)} \leq \theta\}$ holds with probability 1. Also,

$$\Big(\frac{\theta}{\theta + t/n}\Big)^n = \Big(1 + \frac{t/\theta}{n}\Big)^{-n} = \exp\{-t/\theta\} + o_n(1) \text{ as } n \to \infty$$

and

$$\mathbb{I}(X_{(n)} \leq \theta + t/n) = \mathbb{I}(t \geq -T_n) \text{ where } T_n = n(\theta - X_{(n)}).$$

It remains to show that T_n has a limiting exponential distribution. Indeed,

$$\lim_{n\to\infty} \mathbb{P}_\theta(T_n \geq \tau) = \lim_{n\to\infty} \mathbb{P}_\theta(n(\theta - X_{(n)}) \geq \tau)$$

$$= \lim_{n\to\infty} \mathbb{P}_\theta(X_{(n)} \leq \theta - \tau/n) = \lim_{n\to\infty} \Big(\frac{\theta - \tau/n}{\theta}\Big)^n = e^{-\tau/\theta}. \qquad \square$$

A similar argument proves the next proposition (see Exercise 4.26).

Proposition 4.6. *The shifted exponential statistical experiment defined in Example 4.2 is asymptotically exponential with* $\lambda(\theta) = 1$.

4.4. Minimax Rate of Convergence

In accordance with the definition (3.7), the estimators in Examples 4.1 and 4.2 have guaranteed rate of convergence $\psi_n = O(n^{-1})$. Can this rate be improved? That is, are there estimators that converge with faster rates? The answer is negative, and the proof is relatively easy.

Lemma 4.7. *In an asymptotically exponential statistical experiment, there exists a constant $r_* > 0$ not depending on n such that for any estimator $\hat{\theta}_n$, the following lower bound holds:*

$$\liminf_{n\to\infty} \sup_{\theta\in\Theta} \mathbb{E}_\theta\Big[\big(n\,(\hat{\theta}_n - \theta)\big)^2\Big] \geq r_*.$$

Proof. Take $\theta_0 \in \Theta$ and $\theta_1 = \theta_0 + 1/n \in \Theta$. Assume that property (ii) in the definition of an asymptotically exponential model holds. Then, as in the proof of Lemma 3.4, we have

$$\sup_{\theta\in\Theta} \mathbb{E}_\theta\Big[\big(n\,(\hat{\theta}_n - \theta)\big)^2\Big] \geq \max_{\theta\in\{\theta_0,\,\theta_1\}} \mathbb{E}_\theta\Big[\big(n\,(\hat{\theta}_n - \theta)\big)^2\Big]$$

$$\geq \frac{n^2}{2}\,\mathbb{E}_{\theta_0}\Big[(\hat{\theta}_n - \theta_0)^2 + (\hat{\theta}_n - \theta_1)^2\,e^{\lambda(\theta_0)+o_n(1)}\,\mathbb{I}(1 \leq T_n)\Big]$$

$$\geq \frac{n^2}{2}\,\mathbb{E}_{\theta_0}\Big[\Big((\hat{\theta}_n - \theta_0)^2 + (\hat{\theta}_n - \theta_1)^2\Big)\,\mathbb{I}(T_n \geq 1)\Big],$$

since $\lambda(\theta_0) + o_n(1) \geq 0$,

$$\geq \frac{n^2}{2}\,\frac{(\theta_1 - \theta_0)^2}{2}\,\mathbb{P}_{\theta_0}(T_n \geq 1)$$

$$= \frac{1}{4}\,\mathbb{P}_{\theta_0}(T_n \geq 1) \to \frac{1}{4}\,\exp\{-\lambda(\theta_0)\} \text{ as } n\to\infty. \qquad \square$$

Remark 4.8. The rate of convergence may be different from $O(n^{-1})$ for some other irregular statistical experiments, but those models are not asymptotically exponential. For instance, the model described in Exercise 1.8 is not regular (the Fisher information does not exist) if $-1 < \alpha \leq 1$. The rate of convergence in this model depends on α and is, generally speaking, different from $O(n^{-1})$. $\square$

4.5. Sharp Lower Bound

The constant $r_* = \frac{1}{4}\exp\{-\lambda(\theta_0)\}$ in the proof of Lemma 4.7 is far from being sharp. In the theorem that follows, we state a local version of the lower bound with an exact constant for an asymptotically exponential experiment.

Theorem 4.9. *Consider an asymptotically exponential statistical experiment. Assume that it satisfies property (ii) of the definition, and put* $\lambda_0 = \lambda(\theta_0)$. *Then for any* $\theta_0 \in \Theta$, *any loss function* w, *and any estimator* $\hat{\theta}_n$, *the following lower bound holds:*

$$\lim_{\delta \to 0} \liminf_{n \to \infty} \sup_{\theta\,:\,|\theta - \theta_0| < \delta} \mathbb{E}_\theta\Big[w\big(n\,(\hat{\theta}_n - \theta)\big)\Big] \geq \lambda_0 \min_{y \in \mathbb{R}} \int_0^\infty w(u-y)\, e^{-\lambda_0\, u}\, du.$$

Proof. Choose a large positive number b and assume that n is so large that $b < \delta\, n$. Put $w_C(u) = \min(w(u), C)$ where C is an arbitrarily large constant. For any $\hat{\theta}_n$, we estimate the supremum over θ of the normalized risk by the integral

$$\sup_{\theta\,:\,|\theta - \theta_0| < \delta} \mathbb{E}_\theta\Big[w_C\big(n(\hat{\theta}_n - \theta)\big)\Big] \geq \frac{1}{b}\int_0^b \mathbb{E}_{\theta_0 + u/n}\Big[w_C\big(n(\hat{\theta}_n - \theta_0 - u/n)\big)\Big]\, du$$

$$(4.1)\quad = \frac{1}{b}\,\mathbb{E}_{\theta_0}\Big[\int_0^b w_C\big(n\,(\hat{\theta}_n - \theta_0) - u\big)\,\big(e^{\lambda_0 u}\mathbb{I}(u \leq T_n) + o_n(1)\big)\, du\Big].$$

Here we applied the change of measure formula. Now, since w_C is a bounded function,

$$\frac{1}{b}\,\mathbb{E}_{\theta_0}\Big[\int_0^b w_C\big(n\,(\hat{\theta}_n - \theta_0) - u\big)\, o_n(1)\big)\, du\Big] = o_n(1),$$

and, continuing from (4.1), we obtain

$$= \frac{1}{b}\,\mathbb{E}_{\theta_0}\Big[\int_0^b w_C\big(n\,(\hat{\theta}_n - \theta_0) - u\big)\, e^{\lambda_0 u}\mathbb{I}(u \leq T_n)\, du\Big] + o_n(1)$$

$$\geq \frac{1}{b}\,\mathbb{E}_{\theta_0}\Big[\mathbb{I}(\sqrt{b} \leq T_n \leq b) \int_0^{T_n} w_C\big(n\,(\hat{\theta}_n - \theta_0) - u\big)\, e^{\lambda_0 u}\, du\Big] + o_n(1),$$

which, after a substitution $u = t + T_n$, takes the form

$$= \frac{1}{b}\mathbb{E}_{\theta_0}\Big[e^{\lambda_0 T_n}\mathbb{I}(\sqrt{b} \leq T_n \leq b) \int_{-T_n}^0 w_C\big(n(\hat{\theta}_n - \theta_0) - T_n - t\big) e^{\lambda_0 t} dt\Big] + o_n(1).$$

Put $y = n\,(\hat{\theta}_n - \theta_0) - T_n$, and continue:

(4.2)

$$\geq \frac{1}{b}\,\mathbb{E}_{\theta_0}\Big[e^{\lambda_0 T_n}\,\mathbb{I}(\sqrt{b} \leq T_n \leq b)\Big] \min_{y \in \mathbb{R}} \int_{-\sqrt{b}}^0 w_C\big(y - t\big)\, e^{\lambda_0 t}\, dt + o_n(1).$$

Note that

$$\lim_{n \to \infty} \frac{1}{b}\,\mathbb{E}_{\theta_0}\Big[e^{\lambda_0 T_n}\,\mathbb{I}(\sqrt{b} \leq T_n \leq b)\Big] = \frac{1}{b}\int_{\sqrt{b}}^b \lambda_0\, e^{-\lambda_0 t + \lambda_0 t}\, dt$$

$$= \frac{\lambda_0\,(b - \sqrt{b})}{b} = \lambda_0\,\big(1 - \frac{1}{\sqrt{b}}\big).$$

This provides the asymptotic lower bound for (4.2)

$$\lambda_0 \left(1 - \frac{1}{\sqrt{b}}\right) \min_{y \in \mathbb{R}} \int_{-\sqrt{b}}^{0} w_C(y - u)\, e^{\lambda_0 u}\, du$$

$$= \lambda_0 \left(1 - \frac{1}{\sqrt{b}}\right) \min_{y \in \mathbb{R}} \int_{0}^{\sqrt{b}} w_C(u - y)\, e^{-\lambda_0 u}\, du$$

where b and C can be taken however large, which proves the theorem. □

For the quadratic risk function, the lower bound in Theorem 4.9 can be found explicitly.

Example 4.10. If $w(u) = u^2$, then

$$\lambda_0 \min_{y \in \mathbb{R}} \int_0^{\infty} (u - y)^2 e^{-\lambda_0 u}\, du = \min_{y \in \mathbb{R}} \left[y^2 - 2y/\lambda_0 + 2/\lambda_0^2 \right]$$

$$= \min_{y \in \mathbb{R}} \left[(y - 1/\lambda_0)^2 + 1/\lambda_0^2 \right] = 1/\lambda_0^2.$$

By Proposition 4.5, for the uniform model, $\lambda_0 = 1/\theta_0$, hence, the exact lower bound equals θ_0^2. For the shifted exponential experiment, according to Proposition 4.6, $\lambda_0 = 1$ and thus, the lower bound is 1. □

Remark 4.11. In Exercise 4.27 we ask the reader to show that the lower bound limiting constant of Theorem 4.9 is attainable in the uniform and shifted exponential models under the quadratic risk. The sharpness of the bound holds in general, for all asymptotically exponential models, but under some additional conditions. □

Exercises

Exercise 4.22. Show that if $X_1, \dots, X_n$ are independent uniform$(0, \theta)$ random variables, then (i) the MLE of θ is $\hat{\theta}_n = X_{(n)} = \max(X_1, \dots, X_n)$, (ii) $\theta_n^* = (n+1)X_{(n)}/n$ is an unbiased estimator of θ, and (iii) $\mathbb{V}ar_\theta\left[\theta_n^*\right] = \theta^2/[n(n+2)]$.

Exercise 4.23. Consider independent observations $X_1, \dots, X_n$ from a shifted exponential distribution with the density

$$p(x, \theta) = \exp\{-(x - \theta)\}, \quad x \geq \theta, \quad \theta \in \mathbb{R}.$$

Verify that (i) $\hat{\theta}_n = X_{(1)} = \min(X_1, \dots, X_n)$ is the MLE of θ, (ii) $\theta_n^* = X_{(1)} - 1/n$ is an unbiased estimator of θ, and (iii) $\mathbb{V}ar_\theta\left[\theta_n^*\right] = 1/n^2$.

Exercise 4.24. Refer to Exercise 4.23. Prove that in the shifted exponential model the Fisher information does not exist.

Exercise 4.25. Let $p(x, \theta) = c_-$ if $-1 < x < 0$ and $p(x, \theta) = c_+$ if $0 < x < 1$. Assume that $p(x, \theta) = 0$ if x is outside of the interval $(-1, 1)$, and that the jump of the density at the origin equals θ, i.e., $c_+ - c_- = \theta$, $\theta \in \Theta \subset (0, 1)$. Use the formula in Theorem 4.3 to compute the Fisher information.

Exercise 4.26. Prove Proposition 4.6.

Exercise 4.27. Show that (i) in the uniform model (see Exercise 4.22),

$$\lim_{n\to\infty} \mathbb{E}_{\theta_0}\Big[\big(n(\theta_n^* - \theta_0)\big)^2 \Big] = \theta_0^2.$$

(ii) in the shifted exponential model (see Exercise 4.23),

$$\mathbb{E}_{\theta_0}\Big[\big(n(\theta_n^* - \theta_0)\big)^2 \Big] = 1.$$

Exercise 4.28. Compute explicitly the lower bound in Theorem 4.9 for the absolute loss function $w(u) = |u|$.

Exercise 4.29. Suppose n independent observations have the shifted exponential distribution with the location parameter θ. Using an argument involving the Bayes risk, show that for any estimator $\hat{\theta}_n$, the quadratic minimax risk is bounded from below (cf. Example 4.10),

$$\inf_{\hat{\theta}_n} \sup_{\theta\in\mathbb{R}} \mathbb{E}_\theta\Big[\big(n\,(\hat{\theta}_n - \theta) \big)^2 \Big] \geq 1.$$

(i) Take a uniform$(0, b)$ prior density and let $Y = \min(X_{(1)}, b)$. Verify that the posterior density is defined only if $X_{(1)} > 0$, and is given by the formula

$$f_b(\theta \,|\, X_1, \dots, X_n) = \frac{n\,\exp\{n\,\theta\}}{\exp\{n\,Y\} - 1}, \quad 0 \leq \theta \leq Y.$$

(ii) Check that the posterior mean is equal to

$$\theta_n^*(b) = Y - \frac{1}{n} + \frac{Y}{\exp\{n\,Y\} - 1}.$$

(iii) Argue that for any θ, $\sqrt{b} \leq \theta \leq b - \sqrt{b}$, the normalized quadratic risk of the estimator $\theta_n^*(b)$ has the limit

$$\lim_{b\to\infty} \mathbb{E}_\theta\Big[\big(n\,(\theta_n^*(b) - \theta) \big)^2 \Big] = 1.$$

(iv) Show that

$$\sup_{\theta\in\mathbb{R}} \mathbb{E}_\theta\Big[\big(n\,(\hat{\theta}_n - \theta) \big)^2 \Big] \geq \frac{b - 2\sqrt{b}}{b} \inf_{\sqrt{b}\leq\theta\leq b-\sqrt{b}} \mathbb{E}_\theta\Big[\big(n\,(\theta_n^*(b) - \theta) \big)^2 \Big]$$

where the right-hand side is arbitrarily close to 1 for sufficiently large b.

Chapter 5

Change-Point Problem

5.1. Model of Normal Observations

Consider a statistical model with normal observations $X_1, \dots, X_n$ where $X_i \sim \mathcal{N}(0, \sigma^2)$ if $i = 1, \dots, \theta$, and $X_i \sim \mathcal{N}(\mu, \sigma^2)$ if $i = \theta + 1, \dots, n$. An integer parameter θ belongs to a subset Θ of all positive integers $\mathbb{Z}^+$,

$$\Theta = \Theta_\alpha = \big\{ \theta : \alpha n \leq \theta \leq (1 - \alpha) n, \ \ \theta \in \mathbb{Z}^+ \big\}$$

where α is a given number, $0 < \alpha < 1/2$. We assume that the standard deviation σ and the expectation μ are known. Put $c = \mu/\sigma$. This ratio is called a *signal-to-noise ratio.*

The objective is to estimate θ from observations $X_1, \dots, X_n$. The parameter θ is called the *change point*, and the problem of its estimation is termed the *change-point problem.* Note that it is assumed that there are at least $\alpha\, n$ observations obtained before and after the change point θ, that is, the number of observations of both kinds are of the same order $O(n)$.

In the context of the change-point problem, the index i may be associated with the time at which the observation X_i becomes available. This statistical model differs from the models of the previous chapters in the respect that it deals with non-homogeneous observations since the expected value of the observations suffers a jump at θ.

The joint probability density of the observations has the form

$$p(x_1, \dots, x_n, \theta) = \prod_{i=1}^{\theta} \frac{\exp\big\{ -x_i^2/(2\sigma^2)\big\}}{\sqrt{2\pi\sigma^2}} \prod_{i=\theta+1}^{n} \frac{\exp\big\{ -(x_i - \mu)^2/(2\sigma^2)\big\}}{\sqrt{2\pi\sigma^2}}.$$

Denote by θ_0 the true value of the parameter θ. We want to study the log-likelihood ratio

$$L_n(\theta) - L_n(\theta_0) = \ln \frac{p(X_1, \dots, X_n, \theta)}{p(X_1, \dots, X_n, \theta_0)} = \sum_{i=1}^{n} \ln \frac{p(X_i, \theta)}{p(X_i, \theta_0)}.$$

Introduce a set of random variables

$$\varepsilon_i = \begin{cases} X_i/\sigma & \text{if } 1 \le i \le \theta_0, \\ -(X_i - \mu)/\sigma & \text{if } \theta_0 < i \le n. \end{cases} \tag{5.1}$$

These random variables are independent and have $\mathcal{N}(0,\, 1)$ distribution with respect to $\mathbb{P}_{\theta_0}$-probability.

Define a stochastic process $W(j)$ for integer-valued j's by

$$W(j) = \begin{cases} \displaystyle\sum_{\theta_0 < i \le \theta_0 + j} \varepsilon_i & \text{if } j > 0, \\ \displaystyle\sum_{\theta_0 + j < i \le \theta_0} \varepsilon_i & \text{if } j < 0, \\ 0 & \text{if } j = 0, \end{cases} \tag{5.2}$$

where the ε_i's are as in (5.1) for $1 \le i \le n$; and for i outside of the interval $[1, n]$, the ε_i's are understood as supplemental independent standard normal random variables. The process $W(j)$ is called the *two-sided Gaussian random walk* (see Figure 2).

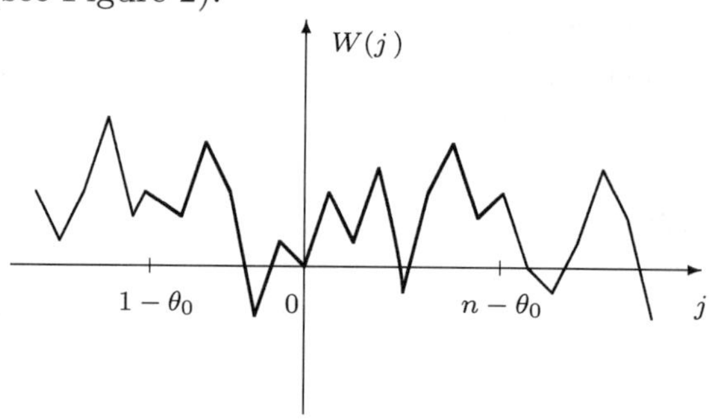

Figure 2. A sample path of the two-sided Gaussian random walk W.

The next theorem suggests an explicit form of the likelihood ratio in terms of the process W and the signal-to-noise ratio c.

Theorem 5.1. *For any θ, $1 \le \theta \le n$, and any $\theta_0 \in \Theta_\alpha$, the log-likelihood ratio has the representation*

$$L_n(\theta) - L_n(\theta_0) = c\, W(\theta - \theta_0) - \frac{c^2}{2}\, |\, \theta - \theta_0\, |$$

where $W(\theta - \theta_0)$ *is the two-sided Gaussian random walk in* $\mathbb{P}_{\theta_0}$*-probability defined by* (5.2).

Proof. By definition, for any $\theta > \theta_0$, the log-likelihood ratio is expressed as

$$L_n(\theta) - L_n(\theta_0) = -\sum_{i=1}^{\theta} \frac{X_i^2}{2\sigma^2} - \sum_{i=\theta+1}^{n} \frac{(X_i - \mu)^2}{2\sigma^2}$$

$$+\sum_{i=1}^{\theta_0} \frac{X_i^2}{2\sigma^2} + \sum_{i=\theta_0+1}^{n} \frac{(X_i-\mu)^2}{2\sigma^2} = -\sum_{i=\theta_0+1}^{\theta} \frac{X_i^2}{2\sigma^2} + \sum_{i=\theta_0+1}^{\theta} \frac{(X_i-\mu)^2}{2\sigma^2}$$

$$= \frac{\mu}{\sigma} \sum_{i=\theta_0+1}^{\theta} \Big(-\frac{X_i}{\sigma} + \frac{\mu}{2\sigma} \Big) = \frac{\mu}{\sigma} \sum_{i=\theta_0+1}^{\theta} \Big[-\Big(\frac{X_i - \mu}{\sigma} \Big) - \frac{\mu}{2\sigma} \Big]$$

$$= \frac{\mu}{\sigma} \sum_{i=\theta_0+1}^{\theta} \varepsilon_i - \frac{\mu^2}{2\sigma^2}(\theta - \theta_0) = c\,W(\theta - \theta_0) - \frac{c^2}{2}(\theta - \theta_0)$$

with $c = \mu/\sigma$. For $\theta < \theta_0$, we get a similar formula,

$$L_n(\theta) - L_n(\theta_0) = \frac{\mu}{\sigma} \sum_{i=\theta+1}^{\theta_0} \Big(\frac{X_i}{\sigma} - \frac{\mu}{2\sigma} \Big)$$

$$= c \sum_{i=\theta+1}^{\theta_0} \varepsilon_i - \frac{c^2}{2}(\theta_0 - \theta) = c\,W(\theta - \theta_0) - \frac{c^2}{2}|\theta - \theta_0|. \qquad \square$$

Remark 5.2. The two-sided Gaussian random walk W plays a similar role as a standard normal random variable Z in the regular statistical models under the LAN condition (see Section 3.4.) The essential difference is that the dimension of W grows as $n \to \infty$. $\square$

The next result establishes a rough minimax lower bound.

Lemma 5.3. *There exists a positive constant* r_* *independent of* n *such that*

$$\inf_{\hat{\theta}_n} \max_{\theta \in \Theta_\alpha} \mathbb{E}_\theta\big[(\hat{\theta}_n - \theta)^2 \big] \geq r_*.$$

Proof. Take $\theta_0, \theta_1 \in \Theta_\alpha$ such that $\theta_1 - \theta_0 = 1$. From Theorem 5.1, we have that

$$L_n(\theta_1) - L_n(\theta_0) = c\,W(1) - \frac{c^2}{2}$$

where $W(1)$ is a standard normal random variable in $\mathbb{P}_{\theta_0}$-probability. Thus,

$$\mathbb{P}_{\theta_0}\big(L_n(\theta_1) - L_n(\theta_0) \geq 0 \big) = \mathbb{P}_{\theta_0}\big(W(1) \geq c/2 \big) = p_0$$

with a positive constant p_0 independent of n. Taking the same steps as in the proof of Lemma 3.4, we obtain the result with $r_* = p_0/4$. $\square$

Remark 5.4. Lemma 5.3 is very intuitive. Any estimator $\hat{\theta}_n$ misses the true change point θ_0 by at least 1 with a positive probability, which is not a surprise due to the stochastic nature of observations. Thus, the anticipated minimax rate of convergence in the change-point problem should be $O(1)$ as $n \to \infty$. □

Remark 5.5. We can define a change-point problem on the interval $[0, 1]$ by $X_i \sim \mathcal{N}(0, \sigma^2)$ if $i/n \leq \theta$, and $X_i \sim \mathcal{N}(\mu, \sigma^2)$ if $i/n > \theta$. On this scale, the anticipated minimax rate of convergence is $O(n^{-1})$ for $n \to \infty$. Note that the convergence in this model is faster than that in regular models, and, though unrelated, is on the order of that in the asymptotically exponential experiments. □

5.2. Maximum Likelihood Estimator of Change Point

The log-likelihood function $L_n(\theta)$ in the change-point problem can be written as

$$L_n(\theta) = \sum_{i=1}^{n} \Big[- \frac{1}{2\sigma^2} \Big(X_i - \mu\, \mathbb{I}\,(i > \theta) \Big)^2 - \frac{1}{2} \ln (2\pi\sigma^2) \Big], \quad 1 \leq \theta \leq n.$$

The maximum likelihood estimator (MLE) of θ exists and is unique with probability 1 for any true value θ_0. We denote the MLE of θ by

$$\tilde{\theta}_n = \underset{1 \leq \theta \leq n}{\operatorname{argmax}}\, L_n(\theta).$$

The goal of this section is to describe the exact large sample performance of the MLE.

Introduce a stochastic process

$$L_\infty(j) = c\, W(j) - \frac{c^2}{2}\, |\, j\, |, \quad j \in \mathbb{Z},$$

where the subscript "∞" indicates that j is unbounded in both directions. Define j_∞^* as the point of maximum of the process $L_\infty(j)$,

$$j_\infty^* = \underset{j \in \mathbb{Z}}{\operatorname{argmax}}\, L_\infty(j),$$

and put

$$p_j = \mathbb{P}_{\theta_0}\big(j_\infty^* = j \big), \quad j \in \mathbb{Z}.$$

Note that the distribution of j_∞^* is independent of θ_0.

Theorem 5.6. *For any $\theta_0 \in \Theta_\alpha$, and any loss function w, the risk of the MLE $\tilde{\theta}_n$ has a limit as $n \to \infty$, independent of θ_0,*

$$\lim_{n \to \infty} \mathbb{E}_{\theta_0}\Big[w(\tilde{\theta}_n - \theta_0) \Big] = \sum_{j \in \mathbb{Z}} w(j)\, p_j.$$

Before we turn to proving the theorem, we state two technical lemmas, the proofs of which are postponed until the final section of this chapter.

Lemma 5.7. *Let* $\tilde{p}_j = \mathbb{P}_{\theta_0}\big(\tilde{\theta}_n - \theta_0 = j\big)$. *There exist positive constants* b_1 *and* b_2 *such that for any* $\theta_0 \in \Theta_\alpha$, *and any* j *satisfying* $1 \le \theta_0 + j \le n$, *the following bounds hold:*

$$p_j \le \tilde{p}_j \le p_j + b_1 e^{-b_2 n}.$$

Lemma 5.8. *There exist positive constants* b_3 *and* b_4 *such that*

$$p_j \le b_3 e^{-b_4 |j|}, \quad j \in \mathbb{Z}.$$

Proof of Theorem 5.6. Applying Lemma 5.7, we find that

$$\mathbb{E}_{\theta_0}\Big[w(\tilde{\theta}_n - \theta_0)\Big] = \sum_{j:1 \le \theta_0 + j \le n} w(j)\,\tilde{p}_j \ge \sum_{j:1 \le \theta_0 + j \le n} w(j)\,p_j$$

$$= \sum_{j \in \mathbb{Z}} w(j)\,p_j - \sum_{j \in \mathbb{Z} \setminus [1-\theta_0,\, n-\theta_0]} w(j)\,p_j$$

where the latter sum is taken over integers j that do not belong to the set $1 \le \theta_0 + j \le n$. As a loss function, $w(j)$ does not increase faster than a polynomial function in $|j|$, while p_j, in accordance with Lemma 5.8, decreases exponentially fast. Besides, the absolute value of any $j \in \mathbb{Z} \setminus [1-\theta_0\,,\, n-\theta_0]$ is no less than $\alpha\, n$. Thus,

$$\lim_{n \to \infty} \sum_{j \in \mathbb{Z} \setminus [1-\theta_0,\, n-\theta_0]} w(j)\,p_j = 0$$

and

$$\liminf_{n \to \infty} \mathbb{E}_{\theta_0}\Big[w(\tilde{\theta}_n - \theta_0)\Big] \ge \sum_{j \in \mathbb{Z}} w(j)\,p_j.$$

Similarly, we get the upper bound

$$\mathbb{E}_{\theta_0}\Big[w(\tilde{\theta}_n - \theta_0)\Big] = \sum_{j:1 \le \theta_0 + j \le n} w(j)\tilde{p}_j \le \sum_{j:1 \le \theta_0 + j \le n} w(j)\Big[p_j + b_1 e^{-b_2 n}\Big]$$

$$\le \sum_{j \in \mathbb{Z}} w(j)\,p_j + b_1 e^{-b_2 n}\Big[\max_{j:1 \le \theta_0 + j \le n} w(j)\Big].$$

The maximum of $w(j)$ does not grow faster than a polynomial function in n. That is why the latter term is vanishing as $n \to \infty$, and

$$\limsup_{n \to \infty} \mathbb{E}_{\theta_0}\Big[w(\tilde{\theta}_n - \theta_0)\Big] \le \sum_{j \in \mathbb{Z}} w(j)\,p_j. \qquad \square$$

5.3. Minimax Limiting Constant

In this section we will find the minimax limiting constant for the quadratic risk function.

Lemma 5.9. *For any θ_0, the Bayes estimator θ_n^* with respect to the uniform prior $\pi_n(\theta) = 1/n$, $1 \leq \theta \leq n$, satisfies the formula*

$$\theta_n^* = \theta_0 + \frac{\sum_{j:1 \leq j+\theta_0 \leq n} j \exp\{ c\,W(j) - c^2\,|\,j\,|/2 \}}{\sum_{j:1 \leq j+\theta_0 \leq n} \exp\{ c\,W(j) - c^2\,|\,j\,|/2 \}}.$$

Proof. The proof is left as an exercise (see Exercise 5.30). □

Introduce a new random variable,

$$\xi = \frac{\sum_{j \in \mathbb{Z}} j \exp\{ c\,W(j) - c^2\,|\,j\,|/2 \}}{\sum_{j \in \mathbb{Z}} \exp\{ c\,W(j) - c^2\,|\,j\,|/2 \}}. \tag{5.3}$$

The next lemma asserts that $\theta_n^* - \theta_0$ converges to ξ in the quadratic sense. The proof of the lemma is deferred until Section 5.5.

Lemma 5.10. *There exists a finite second moment $r_* = \mathbb{E}_{\theta_0}\big[\,\xi^2\,\big]$ and, uniformly in $\theta_0 \in \Theta_\alpha$, the Bayes estimator θ_n^* satisfies the identity*

$$\lim_{n\to\infty} \mathbb{E}_{\theta_0}\Big[\,(\,\theta_n^* - \theta_0\,)^2\,\Big] = r_*.$$

Now we can show that the minimax quadratic risk of any estimator of θ_0 is bounded from below by r_*.

Theorem 5.11. *Let r_* be the constant defined in Lemma 5.10. For all large enough n, and for any estimator $\hat{\theta}_n$, the following inequality takes place:*

$$\liminf_{n\to\infty} \max_{\theta_0 \in \Theta_\alpha} \mathbb{E}_{\theta_0}\Big[\,(\,\hat{\theta}_n - \theta_0\,)^2\,\Big] \geq r_*.$$

Proof. Assume, without loss of generality, that αn is an integer, and so is $N = n - 2\alpha n$ where N is the number of points in the parameter set Θ_α. As we typically deal with lower bounds, we estimate the maximum over Θ_α from below by the mean value over the same set,

$$\max_{\theta_0 \in \Theta_\alpha} \mathbb{E}_{\theta_0}\Big[(\hat{\theta}_n - \theta_0)^2\Big] \geq N^{-1} \sum_{\theta_0 \in \Theta_\alpha} \mathbb{E}_{\theta_0}\Big[(\hat{\theta}_n - \theta_0)^2\Big]$$

$$\geq N^{-1} \sum_{\theta_0 \in \Theta_\alpha} \mathbb{E}_{\theta_0}\Big[(\tilde{\theta}_N - \theta_0)^2\Big] \tag{5.4}$$

where $\tilde{\theta}_N$ is the Bayes estimator with respect to the uniform prior distribution $\pi_N(\theta) = 1/N$, $\alpha n \leq \theta \leq (1-\alpha)n$.

Take an arbitrarily small positive β and define a set of integers $\Theta_{\alpha,\beta}$ by

$$\Theta_{\alpha,\beta} = \big\{ \theta : \alpha n + \beta N \le \theta \le n - (\alpha n + \beta N) \big\}.$$

The set $\Theta_{\alpha,\beta}$ contains no less than $n - 2(\alpha n + \beta N) = N - 2\beta N$ points. This set plays the same role of the "inner points" for Θ_α as Θ_α does for the original set $\{\theta : 1 \le \theta \le n\}$. Applying Lemma 5.10 to $\tilde{\theta}_N$, we have that, uniformly in $\theta_0 \in \Theta_{\alpha,\beta}$,

$$\lim_{n\to\infty} \mathbb{E}_{\theta_0}\Big[(\tilde{\theta}_N - \theta_0)^2 \Big] = r_*.$$

Substituting this limit into (5.4), we obtain that for any estimator $\hat{\theta}_n$,

$$\liminf_{n\to\infty} \max_{\theta_0\in\Theta_\alpha} \mathbb{E}_{\theta_0}\big[(\hat{\theta}_n - \theta_0)^2\big] \ge \liminf_{n\to\infty} N^{-1} \sum_{\theta_0\in\Theta_\alpha} \mathbb{E}_{\theta_0}\Big[(\tilde{\theta}_N - \theta_0)^2\Big]$$

$$\ge \liminf_{n\to\infty} N^{-1} \sum_{\theta_0\in\Theta_{\alpha,\beta}} \mathbb{E}_{\theta_0}\Big[(\tilde{\theta}_N - \theta_0)^2\Big]$$

$$\ge \liminf_{n\to\infty} \Big\{ \frac{N - 2\beta N}{N} \min_{\theta_0\in\Theta_{\alpha,\beta}} \mathbb{E}_{\theta_0}\Big[(\tilde{\theta}_N - \theta_0)^2\Big]\Big\} = (1 - 2\beta) r_*.$$

Since β is arbitrarily small, the result follows. □

Remark 5.12. Combining the results of Lemma 5.10 and Theorem 5.11, we obtain the sharp limit of the quadratic minimax risk. The Bayes estimator corresponding to the uniform prior distribution is asymptotically minimax. To the best of our knowledge, the exact value of r_* is unknown. Nevertheless, it is possible to show that the MLE estimator discussed in Section 5.2 is not asymptotically minimax, since the limit of its quadratic risk determined in Theorem 5.6 is strictly larger than r_*. □

5.4. Model of Non-Gaussian Observations

Let $p_0(x)$ be a probability density which is positive on the whole real line, $x \in \mathbb{R}$, and let $\mu \ne 0$, be a fixed number. Assume that observations X_i's have a distribution with the density $p_0(x)$ if $1 \le i \le \theta$, and $p_0(x - \mu)$ if $\theta < i \le n$. As in the previous sections, an integer θ is an unknown change point which belongs to Θ_α. In this section, we will describe the structure of the likelihood ratio to understand the difference from the case of normal observations.

Denote by l_i the log-likelihood ratio associated with a single observation X_i,

$$l_i = \ln \frac{p_0(X_i - \mu)}{p_0(X_i)}, \quad 1 \le i \le n. \tag{5.5}$$

The two quantities

$$K_{\pm} = - \int_{-\infty}^{\infty} \Big[\ln \frac{p_0(x \pm \mu)}{p_0(x)} \Big] p_0(x)\, dx$$

are called the *Kullback-Leibler information numbers.*

Let θ_0 denote the true value of the parameter θ. Consider the random variables

$$\varepsilon_i = \begin{cases} l_i + K_- & \text{if } 1 \le i \le \theta_0, \\ -l_i + K_+ & \text{if } \theta_0 < i \le n. \end{cases} \tag{5.6}$$

Lemma 5.13. *For any integer i, $1 \le i \le n$, the random variables ε_i defined in (5.6) have an expected value of zero under P_{θ_0}-distribution.*

Proof. If $1 \le i \le \theta_0$, then by definition of K_-,

$$\mathbb{E}_{\theta_0}[\varepsilon_i] = \mathbb{E}_{\theta_0}[l_i] + K_- = (-K_-) + K_- = 0.$$

If $\theta_0 < i \le n$, then

$$\mathbb{E}_{\theta_0}[\varepsilon_i] = - \int_{-\infty}^{\infty} \Big[\ln \frac{p_0(x - \mu)}{p_0(x)} \Big] p_0(x-\mu)\, dx + K_+$$

$$= - \int_{-\infty}^{\infty} \Big[\ln \frac{p_0(x)}{p_0(x+\mu)} \Big] p_0(x)\, dx + K_+$$

$$= \int_{-\infty}^{\infty} \Big[\ln \frac{p_0(x + \mu)}{p_0(x)} \Big] p_0(x)\, dx + K_+ = -K_+ + K_+ = 0. \qquad \square$$

For any $j \in \mathbb{Z}$, analogously to the Gaussian case (cf. (5.2)), introduce a stochastic process $W(j)$ by $W(0) = 0$,

$$W(j) = \sum_{\theta_0 < i \le \theta_0 + j} \varepsilon_i \text{ if } j > 0, \text{ and } W(j) = \sum_{\theta_0 + j < i \le \theta_0} \varepsilon_i \text{ if } j < 0$$

where for $1 \le i \le n$, ε_i's are the random variables from Lemma 5.13 that have mean zero with respect to the P_{θ_0}-distribution. For all other values of i, the random variables ε_i's are assumed independent with the zero expected value. Note that $W(j)$ is a two-sided random walk, which in general is not symmetric. Indeed, the distributions of ε_i's may be different for $i \le \theta_0$ and $i > \theta_0$.

Define a constant $K_{\text{sgn}(i)}$ as K_+ for $i > 0$, and K_- for $i < 0$.

Theorem 5.14. *For any integer θ, $1 \le \theta \le n$, and any true change point $\theta_0 \in \Theta_\alpha$, the log-likelihood ratio has the form*

$$L_n(\theta) - L_n(\theta_0) = \ln \frac{p(X_1, \dots, X_n, \theta)}{p(X_1, \dots, X_n, \theta_0)} = W(\theta - \theta_0) - K_{\text{sgn}(\theta-\theta_0)} \, |\, \theta - \theta_0 \,|.$$

Proof. The joint density is computed according to the formula

$$p(X_1, \dots, X_n, \theta) = \prod_{1 \le i \le \theta} p_0(X_i) \prod_{\theta < i \le n} p_0(X_i - \mu).$$

Therefore, if $\theta > \theta_0$, the likelihood ratio satisfies

$$L_n(\theta) - L_n(\theta_0) = \ln \prod_{\theta_0 < i \le \theta} \frac{p_0(X_i)}{p_0(X_i - \mu)} = \sum_{\theta_0 < i \le \theta} (-l_i)$$

$$= \sum_{\theta_0 < i \le \theta} (\varepsilon_i - K_+) = W(\theta - \theta_0) - K_+ (\theta - \theta_0).$$

In the case $\theta < \theta_0$, we write

$$L_n(\theta) - L_n(\theta_0) = \ln \prod_{\theta < i \le \theta_0} \frac{p_0(X_i - \mu)}{p_0(X_i)} = \sum_{\theta < i \le \theta_0} l_i$$

$$= \sum_{\theta < i \le \theta_0} (\varepsilon_i - K_-) = W(\theta - \theta_0) - K_- |\theta - \theta_0|. \qquad \square$$

From Theorem 5.14 we can expect that the MLE of θ_0 in the non-Gaussian case possesses properties similar to that in the normal case. This is true under some restrictions on p_0 (see Exercise 5.33).

5.5. Proofs of Lemmas

Proof of Lemma 5.7. Note that if $j_\infty^* = j$, and j is such that $1 \le \theta_0 + j \le n$, then $\tilde{\theta}_n - \theta_0 = j$. Therefore, for any j satisfying $1 \le \theta_0 + j \le n$, we have

$$p_j = \mathbb{P}_{\theta_0}(j_\infty^* = j) \le \mathbb{P}_{\theta_0}(\tilde{\theta}_n - \theta_0 = j) = \tilde{p}_j$$

$$= \mathbb{P}_{\theta_0}\big(\tilde{\theta}_n - \theta_0 = j, 1 \le j_\infty^* + \theta_0 \le n\big) + \mathbb{P}_{\theta_0}\big(\tilde{\theta}_n - \theta_0 = j, j_\infty^* + \theta_0 \notin [1, n]\big)$$

$$\le p_j + \mathbb{P}_{\theta_0}\big(j_\infty^* + \theta_0 \notin [1, n]\big) = p_j + \mathbb{P}_{\theta_0}\big(j_\infty^* \le -\theta_0\big) + \mathbb{P}_{\theta_0}\big(j_\infty^* \ge n + 1 - \theta_0\big)$$

$$\le p_j + \mathbb{P}_{\theta_0}\big(j_\infty^* \le -\alpha n\big) + \mathbb{P}_{\theta_0}\big(j_\infty^* \ge \alpha n\big) \le p_j + \sum_{k \le -\alpha n} p_k + \sum_{k \ge \alpha n} p_k$$

$$= p_j + 2 \sum_{k \ge \alpha n} p_k \le p_j + 2 b_3 \sum_{k \ge \alpha n} e^{-b_4 k} = p_j + b_1 e^{-b_2 n}$$

where we have applied Lemma 5.8. The constants b_1 and b_2 are independent of n, $b_1 = 2\, b_3 / (1 - \exp\{-b_4\})$ and $b_2 = b_4\, \alpha$. $\square$

Proof of Lemma 5.8. For a positive integer j we can write

$$p_j = \mathbb{P}_{\theta_0}\big(j_\infty^* = j\big) \le \mathbb{P}_{\theta_0}\big(j_\infty^* \ge j\big) \le \mathbb{P}_{\theta_0}\big(\max_{k \ge j} \big[c\, W(k) - c^2 k/2\big] \ge 0\big)$$

$$\le \sum_{k \ge j} \mathbb{P}_{\theta_0}\big(c\, W(k) - c^2 k/2 \ge 0\big) = \sum_{k \ge j} \mathbb{P}_{\theta_0}\big(W(k)/\sqrt{k} \ge c\sqrt{k}/2\big)$$

$$= \sum_{k \geq j} \left[1 - \Phi(c\sqrt{k}/2) \right] \leq \sum_{k \geq j} \exp\{ -c^2 k/8 \} \leq b_3 \exp\{ -b_4 j \}$$

with the positive constants $b_3 = 1/(1 - \exp\{-c^2/8\})$ and $b_4 = c^2/8$.

In the above, we have estimated the probability $\mathbb{P}_{\theta_0}\big(j_\infty^* \geq j \big)$ using the fact that $c\,W(j) - c^2 j/2 = 0$ at $j = 0$, which implies that at the point of the maximum $c\,W(j_\infty^*) - c^2 j_\infty^*/2 \geq 0$. Next we applied the fact that $W(k)/\sqrt{k}$ has the standard normal distribution with the c.d.f. $\Phi(x)$. In the last step, the inequality $1 - \Phi(x) \leq \exp\{-x^2/2\}$ is used.

A similar upper bound for the probabilities p_j holds for negative j, so that

$$p_j \leq b_3 \exp\{ -b_4 \,|\, j \,|\, \}, \quad j \in \mathbb{Z}. \qquad \square$$

Proof of Lemma 5.10. First we show that uniformly in $\theta_0 \in \Theta_\alpha$, there exists a limit in $\mathbb{P}_{\theta_0}$-probability

$$\lim_{n \to \infty} \left(\theta_n^* - \theta_0 \right) = \xi.$$

To see that ξ is well-defined, it suffices to show that the denominator in the definition (5.3) of ξ is separated away from zero. Indeed, at $j = 0$ the contributing term equals 1. We want to demonstrate that the infinite sums in the numerator and denominator converge in probability. That is, for however small the positive ε, uniformly in $\mathbb{P}_{\theta_0}$-probability, the following limits exist:

$$\text{(5.7)} \qquad \lim_{n \to \infty} \mathbb{P}_{\theta_0} \Big(\sum_{j\,:\, j + \theta_0 \notin [1, n]} |\, j \,| \exp\big\{ c\,W(j) - c^2 \,|\, j \,|\, / \,2 \big\} > \varepsilon \Big) = 0$$

and

$$\text{(5.8)} \qquad \lim_{n \to \infty} \mathbb{P}_{\theta_0} \Big(\sum_{j\,:\, j + \theta_0 \notin [1, n]} \exp\big\{ c\,W(j) - c^2 \,|\, j \,|\, / \,2 \big\} > \varepsilon \Big) = 0.$$

Introduce a random variable:

$$\text{(5.9)} \qquad \zeta = \min\big\{ k \,:\, W(|j|) \leq c\,|\,j\,|/\,4 \text{ for all } j\,,\, |\,j\,| \geq k \big\}.$$

Starting from ζ, the random walk $W(j)$ does not exceed $c\,|\,j\,|\,/\,4$. First of all, note that the tail probabilities for ζ are decreasing exponentially fast. In fact,

$$\mathbb{P}_{\theta_0}(\zeta \geq k) \leq \mathbb{P}_{\theta_0}\Big(\bigcup_{j \geq k} \{ W(j) \geq cj/4 \} \Big) + \mathbb{P}_{\theta_0}\Big(\bigcup_{j \leq -k} \{ W(j) \geq c|j|/4 \} \Big)$$

$$\leq 2 \sum_{j \geq k} \mathbb{P}_{\theta_0}\Big(W(j) \geq cj/4 \Big) = 2 \sum_{j \geq k} \mathbb{P}_{\theta_0}\Big(W(j)/\sqrt{j} \geq c\sqrt{j}/4 \Big)$$

$$\leq 2 \sum_{j \geq k} \exp\big\{ -c^2 j/32 \big\} \leq a_1 \exp\big\{ -c^2 k/32 \big\} = a_1 \exp\big\{ -a_2 k \big\}$$

with $a_1 = 2/(1 - \exp\{-c^2/32\})$ and $a_2 = c^2/32$.

Next we verify (5.7). The limit (5.8) is shown analogously. We have

$$\mathbb{P}_{\theta_0}\Big(\sum_{j+\theta_0 \notin [1,n]} |j| \exp\big\{cW(j) - c^2|j|/2\big\} > \varepsilon\Big)$$

$$\leq \mathbb{P}_{\theta_0}\Big(\sum_{|j| \geq \alpha n} |j| \exp\big\{cW(j) - c^2|j|/2\big\} > \varepsilon\Big)$$

$$\leq \mathbb{P}_{\theta_0}\Big(\sum_{|j| \geq \alpha n} |j| \exp\big\{cW(j) - c^2|j|/2\big\} > \varepsilon; \zeta \leq \alpha n\Big) + \mathbb{P}_{\theta_0}\big(\zeta > \alpha n\big)$$

$$\leq \mathbb{P}_{\theta_0}\Big(\sum_{|j| \geq \alpha n} |j| \exp\big\{ - c^2|j|/4\big\} > \varepsilon; \zeta \leq \alpha n\Big) + \mathbb{P}_{\theta_0}\big(\zeta > \alpha n\big)$$

$$\leq \mathbb{P}_{\theta_0}\Big(\sum_{|j| \geq \alpha n} |j| \exp\big\{ - c^2|j|/4\big\} > \varepsilon\Big) + \mathbb{P}_{\theta_0}\big(\zeta > \alpha n\big)$$

where we have applied the definition of the random variable ζ. In the latter sum, the first probability is zero for all sufficiently large n as the probability of a non-random event. The second probability is decreasing exponentially fast as $n \to \infty$, which proves (5.7).

Further, we check that the second moment of ξ is finite despite the fact that neither the numerator nor denominator is integrable (see Exercise 5.31). Thus, we prove that there exists a finite second moment

$$r_* = \mathbb{E}_{\theta_0}[\,\xi^2\,] < \infty.$$

Introduce the notation for the denominator in the formula (5.3) for the random variable ξ,

$$D = \sum_{j \in \mathbb{Z}} \exp\big\{\, c\,W(j) - c^2\,|\,j\,|/2\,\big\}.$$

Involving the random variable ζ defined in (5.9), we write

$$|\,\xi\,| = \sum_{|\,j\,| \leq \zeta} |\,j\,|\,D^{-1} \exp\big\{\, c\,W(j) - c^2\,|\,j\,|/2\,\big\}$$

$$+ \sum_{|\,j\,| > \zeta} |\,j\,|\,D^{-1} \exp\big\{\, c\,W(j) - c^2\,|\,j\,|/2\,\big\}.$$

Note that for any j, $D^{-1} \exp\{\, c\,W(j) - c^2\,|\,j\,|/2\,\} \leq 1$. We substitute this inequality into the first sum. In the second sum, we use the obvious fact that $D > \exp\{\, c\,W(0)\,\} = 1$. Hence, we arrive at the following inequalities:

$$|\,\xi\,| \leq \zeta^2 + \sum_{|\,j\,| > \zeta} |\,j\,| \exp\{\, c\,W(j) - c^2\,|\,j\,|/2\,\}.$$

If $|\,j\,|$ is larger than ζ, then we can bound $W(j)$ from above by $c\,|\,j\,|\,/\,4$ and find that

$$|\,\xi\,| \le \zeta^2 + 2 \sum_{j>\zeta} j \exp\{\,-c^2 j\,/\,4\,\}$$

$$\le \zeta^2 + 2 \sum_{j \ge 1} j \exp\{\,-c^2 j\,/\,4\,\} = \zeta^2 + a_3 \tag{5.10}$$

with $a_3 = \sum_{j \ge 1} j \exp\{-c^2 j/4\}$. Because the tail probabilities of ζ decrease exponentially fast, any power moment of ξ is finite, in particular, $r_* = \mathbb{E}_{\theta_0}\left[\,\xi^2\,\right] < \infty$.

Finally, we verify that $\theta_n^* - \theta_0$ converges to ξ in the L_2 sense, that is, uniformly in $\theta_0 \in \Theta_\alpha$,

$$\lim_{n\to\infty} \mathbb{E}_{\theta_0}\left[\,\left(\,\theta_n^* - \theta_0 - \xi\,\right)^2\,\right] = 0.$$

Apply the representation for the difference $\theta_n^* - \theta_0$ from Lemma 5.9. Similarly to the argument used to derive (5.10), we obtain that

$$\left|\,\theta_n^* - \theta_0\,\right| \le \zeta^2 + a_3$$

with the same definitions of the entries on the right-hand side. Thus, the difference

$$\left|\theta_n^* - \theta_n - \xi\,\right| \le \left|\theta_n^* - \theta_0\,\right| + |\xi| \le 2\,(\zeta^2 + a_3)$$

where the random variable $\zeta^2 + a_3$ is square integrable and independent of n. As shown above, the difference $\theta_n^* - \theta_n - \xi$ converges to 0 in $\mathbb{P}_{\theta_0}$-probability as $n \to \infty$. By the dominated convergence theorem, this difference converges to zero in the quadratic sense. □

Exercises

Exercise 5.30. Prove Lemma 5.9.

Exercise 5.31. Show that $\mathbb{E}_{\theta_0}\Big[\,\exp\big\{\,c\,W(j) - c^2\,|\,j\,|\,/\,2\,\big\}\Big] = 1$, for any integer j. Deduce from here that the numerator and denominator in (5.3) have infinite expected values.

Exercise 5.32. Show that the Kullback-Leibler information numbers $K_\pm$ are positive. Hint: Check that $-\,K_\pm < 0$. Use the inequality $\ln(1+x) < x$, for any $x \ne 0$.

Exercise 5.33. Suppose that $\mathbb{E}_{\theta_0}\big[\,|\,l_i\,|^{5+\delta}\,\big] < \infty$ for a small $\delta > 0$, where l_i's are the log-likelihood ratios defined in (5.5), $i = 1, \dots, n$, and θ_0 denotes the true value of the change point. Show that uniformly in $\theta_0 \in \Theta_\alpha$, the quadratic risk of the MLE $\tilde{\theta}_n$ of θ_0 is finite for any n, that is,

$$\mathbb{E}_{\theta_0}\Big[\,\big(\tilde{\theta}_n - \theta_0\,\big)^2\,\Big] < \infty\,.$$

Hint: Argue that if $\tilde{\theta}_n > \theta_0$,

$$\mathbb{P}_{\theta_0}\big(\,\tilde{\theta}_n - \theta_0 = m\,\big) \le \sum_{l=m}^{\infty} \mathbb{P}_{\theta_0}\Big(\,\sum_{i=1}^{l} \varepsilon_i \ge K_+\, l\,\Big)$$

where ε_i's are as in (5.6). Now use the fact that if $\mathbb{E}_{\theta_0}\big[\,|\,\varepsilon_i\,|^{5+\delta}\big] < \infty$, then there exists a positive constant C with the property

$$\mathbb{P}_{\theta_0}\Big(\,\sum_{i=1}^{l} \varepsilon_i \ge K_+\, l\,\Big) \le C\, l^{-(4+\delta)}.$$

This fact can be found, for example, in Petrov [**Pet75**], Chapter IX, Theorem 28.

Exercise 5.34. Consider the Bernoulli model with 30 independent observations X_i that take on values 1 or 0 with the respective probabilities p_i and $1 - p_i$, $i = 1, \dots, 30$. Suppose that $p_i = 0.4$ if $1 \le i \le \theta$, and $p_i = 0.7$ if $\theta < i \le 30$, where θ is an integer change point, $1 \le \theta \le 30$. Estimate θ from the following set of data:

i	X_i	i	X_i	i	X_i
1	0	11	0	21	1
2	0	12	1	22	1
3	1	13	1	23	1
4	0	14	0	24	0
5	1	15	1	25	1
6	0	16	0	26	0
7	0	17	0	27	1
8	1	18	1	28	1
9	1	19	1	29	1
10	0	20	0	30	0

Exercise 5.35. Suppose that in the change-point problem, observations X_i have a known c.d.f. $F_1(x)$, $x \in \mathbb{R}$, for $1 \le i \le \theta$, and another known c.d.f. $F_2(x)$, $x \in \mathbb{R}$, for $\theta < i \le n$. Assume that the two c.d.f.'s are not identically equal. Suggest an estimator of the true change point $\theta_0 \in \Theta_\alpha$. Hint: Consider a set $\mathfrak{X}$ such that $\int_{\mathfrak{X}} dF_1(x) \ne \int_{\mathfrak{X}} dF_2(x)$, and introduce indicators $\mathbb{I}(X_i \in \mathfrak{X})$.

Chapter 6

Sequential Estimators

6.1. The Markov Stopping Time

Sequential estimation is a method in which the size of the sample is not predetermined, but instead parameters are estimated as new observations become available. The data collection is terminated in accordance with a predefined stopping rule.

In this chapter we consider only the model of Gaussian observations. We address two statistical problems, using the sequential estimation approach. First, we revisit the change-point problem discussed in Chapter 5, and, second, we study the parameter estimation problem from a sample of a random size in an autoregressive model. Solutions to both problems are based on the concept of the Markov stopping time.

Let X_i, $i = 1, 2, \dots$, be a sequence of real-valued random variables. For any integer $t \geq 1$, and any real numbers a_i and b_i such that $a_i \leq b_i$, consider the random events

$$\left\{ X_i \in [a_i\,, b_i] \right\}, \quad i = 1, \dots, t.$$

All countable intersections, unions and complements of these random events form a *σ-algebra* generated by the random variables $X_1, \dots, X_t$. Denote this σ-algebra by $\mathcal{F}_t$, that is,

$$\mathcal{F}_t \, = \, \sigma\left\{ X_i \in [a_i\,, \, b_i]\,, \; i = 1, \dots, t \right\}.$$

All the random events that belong to $\mathcal{F}_t$ are called *$\mathcal{F}_t$-measurable.* We interpret the integer t as *time*, and we call $\mathcal{F}_t$ a *σ-algebra generated by the observations X_i up to time t.*

It is easily seen that these inclusions are true:

$$\mathcal{F}_1 \subseteq \mathcal{F}_2 \subseteq \cdots \subseteq \mathcal{F}$$

where $\mathcal{F}$ denotes the σ-algebra that contains all the σ-algebras $\mathcal{F}_t$. The set of the ordered σ-algebras $\{ \mathcal{F}_t\,,\ t \geq 1 \}$ is called a *filter*.

Example 6.1. The random event $\{ X_1^2 + X_2^2 < 1 \}$ is $\mathcal{F}_2$-measurable. Indeed, this random event can be presented as the union of intersections:

$$\bigcup_{m=1}^{\infty} \bigcup_{i^2+j^2 < m^2} \Big(\{ |X_1| < i/m \} \cap \{ |X_2| < j/m \} \Big). \qquad \square$$

An integer-valued random variable τ, $\tau \in \{1, 2, \dots \}$, is called a *Markov stopping time* (or, simply, *stopping time*) with respect to the filter $\{ \mathcal{F}_t, t \geq 1 \}$ if for any integer t, the random event $\{ \tau = t \}$ is $\mathcal{F}_t$-measurable.

Example 6.2. The following are examples of the Markov stopping times (for the proof see Exercise 6.37):
(i) A non-random variable $\tau = T$ where T is a given positive integer number.
(ii) The first time when the sequence X_i hits a given interval $[a, b]$, that is, $\tau = \min\{ i \,:\, X_i \in [a, b] \}$.
(iii) The minimum or maximum of two given Markov stopping times τ_1 and τ_2, $\tau = \min(\tau_1, \tau_2)$ or $\tau = \max(\tau_1, \tau_2)$.
(iv) The time $\tau = \tau_1 + s$ for any positive integer s, where τ_1 is a given Markov stopping time. $\square$

Example 6.3. Some random times are not examples of Markov stopping times (for the proof see Exercise 6.38):
(i) The last time when the sequence X_i, $1 \leq i \leq n$, hits a given interval $[a, b]$, that is, $\tau = \max\{ i \,:\, X_i \in [a, b],\ 1 \leq i \leq n \}$.
(ii) The time $\tau = \tau_1 - s$ for any positive integer s, where τ_1 is a given stopping time. $\square$

Lemma 6.4. *If τ is a stopping time, then the random events: (i) $\{\tau \leq t\}$ is $\mathcal{F}_t$-measurable, and (ii) $\{\tau \geq t\}$ is $\mathcal{F}_{t-1}$-measurable.*

Proof. (i) We write $\{\tau \leq t\} = \bigcup_{s=1}^{t} \{\tau = s\}$, where each event $\{\tau = s\}$ is $\mathcal{F}_s$-measurable, and since $\mathcal{F}_s \subseteq \mathcal{F}_t$, it is $\mathcal{F}_t$-measurable as well. Thus, $\{\tau \leq t\}$ is $\mathcal{F}_t$-measurable as the union of $\mathcal{F}_t$-measurable events.
(ii) The random event $\{\tau \geq t\}$ is $\mathcal{F}_{t-1}$-measurable as the complement of $\{\tau < t\} = \{\tau \leq t - 1\}$, an $\mathcal{F}_{t-1}$-measurable event. $\square$

The next important result is known as *Wald's first identity*.

Theorem 6.5. *Let $X_1, X_2, \dots$, be a sequence of independent identically distributed random variables with $\mathbb{E}[\,X_1\,] < \infty$. Then for any Markov stopping time τ such that $\mathbb{E}[\,\tau\,] < \infty$, the following identity holds:*

$$\mathbb{E}\big[\,X_1 + \cdots + X_\tau\,\big] \;=\; \mathbb{E}[\,X_1\,]\,\mathbb{E}[\,\tau\,].$$

Proof. By definition,

$$\mathbb{E}\big[\,X_1 + \cdots + X_\tau\,\big] \;=\; \mathbb{E}\Big[\,\sum_{t=1}^{\infty}(X_1 + \cdots + X_t)\,\mathbb{I}(\tau = t)\Big]$$

$$= \mathbb{E}\big[\,X_1\,\mathbb{I}(\tau \geq 1) + X_2\,\mathbb{I}(\tau \geq 2) + \cdots + X_t\,\mathbb{I}(\tau \geq t) + \dots\,\big].$$

For a Markov stopping time τ, the random event $\{\tau \geq t\}$ is $\mathcal{F}_{t-1}$-measurable by Lemma 6.4, that is, it is predictable from the observations up to time $t-1$, $X_1, \dots, X_{t-1}$, and is independent of the future observations $X_t, X_{t+1}, \dots$. In particular, X_t and $\mathbb{I}(\tau \geq t)$ are independent, and hence, $\mathbb{E}\big[\,X_t\,\mathbb{I}(\tau \geq t)\big] = \mathbb{E}[\,X_1\,]\,\mathbb{P}(\,\tau \geq t\,)$. Consequently,

$$\mathbb{E}\big[\,X_1 + \cdots + X_\tau\,\big] \;=\; \mathbb{E}[\,X_1\,]\sum_{t=1}^{\infty}\mathbb{P}(\tau \geq t) \;=\; \mathbb{E}[\,X_1\,]\,\mathbb{E}[\,\tau\,].$$

Here we used the straightforward fact that

$$\sum_{t=1}^{\infty}\mathbb{P}(\,\tau \geq t\,) \;=\; \mathbb{P}(\tau = 1) \;+\; 2\,\mathbb{P}(\tau = 2) \;+\; 3\,\mathbb{P}(\tau = 3) \;+\; \dots$$

$$= \sum_{t=1}^{\infty} t\,\mathbb{P}(\tau = t) \;=\; \mathbb{E}[\,\tau\,]. \qquad \square$$

Let τ be a Markov stopping time. Introduce a set of random events:

$$\mathcal{F}_\tau \;=\; \big\{\,A \in \mathcal{F}\,:\, A \cap \{\tau = t\} \in \mathcal{F}_t \;\text{ for all }\; t,\; t \geq 1\,\big\}.$$

Lemma 6.6. *The set $\mathcal{F}_\tau$ is a σ-algebra, that is, this set is closed under countable intersections, unions, and complements.*

Proof. Suppose events A_1 and A_2 belong to $\mathcal{F}_\tau$. To show that $\mathcal{F}_\tau$ is a σ-algebra, it suffices to show that the intersection $A_1 \cap A_2$, union $A_1 \cup A_2$, and complement $\overline{A}_1$ belong to $\mathcal{F}_\tau$. The same proof extends to countably many random events.

Denote by $B_1 = A_1 \cap \{\tau = t\} \in \mathcal{F}_t$ and $B_2 = A_2 \cap \{\tau = t\} \in \mathcal{F}_t$. The intersection $A_1 \cap A_2$ satisfies

$$(A_1 \cap A_2) \cap \{\tau = t\} \;=\; B_1 \cap B_2 \in \mathcal{F}_t,$$

as the intersection of two $\mathcal{F}_t$-measurable events. Also, the union $A_1 \cup A_2$ is such that

$$(A_1 \cup A_2) \cap \{\tau = t\} \;=\; B_1 \cup B_2 \in \mathcal{F}_t,$$

as the union of two $\mathcal{F}_t$-measurable events. As for the complement $\overline{A}_1$, note first that both events $\{\tau = t\}$ and $\overline{A_1 \cap \{\tau = t\}}$ belong to $\mathcal{F}_t$, therefore,

$$\begin{aligned}\overline{A}_1 \cap \{\tau = t\} &= \{\tau = t\} \setminus (A_1 \cap \{\tau = t\}) \\ &= \{\tau = t\} \cap \overline{A_1 \cap \{\tau = t\}} \in \mathcal{F}_t,\end{aligned}$$

as an intersection of two $\mathcal{F}_t$-measurable events. □

The σ-algebra $\mathcal{F}_\tau$ is referred to as a σ-algebra of random events *measurable up to the random time* τ.

Lemma 6.7. *The Markov stopping time* τ *is* $\mathcal{F}_\tau$*-measurable.*

Proof. For any positive integer s, put $A = \{\tau = s\}$. We need to show that $A \in \mathcal{F}_\tau$. For all t we find that

$$A \cap \{\tau = t\} = \{\tau = s\} \cap \{\tau = t\} = \{\tau = t\} \text{ if } s = t,$$

and is the empty set, otherwise. The set $\{\tau = t\}$ belongs to $\mathcal{F}_t$ by the definition of a stopping time. The empty set is $\mathcal{F}_t$ - measurable as well (refer to Exercise 6.36). Thus, by the definition of $\mathcal{F}_\tau$, the event A belongs to $\mathcal{F}_\tau$. □

Recall that we defined $\mathcal{F}_t$ as a σ-algebra generated by the random variables X_i up to time t.

Lemma 6.8. *The random variable* X_τ *is* $\mathcal{F}_\tau$*-measurable.*

Proof. Take any interval $[a, b]$ and define $A = \{ X_\tau \in [a, b] \}$. Note that

$$A = \bigcup_{s=1}^{\infty} \Big(\{ X_s \in [a, b] \} \cap \{\tau = s\} \Big).$$

Then for all t, we have that

$$\begin{aligned}A \cap \{\tau = t\} &= \bigcup_{s=1}^{\infty} \Big(\{ X_s \in [a, b] \} \cap \{\tau = s\} \Big) \bigcap \{\tau = t\} \\ &= \{ X_t \in [a, b] \} \cap \{\tau = t\}.\end{aligned}$$

The latter intersection belongs to $\mathcal{F}_t$ because both random events belong to $\mathcal{F}_t$. Hence A is $\mathcal{F}_\tau$-measurable. □

Remark 6.9. The concept of the σ-algebra $\mathcal{F}_\tau$ is essential in the sequential analysis. All parameter estimators constructed from sequential observations are $\mathcal{F}_\tau$-measurable, that is, are based on observations $X_1, \dots, X_\tau$ obtained up to a random stopping time τ. □

6.2. Change-Point Problem. Rate of Detection

In this section we return to the change-point problem studied in Chapter 5, and look at it from the sequential estimation point of view. The statistical setting of the problem is modified. If previously all n observations were available for estimation of the true change point $\theta_0 \in \Theta_\alpha$, in this section we assume that observations X_i's arrive sequentially one at a time at moments $t_i = i$ where $i = 1, \dots, n$.

Define a filter $\{ \mathcal{F}_t,\, 1 \le t \le n \}$ of σ-algebras $\mathcal{F}_t$ generated by the observations $X_1, \dots, X_t$ up to time t, $1 \le t \le n$. Introduce $\mathcal{T}$ as a set of all Markov stopping times with respect to this filter.

If a sequential estimator $\hat{\tau}_n$ of the change point θ_0 belongs to $\mathcal{T}$, that is, if $\hat{\tau}_n$ is a Markov stopping time, then we call this estimator an *on-line detector* (or just *detector*) and the estimation problem itself, the *on-line detection problem* (or, simply, *detection*).

In the on-line detection problem, we use the same loss functions as in the regular estimation problems studied so far. For example, for the quadratic loss, the *minimax risk of detection* is defined as

$$r_n^D = \inf_{\hat{\tau}_n \in \mathcal{T}} \max_{\theta_0 \in \Theta_\alpha} \mathbb{E}_{\theta_0}\big[(\hat{\tau}_n - \theta_0)^2 \big].$$

The crucial difference between the minimax risk r_n in the previous chapters and r_n^D consists of restrictions on the set of admissible estimators. In the on-line detection, we cannot use an arbitrary function of observations.

Remark 6.10. In this section we focus our attention on the quadratic loss, even though, sometimes in practice, other loss functions are used. For instance, we can restrict the class of admissible detectors to a class $\mathcal{T}_\gamma$ defined by

$$\mathcal{T}_\gamma = \big\{ \hat{\tau}_n : \hat{\tau}_n \in \mathcal{T} \text{ and } \max_{\theta_0 \in \Theta_\alpha} \mathbb{P}_{\theta_0}\big(\hat{\tau}_n \le \theta_0 \big) \le \gamma \big\} \tag{6.1}$$

where γ is a given small positive number. The probability $\mathbb{P}_{\theta_0}\big(\hat{\tau}_n \le \theta_0 \big)$ is called the *false alarm probability*. The name is inherited from the military air defense problems where θ is associated with the time of a target appearance, so that any detection of the target before it actually appears on the radar screen is, indeed, a false alarm. The condition on detectors in $\mathcal{T}_\gamma$ requires that the false alarm probability is small, uniformly in $\theta_0 \in \Theta_\alpha$. Another natural criterion in detection, also rooted in the military concerns, is a so-called *expected detection delay*,

$$\mathbb{E}_{\theta_0}\big[(\hat{\tau}_n - \theta_0)_+ \big] = \mathbb{E}_{\theta_0}\big[(\hat{\tau}_n - \theta_0)\, \mathbb{I}(\hat{\tau}_n > \theta_0) \big]. \tag{6.2}$$

The expected detection delay generates a minimax risk

$$\inf_{\hat{\tau}_n \in \mathcal{T}_\gamma} \max_{\theta_0 \in \Theta_\alpha} \mathbb{E}_{\theta_0}\big[\, (\, \hat{\tau}_n \, - \, \theta_0 \,)_+ \big].$$

Clearly, any additional constraints on the admissible detectors increase the value of the minimax risk. And every additional restriction makes the problem more difficult. □

Below we find the rate of convergence for the minimax quadratic risk of detection r_n^D for the Gaussian model, and define the rate-optimal detectors.

Assume that $X_i \sim \mathcal{N}(0, \sigma^2)$ if $1 \leq i \leq \theta_0$, and $X_i \sim \mathcal{N}(\mu, \sigma^2)$ if $\theta_0 < i \leq n$, where $\mu > 0$ is known. Our goal is to show that there exists a Markov stopping time τ_n^* such that its deviation away from the true value of θ_0 has the magnitude $O(\ln n)$ as $n \to \infty$. It indicates a slower rate of convergence for the on-line detection as opposed to the estimation based on the entire sample. Recall that in the latter case, the rate is $O(1)$.

Remark 6.11. Note that on the integer scale, the convergence with the rate $O(\ln n)$ is not a convergence at all. This should not be surprising since the convergence rate of $O(1)$ means no convergence as well. If we compress the scale and consider the on-line detection problem on the unit interval $[0, 1]$ with the frequency of observations n (see Remark 5.5), then the rate of convergence guaranteed by the Markov stopping time detectors becomes $(\ln n)/n$. □

Theorem 6.12. *In the on-line detection problem with n Gaussian observations, there exists a Markov stopping time τ_n^* and a constant r^* independent of n such that the following upper bound holds:*

$$\max_{\theta_0 \in \Theta_\alpha} \mathbb{E}_{\theta_0}\Big[\Big(\frac{\tau_n^* - \theta_0}{\ln n} \Big)^2 \Big] \leq r^*.$$

Proof. The construction of the stopping time τ_n^* is based on the idea of averaging. Roughly speaking, we partition the interval $[1, n]$ and compute the sample means in each of the subintervals. At the lower end of the interval, the averages are close to zero. At the upper end, they are close to the known number μ, while in the subinterval that captures the true change point, the sample mean is something in-between.

Put $N = b \ln n$ where b is a positive constant independent of n that will be chosen later. Define $M = n/N$. Without loss of generality, we assume that N and M are integer numbers. Introduce the normalized mean values of observations in subintervals of length N by

$$\bar{X}_m = \frac{1}{\mu N} \sum_{i=1}^{N} X_{(m-1)N+i}\,, \quad m = 1, \dots, M.$$

Let m_0 be an integer such that $(m_0 - 1)N + 1 \le \theta_0 \le m_0 N$. Note that

$$\bar{X}_m = \frac{1}{N} \sum_{i=1}^{N} \mathbb{I}\big((m-1)\, N + i > \theta_0 \big) + \frac{1}{c\, N} \sum_{i=1}^{N} \varepsilon_{(m-1)\, N+i}$$

where ε_i's are independent standard normal random variables and c is the signal-to-noise ratio, $c = \mu/\sigma$. For any $m < m_0$ the first sum equals zero, and for $m > m_0$ this sum is 1. The value of the first sum at $m = m_0$ is a number between 0 and 1, and it depends on the specific location of the true change point θ_0. The second sum can be shown to be

$$\frac{1}{c\, N} \sum_{i=1}^{N} \varepsilon_{(m-1)N+i} = \frac{1}{c\sqrt{N}}\, Z_m, \quad m = 1, \dots, M,$$

where the Z_m's are independent standard normal random variables under $\mathbb{P}_{\theta_0}$-probability.

Next we show that for sufficiently large n,

$$\mathbb{P}_{\theta_0}\Big(\max_{1 \le m \le M} |\, Z_m \,| \ge \sqrt{10 \ln M} \,\Big) \le n^{-3}.$$

Put $y = \sqrt{10 \ln M} > 1$. Notice that the probability that the maximum is not less than y equals the probability that at least one of the random variables is not less than y, therefore, we estimate

$$\mathbb{P}\big(\max_{1 \le m \le M} |\, Z_m \,| \ge y \big) = \mathbb{P}\big(\bigcup_{m=1}^{M} \{ |\, Z_m \,| \ge y \} \big) \le \sum_{m=1}^{M} \mathbb{P}\big(|\, Z_m \,| \ge y \big)$$

$$= 2\, M \,(\, 1 - \Phi(y) \,) \le \frac{2}{\sqrt{2\pi y^2}}\, M \,\exp\{-y^2/2\} \le M \,\exp\{-y^2/2\}$$

where $\Phi(y)$ denotes the cumulative distribution function of a $\mathcal{N}(0,1)$ random variable. In the above we used the standard inequality $1 - \Phi(y) \le \exp\{-y^2/2\}/\sqrt{2\pi y^2}$ if $y > 1$. Thus, we have

$$\mathbb{P}_{\theta_0}\Big(\max_{1 \le m \le M} |\, Z_m \,| \ge \sqrt{10 \ln M} \,\Big) \le M \,\exp\{-10 \ln M/2\}$$

$$= M^{-4} = \big(n/(b \ln n)\big)^{-4} \le n^{-3}.$$

Consider the random event

$$\mathcal{A} = \Big\{ \max_{1 \le m \le M} |\, Z_m \,| < \sqrt{10 \ln M} \,\Big\}.$$

We have just shown that, uniformly in $\theta_0 \in \Theta_\alpha$, the probability of $\overline{\mathcal{A}}$, the complement of $\mathcal{A}$, is bounded from above,

$$\mathbb{P}_{\theta_0}\big(\overline{\mathcal{A}} \big) \le n^{-3}. \qquad (6.3)$$

Choose $b = 10^3\, c^{-2}$. If the event $\mathcal{A}$ occurs, we have the inequalities

$$\max_{1 \le m \le M} \left| \frac{1}{c\sqrt{N}}\, Z_m \right| < \frac{1}{c} \sqrt{10 \ln M/N}$$

$$= \frac{1}{c} \sqrt{10\,(\ln n \;-\; \ln\ln n \;-\; \ln b)\,/\,(b\, \ln n)} \;\le\; \sqrt{10/(b\, c^2)} \;=\; 0.1.$$

Now we can finalize the description of the averaged observations $\bar{X}_m = B_m + \xi_m$ where B_m's are deterministic with the property that $B_m = 0$ if $m < m_0$, and $B_m = 1$ if $m > m_0$. The random variables $|\,\xi_m\,| = |\,Z_m\,/\,(c\sqrt{N})\,|$ do not exceed 0.1 if the random event $\mathcal{A}$ holds.

We are ready to define the Markov stopping time that estimates the change point θ_0. Define an integer-valued random variable m_* by

$$m_* \;=\; \min\big\{\, m \,:\, \bar{X}_m \ge 0.9\,,\; 1 \le m \le M \,\big\},$$

and formally put $m_* = M$ if $\bar{X}_m < 0.9$ for all m. Under the random event $\mathcal{A}$, the minimal m_* exists and is equal to either m_0 or $m_0 + 1$.

Introduce a random variable

$$\tau_n^* \;=\; m_*\, N. \tag{6.4}$$

If t is an integer divisible by N, then the random event $\{\,\tau_n^* = t\,\}$ is defined in terms of $\bar{X}_1, \dots, \bar{X}_{t/N}$, that is, in terms of $X_1, \dots, X_t$, which means that τ_n^* is $\mathcal{F}_t$-measurable. Thus, τ_n^* is a stopping time. We take τ_n^* as the on-line detector. The next step is to estimate its quadratic risk.

As shown above, the inclusion $\mathcal{A} \subseteq \{\,0 \le m_* - m_0 \le 1\,\}$ is true. The definition of m_0 implies the inequalities $0 \le \tau_n^* - \theta_0 \le 2\,N$. We write

$$\max_{\theta_0 \in \Theta_\alpha} \mathbb{E}_{\theta_0}\Big[\, \big(\, (\tau_n^* - \theta_0)/\ln n \,\big)^2 \,\Big]$$

$$= \max_{\theta_0 \in \Theta_\alpha} \Big(\mathbb{E}_{\theta_0}\Big[\, \big(\, (\tau_n^* - \theta_0)/\ln n \,\big)^2\, \mathbb{I}(\mathcal{A}) \Big] \;+\; \mathbb{E}_{\theta_0}\Big[\, \big(\, (\tau_n^* - \theta_0)/\ln n \,\big)^2\, \mathbb{I}(\overline{\mathcal{A}}) \Big] \Big)$$

$$\le \max_{\theta_0 \in \Theta_\alpha} \Big(\mathbb{E}_{\theta_0}\Big[\, (2\,N/\ln n)^2\, \mathbb{I}(\mathcal{A}) \Big] \;+\; \mathbb{E}_{\theta_0}\Big[\, (n/\ln n)^2\, \mathbb{I}(\overline{\mathcal{A}}) \Big] \Big)$$

$$\le\; \big(2\,N/\ln n\big)^2 \;+\; \big(n/\ln n\big)^2\, n^{-3} \;\le\; 4\,b^2 \;+\; 2$$

where at the final stage we have applied (6.3) and the trivial inequality $1/(n \ln^2 n) \;<\; 2\,,\; n \ge 2$. Thus, the statement of the theorem follows with $r^* \;=\; 4\,b^2 \;+\; 2$. □

6.3. Minimax Limit in the Detection Problem.

The rate $\ln n$ in the on-line detection which is guaranteed by Theorem 6.12 is the minimax rate. We show in this section that it cannot be improved by any other detector.

Recall that $\mathcal{T}$ denotes the class of all Markov stopping times with respect to the filter generated by the observations.

Theorem 6.13. *In the on-line detection problem with n Gaussian observations, there exists a positive constant r_* independent of n such that*

$$\liminf_{n\to\infty} \inf_{\hat{\tau}_n \in \mathcal{T}} \max_{\theta_0 \in \Theta_\alpha} \mathbb{E}_{\theta_0}\Big[\Big(\frac{\hat{\tau}_n - \theta_0}{\ln n}\Big)^2\Big] \geq r_*.$$

Proof. Choose points $t_0, \dots, t_M$ in the parameter set Θ_α such that $t_j - t_{j-1} = 3\,b\,\ln n$, $j = 1, \dots, M$, with a positive constant b independent of n. The exact value of b will be selected later. Here the number of points M is equal to $M = n\,(1 - 2\,\alpha)/(3\,b\,\ln n)$. We assume, without loss of generality, that M and $b \ln n$ are integers.

We proceed by contradiction and assume that the claim of the theorem is false. Then there exists a detector $\tilde{\tau}_n$ such that

$$\lim_{n\to\infty} \max_{0\leq j\leq M} \mathbb{E}_{t_j}\Big[\Big(\frac{\tilde{\tau}_n - t_j}{\ln n}\Big)^2\Big] = 0,$$

which implies that

$$\lim_{n\to\infty} \max_{0\leq j\leq M} \mathbb{P}_{t_j}\Big(\,|\,\tilde{\tau}_n - t_j\,| > b\,\ln n\Big) = 0.$$

Indeed, by the Markov inequality,

$$\mathbb{P}_{t_j}\Big(\,|\,\tilde{\tau}_n - t_j\,| > b\,\ln n\Big) \leq b^{-2}\,\mathbb{E}_{t_j}\Big[\Big(\frac{\tilde{\tau}_n - t_j}{\ln n}\Big)^2\Big].$$

Hence for all large enough n, the following inequalities hold:

$$\mathbb{P}_{t_j}\Big(\,|\,\tilde{\tau}_n - t_j\,| \leq b\,\ln n\Big) \geq 3/4\,,\ \ j = 0, \dots, M. \tag{6.5}$$

Consider the inequality for $j = M$. Then

$$\frac{1}{4} \geq \mathbb{P}_{t_M}\Big(\,|\,\tilde{\tau}_n - t_M\,| > b\,\ln n\Big) \geq \mathbb{P}_{t_M}\Big(\bigcup_{j=0}^{M-1}\{\,|\,\tilde{\tau}_n - t_j\,| \leq b\,\ln n\,\}\Big)$$

$$= \sum_{j=0}^{M-1} \mathbb{P}_{t_M}\Big(\,|\,\tilde{\tau}_n - t_j\,| \leq b\,\ln n\Big)$$

$$= \sum_{j=0}^{M-1} \mathbb{E}_{t_j}\Big[\frac{d\mathbb{P}_{t_M}}{d\mathbb{P}_{t_j}}\,\mathbb{I}\big(\,|\,\tilde{\tau}_n - t_j\,| \leq b\,\ln n\big)\Big]. \tag{6.6}$$

Indeed, if $\tilde{\tau}_n$ is close to one of t_j, $j = 0, \dots, M-1$, then $\tilde{\tau}_n$ is distant from t_M, and the random events $\{\, |\, \tilde{\tau}_n - t_j\,| \leq b \ln n \,\}$ are mutually exclusive. The likelihood ratio has the form

$$\frac{d\mathbb{P}_{t_M}}{d\mathbb{P}_{t_j}} = \exp\Big\{ \frac{\mu}{\sigma} \sum_{i=t_j+1}^{t_M} \Big[-\Big(\frac{X_i - \mu}{\sigma}\Big) - \frac{\mu}{2\sigma} \Big] \Big\}$$

$$= \exp\Big\{ c \sum_{i=t_j+1}^{t_M} \varepsilon_i - \frac{c^2}{2}(t_M - t_j) \Big\}$$

where $c = \mu/\sigma$ is the signal-to-noise ratio, and $\varepsilon_i = -(X_i - \mu)/\sigma$ have the standard normal distribution with respect to the $\mathbb{P}_{t_j}$-probability. Note that the number of terms in the sum from $t_j + 1$ to t_M can be as large as $O(n)$. Further, let

$$\mathcal{B}_j = \big\{\, |\, \tilde{\tau}_n - t_j\,| \leq b \ln n \,\big\}.$$

Thus, each expectation in (6.6) can be written as

$$\mathbb{E}_{t_j}\Big[\frac{d\mathbb{P}_{t_M}}{d\mathbb{P}_{t_j}} \mathbb{I}\big(\, |\, \tilde{\tau}_n - t_j\,| \leq b \ln n \big) \Big]$$

$$= \mathbb{E}_{t_j}\Big[\exp\Big\{ c \sum_{i=t_j+1}^{t_M} \varepsilon_i - \frac{c^2}{2}(t_M - t_j) \Big\} \mathbb{I}(\mathcal{B}_j) \Big].$$

Put $u_j = t_j + b \ln n$. The event $\mathcal{B}_j$ is $\mathcal{F}_{u_j}$-measurable because $\tilde{\tau}_n$ is a Markov stopping time. Hence $\mathcal{B}_j$ is independent of the observations $X_{u_j+1}, \dots, X_{t_M}$. Equivalently, $\mathbb{I}(\mathcal{B}_j)$ is independent of ε_i for $i = u_j + 1, \dots, t_M$. Note also that

$$\mathbb{E}_{t_j}\Big[\exp\Big\{ c \sum_{i=u_j+1}^{t_M} \varepsilon_i - \frac{c^2}{2}(t_M - u_j) \Big\} \Big] = \exp\Big\{ \sum_{i=u_j+1}^{t_M} \frac{c^2}{2} - \frac{c^2}{2}(t_M - u_j) \Big\} = 1.$$

We write

$$\mathbb{E}_{t_j}\Big[\frac{d\mathbb{P}_{t_M}}{d\mathbb{P}_{t_j}} \mathbb{I}\big(\, |\, \tilde{\tau}_n - t_j\,| \leq b \ln n \big) \Big]$$

$$= \mathbb{E}_{t_j}\Big[\exp\Big\{ c \sum_{i=t_j+1}^{u_j} \varepsilon_i - \frac{c^2}{2}(u_j - t_j) \Big\} \mathbb{I}(\mathcal{B}_j) \Big]$$

$$= \mathbb{E}_{t_j}\Big[\exp\Big\{ c \sqrt{b \ln n}\, Z_j - \frac{c^2}{2} b \ln n \Big\} \mathbb{I}(\mathcal{B}_j) \Big]$$

where $Z_j = \sum_{i=t_j+1}^{u_j} \varepsilon_i / \sqrt{b \ln n}$ is a standard normal random variable with respect to the $\mathbb{P}_{t_j}$-probability,

$$\geq \mathbb{E}_{t_j}\Big[\exp\Big\{ c \sqrt{b \ln n}\, Z_j - \frac{c^2}{2} b \ln n \Big\} \mathbb{I}(\mathcal{B}_j)\, \mathbb{I}(Z_j \geq 0) \Big]$$

$$\geq \exp\Big\{ -\frac{c^2}{2}\, b \ln n \Big\}\, \mathbb{P}_{t_j}\big(\mathcal{B}_j \cap \{Z_j \geq 0\} \big).$$

Further, the probability of the intersection

$$\mathbb{P}_{t_j}\big(\mathcal{B}_j \cap \{Z_j \geq 0\} \big) = \mathbb{P}_{t_j}(\mathcal{B}_j) + \mathbb{P}_{t_j}(Z_j \geq 0) - \mathbb{P}_{t_j}\big(\mathcal{B}_j \cup \{Z_j \geq 0\} \big)$$

$$\geq \mathbb{P}_{t_j}(\mathcal{B}_j) + \mathbb{P}_{t_j}(Z_j \geq 0) - 1 \geq \frac{3}{4} + \frac{1}{2} - 1 = \frac{1}{4}.$$

In the last step we used the inequality (6.5) and the fact that $\mathbb{P}_{t_j}(Z_j \geq 0) = 1/2$.

Thus, if we choose $b = c^{-2}$, then the following lower bound holds:

$$\mathbb{E}_{t_j}\Big[\frac{d\mathbb{P}_{t_M}}{d\mathbb{P}_{t_j}}\, \mathbb{I}\big(|\,\tilde{\tau}_n - t_j\,| \leq b \ln n \big) \Big]$$

$$\geq \frac{1}{4} \exp\Big\{ -\frac{c^2}{2}\, b \ln n \Big\} = \frac{1}{4\sqrt{n}}.$$

Substituting this inequality into (6.6), we arrive at a contradiction,

$$\frac{1}{4} \geq \sum_{j=0}^{M-1} \frac{1}{4\sqrt{n}} = \frac{M}{4\sqrt{n}} = \frac{n(1-2\alpha)}{3b \ln n\, 4\sqrt{n}} = \frac{1-2\alpha}{12b} \frac{\sqrt{n}}{\ln n} \to \infty \text{ as } n \to \infty.$$

This implies that the statement of the theorem is true. □

6.4. Sequential Estimation in the Autoregressive Model

In the previous two sections we applied the sequential estimation method to the on-line detection problem. In this section, we demonstrate this technique with another example, the *first-order autoregressive model* (also, termed *autoregression*). Assume that the observations X_i satisfy the equation

$$X_i = \theta X_{i-1} + \varepsilon_i, \quad i = 1, 2, \dots \tag{6.7}$$

with the zero initial condition, $X_0 = 0$. Here ε_i's are independent normal random variables with mean zero and variance σ^2. The *autoregression coefficient* θ is assumed bounded, $-1 < \theta < 1$. Moreover, the true value of this parameter is strictly less than 1, $\theta_0 \in \Theta_\alpha = \{\theta : \,|\,\theta_0\,| \leq 1 - \alpha\,\}$ with a given small positive number α.

The following lemma describes the asymptotic behavior of autoregression. The proof of the lemma is moved to Exercise 6.42.

Lemma 6.14. *(i) The autoregressive model admits the representation*

$$X_i = \varepsilon_i + \theta\, \varepsilon_{i-1} + \theta^2\, \varepsilon_{i-2} + \dots + \theta^{\,i-2} \varepsilon_2 + \theta^{\,i-1} \varepsilon_1, \quad i = 1, 2, \dots.$$

(ii) The random variable X_i is normal with the zero mean and variance

$$\sigma_i^2 \;=\; \mathbb{V}ar[\,X_i\,] \;=\; \sigma^2\,\frac{1-\theta^{2i}}{1-\theta^2}.$$

(iii) The variance of X_i has the limit

$$\lim_{i\to\infty} \sigma_i^2 \;=\; \sigma_\infty^2 \;=\; \frac{\sigma^2}{1-\theta^2}.$$

(iv) The covariance between X_i and X_{i+j}, $j \geq 0$, is equal to

$$\mathbb{C}ov[\,X_i, X_{i+j}\,] \;=\; \sigma^2\,\theta^j\,\frac{1-\theta^{2i}}{1-\theta^2}.$$

Our objective is to find an on-line estimator of the parameter θ. Before we do this, we first study the maximum likelihood estimator (MLE).

6.4.1. Heuristic Remarks on MLE. Assume that only n observations are available, $X_1, \dots, X_n$. Then the log-likelihood function has the form

$$L_n(\theta) \;=\; \sum_{i=1}^{n} \Big[\, -\,\frac{(X_i - \theta\,X_{i-1})^2}{2\sigma^2} \;-\; \frac{1}{2}\,\ln(2\pi\sigma^2)\Big].$$

Differentiating with respect to θ, we find the classical MLE θ_n^* of the autoregression coefficient θ:

$$\theta_n^* \;=\; \frac{\sum_{i=1}^n X_{i-1}\,X_i}{\sum_{i=1}^n X_{i-1}^2}.$$

The MLE does not have a normal distribution, which is easy to show for $n = 2$,

$$\theta_2^* \;=\; \frac{X_0\,X_1 \,+\, X_1\,X_2}{X_0^2 \,+\, X_1^2} \;=\; \frac{X_1\,X_2}{X_1^2} \;=\; \frac{\varepsilon_1\,(\theta_0\,\varepsilon_1 \,+\, \varepsilon_2)}{\varepsilon_1^2} \;=\; \theta_0 \,+\, \frac{\varepsilon_2}{\varepsilon_1}$$

where θ_0 is the true value of θ. The ratio $\varepsilon_2/\varepsilon_1$ has the Cauchy distribution (show!). Therefore, the expectation of the difference $\theta_2^* \,-\, \theta_0$ does not exist.

For $n > 2$, the expectation of $\theta_n^* \,-\, \theta_0$ exists but is not zero, so that the MLE is biased. We skip the proofs of these technical and less important facts. What is more important is that θ_n^* is asymptotically normal as $n \to \infty$. We will try to explain this fact at the intuitive level. Note that

$$\theta_n^* \;=\; \frac{\sum_{i=1}^n X_{i-1}\,X_i}{\sum_{i=1}^n X_{i-1}^2}$$

$$=\; \frac{X_0(\theta_0 X_0 + \varepsilon_1) + X_1(\theta_0 X_1 + \varepsilon_2) + \cdots + X_{n-1}(\theta_0 X_{n-1} + \varepsilon_n)}{X_0^2 + X_1^2 + \cdots + X_{n-1}^2}$$

$$=\; \theta_0 + \frac{X_1\varepsilon_2 + \cdots + X_{n-1}\varepsilon_n}{X_1^2 + \cdots + X_{n-1}^2}. \tag{6.8}$$

By Lemma 6.14 (iv), since $|\theta| < 1$, the covariance between two remote terms X_i and X_{i+j} decays exponentially fast as $j \to \infty$. It can be shown that the Law of Large Numbers (LLN) applies to this process exactly as in the case of independent random variables. By the LLN, for all large n, we can substitute the denominator in the latter formula by its expectation

$$\mathbb{E}[\, X_1^2 + \cdots + X_{n-1}^2 \,] = \sum_{i=1}^{n-1} \mathbb{V}ar[\, X_i \,] \sim n\,\sigma_\infty^2 \;=\; n\,\frac{\sigma^2}{1-\theta_0^2}.$$

Thus, on a heuristic level, we may say that

$$\sqrt{n}\,(\theta_n^* \;-\; \theta_0) \;\sim\; \sqrt{n}\,\frac{X_1\,\varepsilon_2 + \cdots + X_{n-1}\,\varepsilon_n}{n\,\sigma^2/(1-\theta_0^2)}$$

$$= \frac{1-\theta_0^2}{\sigma^2}\,\frac{X_1\,\varepsilon_2 + \cdots + X_{n-1}\,\varepsilon_n}{\sqrt{n}}.$$

If the X_i's were independent, then $\big(X_1\varepsilon_2 + \cdots + X_{n-1}\varepsilon_n\big)/\sqrt{n}$ would satisfy the Central Limit Theorem (CLT). It turns out, and it is far from being trivial, that we can work with the X_i's as if they were independent, and the CLT still applies. Thus, the limiting distribution of this quotient is normal with mean zero and the limiting variance

$$\lim_{n\to\infty} \mathbb{V}ar\Big[\,\frac{X_1\,\varepsilon_2 + \cdots + X_{n-1}\,\varepsilon_n}{\sqrt{n}}\,\Big] \;=\; \lim_{n\to\infty}\,\frac{1}{n}\sum_{i=1}^{n-1}\mathbb{E}\big[\,\big(X_i\varepsilon_{i+1}\big)^2\,\big]$$

$$= \lim_{n\to\infty}\,\frac{1}{n}\sum_{i=1}^{n-1}\mathbb{E}[\,X_i^2\,]\mathbb{E}[\,\varepsilon_{i+1}^2\,] \;=\; \frac{\sigma^4}{n(1-\theta_0^2)}\,\lim_{n\to\infty}\sum_{i=1}^{n-1}(1-\theta_0^{2i})$$

$$= \frac{\sigma^4}{n(1-\theta_0^2)}\,\lim_{n\to\infty}\Big(n - \frac{1-\theta_0^{2n}}{1-\theta_0^2}\Big) \;=\; \frac{\sigma^4}{1-\theta_0^2}.$$

It partially explains why the difference $\sqrt{n}\,(\theta_n^* - \theta_0)$ is asymptotically normal with mean zero and variance

$$\Big(\frac{1-\theta_0^2}{\sigma^2}\Big)^2\,\frac{\sigma^4}{1-\theta_0^2} \;=\; 1-\theta_0^2,$$

that is,

$$\sqrt{n}\,(\theta_n^* - \theta_0) \;\to\; \mathcal{N}\big(0, 1-\theta_0^2\big) \;\text{ as }\; n \to \infty.$$

Note that the limiting variance is independent of σ^2, the variance of the noise.

6.4.2. On-Line Estimator. After obtaining a general idea about the MLE and its asymptotic performance, we are ready to try a sequential estimation procedure, termed an *on-line estimation.*

Note that from (6.8) the difference $\theta_n^* - \theta_0$ can be presented in the form $\theta_n^* - \theta_0 = \sum_{i=2}^n \upsilon_{n,i}\,\varepsilon_i$ with the weights $\upsilon_{n,i} = X_{i-1}/(\,X_1^2 + \cdots + X_{n-1}^2\,)$. If the $\upsilon_{n,i}$'s were deterministic, then the variance of the difference $\theta_n^* - \theta_0$ would be

$$\sigma^2 \sum_{i=2}^{n} \upsilon_{n,i}^2 = \sigma^2/(X_1^2 + \cdots + X_{n-1}^2).$$

In a sense, the sum $X_1^2+\cdots+X_{n-1}^2$ plays the role of the information number: the larger it is, the smaller the variance.

The above argument brings us to an understanding of how to construct a sequential estimator of θ, called an *on-line estimator.* Let us stop collecting data at a random time τ when the sum $X_1^2 + \cdots + X_\tau^2$ reaches a prescribed level $H > 0$, that is, define the Markov stopping time τ by (see Exercise 6.39)

$$\tau = \min\big\{\, t \,:\, X_1^2 + \cdots + X_t^2 > H \,\big\}.$$

In the discrete case with normal noise, the overshoot $X_1^2 + \cdots + X_t^2 - H$ is positive with probability 1. The stopping time τ is a random sample size, and the level H controls the magnitude of its expected value, $\mathbb{E}_{\theta_0}[\,\tau\,]$ increases as H grows (see Exercise 6.39). Put

$$\Delta H = H - (\,X_1^2 + \cdots + X_{\tau-1}^2\,) \text{ and } \eta = \frac{\Delta H}{X_\tau}.$$

The definition of η makes sense because the random variable X_τ differs from zero with probability 1.

Define an on-line estimator of θ_0 by

$$\hat\theta_\tau = \frac{1}{H}\Big(\sum_{i=1}^{\tau} X_{i-1}\,X_i + \eta\,X_{\tau+1}\Big). \tag{6.9}$$

This is a sequential version of the MLE (6.8). Apparently, if ΔH (and, respectively, η) were negligible, then $\hat\theta_\tau$ would be the MLE with n substituted by τ. Note that $\hat\theta_\tau$ is not $\mathcal{F}_\tau$-measurable because it depends on one extra observation, $X_{\tau+1}$. This is the tribute to the discrete nature of the model. As shown below, due to this extra term, the estimator (6.9) is unbiased.

Lemma 6.15. *The estimator $\hat\theta_\tau$ given by (6.9) is an unbiased estimator of θ_0, and uniformly over $\theta_0 \in \Theta_\alpha$, its variance does not exceed σ^2/H.*

Proof. First, we show that the estimator is unbiased. Note that

$$\hat{\theta}_\tau = \frac{1}{H}\Big[\sum_{i=1}^{\tau} X_{i-1}(\theta_0 X_{i-1} + \varepsilon_i) + \eta(\theta_0 X_\tau + \varepsilon_{\tau+1})\Big]$$

$$= \frac{1}{H}\Big[\theta_0 \sum_{i=1}^{\tau} X_{i-1}^2 + \sum_{i=1}^{\tau} X_{i-1}\varepsilon_i + \theta_0\,\eta\, X_\tau + \eta\,\varepsilon_{\tau+1}\Big].$$

By definition, $\eta X_\tau = \Delta H$ and $\Delta H + \sum_{i=1}^{\tau} X_{i-1}^2 = H$, hence,

$$\hat{\theta}_\tau = \theta_0 + \frac{1}{H}\Big[\sum_{i=1}^{\tau} X_{i-1}\varepsilon_i + \eta\,\varepsilon_{\tau+1}\Big]. \tag{6.10}$$

Therefore, the bias of $\hat{\theta}_\tau$ is equal to

$$\mathbb{E}_{\theta_0}\big[\hat{\theta}_\tau - \theta_0\big] = \frac{1}{H}\Big(\mathbb{E}_{\theta_0}\Big[\sum_{i=1}^{\tau} X_{i-1}\varepsilon_i\Big] + \mathbb{E}_{\theta_0}\big[\eta\,\varepsilon_{\tau+1}\big]\Big), \tag{6.11}$$

and it suffices to show that both expectations are equal to zero. Start with the first one:

$$\mathbb{E}_{\theta_0}\Big[\sum_{i=1}^{\tau} X_{i-1}\varepsilon_i\Big] = \mathbb{E}_{\theta_0}\Big[X_1\varepsilon_2\,\mathbb{I}(\tau=2) + (X_1\varepsilon_2 + X_2\varepsilon_3)\,\mathbb{I}(\tau=3) + \dots\Big]$$

$$= \mathbb{E}_{\theta_0}\Big[\sum_{i=1}^{\infty} X_{i-1}\varepsilon_i\,\mathbb{I}(\tau\ge i)\Big] = \sum_{i=1}^{\infty}\mathbb{E}_{\theta_0}\Big[X_{i-1}\varepsilon_i\,\mathbb{I}(\tau\ge i)\Big].$$

We already know that the random variable $\mathbb{I}(\tau \ge i)$ is $\mathcal{F}_{i-1}$-measurable, and so is X_{i-1}. The random variable ε_i is independent of $\mathcal{F}_{i-1}$, which yields that each term in this infinite sum is equal to zero,

$$\mathbb{E}_{\theta_0}\Big[X_{i-1}\varepsilon_i\,\mathbb{I}(\tau\ge i)\Big] = \mathbb{E}_{\theta_0}\Big[X_{i-1}\,\mathbb{I}(\tau\ge i)\Big]\,\mathbb{E}_{\theta_0}\big[\varepsilon_i\big] = 0.$$

The second expectation $\mathbb{E}_{\theta_0}\big[\eta\,\varepsilon_{\tau+1}\big]$ requires more attention. Note that η is $\mathcal{F}_\tau$-measurable. Indeed, for any integer t and for any $a \le b$, the intersection of the random events

$$\{a \le \eta \le b\}\cap\{\tau = t\} = \{aX_t \le H - (X_1^2 + \dots + X_{t-1}^2) \le bX_t\}\cap\{\tau = t\}$$

is $\mathcal{F}_t$-measurable, because both random events on the right-hand side are $\mathcal{F}_t$-measurable. Hence for any t, the random variable $\eta\,\mathbb{I}(\tau = t)$ is $\mathcal{F}_t$-measurable. The variable ε_{t+1}, on the other hand, is independent of $\mathcal{F}_t$. Thus,

$$\mathbb{E}_{\theta_0}\big[\eta\,\varepsilon_{\tau+1}\big] = \sum_{t=1}^{\infty}\mathbb{E}_{\theta_0}\big[\eta\,\varepsilon_{t+1}\,\mathbb{I}(\tau=t)\big]$$

$$= \sum_{t=1}^{\infty}\mathbb{E}_{\theta_0}\big[\eta\,\mathbb{I}(\tau=t)\big]\,\mathbb{E}_{\theta_0}\big[\varepsilon_{t+1}\big] = 0.$$

It follows that either sum in (6.11) is equal to zero, which means that the estimator $\hat{\theta}_\tau$ is unbiased.

Next, we want to estimate the variance of $\hat{\theta}_\tau$. Using the representation (6.10) of $\hat{\theta}_\tau$, we need to verify that

$$\mathbb{E}_{\theta_0}\Big[\Big(\sum_{i=1}^{\tau} X_{i-1}\,\varepsilon_i + \eta\,\varepsilon_{\tau+1}\Big)^2\Big] \leq \sigma^2 H.$$

The left-hand side of this inequality is equal to

$$\mathbb{E}_{\theta_0}\Big[\Big(\sum_{i=1}^{\tau} X_{i-1}\,\varepsilon_i\Big)^2 + 2\Big(\sum_{i=1}^{\tau} X_{i-1}\,\varepsilon_i\Big)\,\eta\,\varepsilon_{\tau+1} + \eta^2\,\varepsilon_{\tau+1}^2\Big]. \tag{6.12}$$

Consider the last term. We know that η is $\mathcal{F}_\tau$-measurable. Hence

$$\begin{aligned}\mathbb{E}_{\theta_0}\big[\eta^2\,\varepsilon_{\tau+1}^2\big] &= \sum_{t=1}^{\infty} \mathbb{E}_{\theta_0}\big[\eta^2\,\varepsilon_{t+1}^2\,\mathbb{I}(\tau = t)\big] \\ &= \sum_{t=1}^{\infty} \mathbb{E}_{\theta_0}\big[\eta^2\,\mathbb{I}(\tau=t)\big]\,\mathbb{E}_{\theta_0}\big[\varepsilon_{t+1}^2\big] \\ &= \sigma^2 \sum_{t=1}^{\infty} \mathbb{E}_{\theta_0}\big[\eta^2\,\mathbb{I}(\tau=t)\big] = \sigma^2\,\mathbb{E}_{\theta_0}\big[\eta^2\big].\end{aligned}$$

In a similar way, we can show that the expectation of the cross-term in (6.12) is zero. The analysis of the first term, however, takes more steps. It can be written as

$$\begin{aligned}\mathbb{E}_{\theta_0}\Big[\Big(\sum_{i=1}^{\tau} X_{i-1}\,\varepsilon_i\Big)^2\Big] &= \mathbb{E}_{\theta_0}\Big[(X_1\,\varepsilon_2)^2\,\mathbb{I}(\tau=2) + (X_1\,\varepsilon_2 + X_2\,\varepsilon_3)^2\,\mathbb{I}(\tau=3) \\ &\quad + (X_1\,\varepsilon_2 + X_2\,\varepsilon_3 + X_3\,\varepsilon_4)^2\,\mathbb{I}(\tau=4) + \dots\Big] = \mathbb{E}_{\theta_0}\Big[X_1^2\,\varepsilon_2^2\,\mathbb{I}(\tau=2) \\ &\quad + \big(X_1^2\,\varepsilon_2^2 + X_2^2\,\varepsilon_3^2\big)\,\mathbb{I}(\tau=3) + \big(X_1^2\varepsilon_2^2 + X_2^2\,\varepsilon_3^2 + X_3^2\,\varepsilon_4^2\big)\,\mathbb{I}(\tau=4) + \cdots\Big] \\ &\quad + 2\,\mathbb{E}_{\theta_0}\Big[(X_1\,\varepsilon_2)\,(X_2\,\varepsilon_3)\,\mathbb{I}(\tau \geq 3) + (X_1\,\varepsilon_2 + X_2\,\varepsilon_3)\,(X_3\,\varepsilon_4)\,\mathbb{I}(\tau \geq 4) + \cdots\Big] \\ &= \mathcal{E}_1 + 2\,\mathcal{E}_2\end{aligned}$$

where

$$\begin{aligned}\mathcal{E}_1 &= \mathbb{E}_{\theta_0}\Big[X_1^2\,\varepsilon_2^2\,\mathbb{I}(\tau=2) + \big(X_1^2\,\varepsilon_2^2 + X_2^2\,\varepsilon_3^2\big)\,\mathbb{I}(\tau=3) \\ &\quad + \big(X_1^2\varepsilon_2^2 + X_2^2\,\varepsilon_3^2 + X_3^2\,\varepsilon_4^2\big)\,\mathbb{I}(\tau=4) + \cdots\Big] \\ &= \sigma^2\,\mathbb{E}_{\theta_0}\big[X_1^2\,\mathbb{I}(\tau=2) + (X_1^2+X_2^2)\,\mathbb{I}(\tau=3) + (X_1^2+X_2^2+X_3^2)\,\mathbb{I}(\tau=4) + \cdots\big] \\ &= \sigma^2\,\mathbb{E}_{\theta_0}\Big[\sum_{i=1}^{\tau} X_{i-1}^2\Big]\end{aligned}$$

and

$$\mathcal{E}_2 = \mathbb{E}_{\theta_0}\Big[(X_1\,\varepsilon_2)\,(X_2\,\varepsilon_3)\,\mathbb{I}(\tau \geq 3) + (X_1\,\varepsilon_2 + X_2\,\varepsilon_3)\,(X_3\,\varepsilon_4)\,\mathbb{I}(\tau \geq 4) + \cdots \Big]$$

$$= \mathbb{E}_{\theta_0}\big[\,(X_1\,\varepsilon_2)(X_2)\,\mathbb{I}(\tau \geq 3)\,\big]\;\mathbb{E}_{\theta_0}\big[\,\varepsilon_3\,\big]$$

$$+\,\mathbb{E}_{\theta_0}\big[\,(X_1\,\varepsilon_2 + X_2\,\varepsilon_3)(X_3)\,\mathbb{I}(\tau \geq 3)\,\big]\;\mathbb{E}_{\theta_0}\big[\,\varepsilon_4\,\big] + \cdots = 0.$$

Combining all these estimates, we find that the expectation in (6.12) is equal to

$$\mathbb{E}_{\theta_0}\Big[\,\big(\sum_{i=1}^{\tau} X_{i-1}\,\varepsilon_i\,\big)^2 + 2\,\big(\sum_{i=1}^{\tau} X_{i-1}\,\varepsilon_i\,\big)\,\eta\,\varepsilon_{\tau+1} + \eta^2\,\varepsilon_{\tau+1}^2\,\Big]$$

$$= \sigma^2\,\mathbb{E}_{\theta_0}\big[\,\sum_{i=1}^{\tau} X_{i-1}^2\,\big] + \sigma^2\,\mathbb{E}_{\theta_0}\big[\,\eta^2\,\big].$$

From the definition of ΔH, $\sum_{i=1}^{\tau} X_{i-1}^2 = H - \Delta H$. Also, recall that $\eta = H/X_\tau$. Thus, we continue

$$= \sigma^2\,\mathbb{E}_{\theta_0}\big[\,H - \Delta H + \eta^2\,\big] = \sigma^2\,\Big(H - \mathbb{E}_{\theta_0}\big[\,\Delta H - \eta^2\,\big]\Big)$$

$$= \sigma^2\,\Big(H - \mathbb{E}_{\theta_0}\big[\,\Delta H - (\,\Delta H/X_\tau\,)^2\,\big]\Big)$$

$$= \sigma^2\,\Big(H - \mathbb{E}_{\theta_0}\big[\,\Delta H\,(1 - \Delta H/X_\tau^2\,)\,\big]\Big).$$

Note that at the time $\tau - 1$, the value of the sum $X_1^2 + \cdots + X_{\tau-1}^2$ does not exceed H, which yields the inequality $\Delta H \geq 0$. In addition, by the definition of τ, $\sum_{i=1}^{\tau} X_{i-1}^2 + X_\tau^2 > H$, which implies that

$$\Delta H = H - \sum_{i=1}^{\tau} X_{i-1}^2 < X_\tau^2.$$

Hence, $\Delta H/X_\tau^2 < 1$. Thus, ΔH and $(1 - \Delta H/X_\tau^2\,)$ are positive random variables with probability 1, and therefore,

$$\mathbb{E}_{\theta_0}\Big[\,\big(\sum_{i=1}^{\tau} X_{i-1}\,\varepsilon_i + \eta\,\varepsilon_{\tau+1}\,\big)^2\,\Big]$$

$$(6.13) \qquad = \sigma^2\,\Big(H - \mathbb{E}_{\theta_0}\big[\,\Delta H\,(1 - \Delta H/X_\tau^2\,)\,\big]\Big) \leq \sigma^2\,H. \qquad \square$$

The statement of Lemma 6.15 is true for any continuous distribution of the noise ε_i, if it has the zero mean and variance σ^2. The continuity of the noise guarantees that the distribution of X_i is also continuous, and therefore $\eta = \Delta H/X_\tau$ is properly defined. If we assume additionally that the noise has a bounded distribution, that is, $|\,\varepsilon_i\,| \leq C_0$ for some positive constant C_0, then for any i the random variables $|\,X_i\,|$'s turn out to be bounded as well. Under this additional assumption, we can get a lower bound on the variance of $\hat{\theta}_\tau$.

Theorem 6.16. *If* $|\varepsilon_i| \le C_0$, $\mathbb{E}[\varepsilon_i] = 0$, *and* $\mathbb{V}ar[\varepsilon_i] = \sigma^2$, *then*

$$\mathbb{V}ar_{\theta_0}[\hat{\theta}_\tau] \ge \frac{\sigma^2}{H} - \frac{\sigma^2 C_0^2}{4H^2(1 - |\theta_0|)^2}.$$

Proof. From Lemma 6.14 (i), we find that

$$|X_i| \le |\varepsilon_i| + |\theta||\varepsilon_{i-1}| + |\theta^2||\varepsilon_{i-2}| + \cdots + |\theta^{i-2}||\varepsilon_2| + |\theta^{i-1}||\varepsilon_1|$$
$$\le C_0(1 + |\theta| + |\theta^2| + \cdots + |\theta^{i-2}| + |\theta^{i-1}|) \le C_0/(1 - |\theta_0|).$$

In the proof of Lemma 6.15 we have shown (see (6.9)-(6.13)) that

$$\mathbb{V}ar_{\theta_0}\big[\hat{\theta}_\tau\big] = \frac{\sigma^2}{H^2}\Big(H - \mathbb{E}_{\theta_0}\big[\Delta H(1 - \Delta H/X_\tau^2)\big]\Big)$$

where $0 \le \Delta H \le X_\tau^2$. Now, the parabola $\Delta H(1 - \Delta H/X_\tau^2)$ is maximized at $\Delta H = X_\tau^2/2$, and therefore $\Delta H(1 - \Delta H/X_\tau^2) \le X_\tau^2/4$. Finally, we have that

$$\mathbb{E}_{\theta_0}\Big[\Delta H(1 - \Delta H/X_\tau^2)\Big] \le \frac{1}{4}\mathbb{E}_{\theta_0}\big[X_\tau^2\big] \le \frac{C_0^2}{4(1 - |\theta_0|)^2}.$$

The result of the theorem follows. □

Remark 6.17. Note that the bound for the variance of $\hat{\theta}_\tau$ in Theorem 6.16 is pointwise, that is, the lower bound depends on θ_0. To declare a uniform bound for all $\theta_0 \in \Theta_\alpha = \{\theta : |\theta_0| \le 1 - \alpha\}$, we take the minimum of both sides:

$$\inf_{\theta_0 \in \Theta_\alpha} \mathbb{V}ar_{\theta_0}[\hat{\theta}_\tau] \ge \frac{\sigma^2}{H} - \frac{\sigma^2 C_0^2}{4H^2\alpha^2}.$$

Combining this result with the uniform upper bound in Lemma 6.15, we get that as $H \to \infty$,

$$\inf_{\theta_0 \in \Theta_\alpha} \mathbb{V}ar_{\theta_0}[\hat{\theta}_\tau] = \frac{\sigma^2}{H}(1 + O(H^{-1})).$$ □

Exercises

Exercise 6.36. Show that an empty set is $\mathcal{F}$-measurable.

Exercise 6.37. Check that the random variables τ defined in Example 6.2 are stopping times.

Exercise 6.38. Show that the variables τ specified in Example 6.3 are non-stopping times.

Exercise 6.39. Let X_i's be independent identically distributed random variables, and let τ be defined as the first time when the sum of squared observations hits a given positive level H,

$$\tau = \min\{ i : X_1^2 + \cdots + X_i^2 > H \}.$$

(i) Show that τ is a Markov stopping time.
(ii) Suppose $\mathbb{E}[X_1^2] = \sigma^2$. Prove that $\mathbb{E}[\tau] > H/\sigma^2$. Hint: Use Wald's first identity.

Exercise 6.40. Prove Wald's second identity formulated as follows. Suppose $X_1, X_2, \dots$ are independent identically distributed random variables with finite mean and variance. Then

$$\mathbb{V}ar\Big[X_1 + \cdots + X_\tau - \mathbb{E}[X_1]\,\tau \Big] = \mathbb{V}ar[X_1]\,\mathbb{E}[\tau].$$

Exercise 6.41. Suppose that X_i's are independent random variables, $X_i \sim \mathcal{N}(\theta, \sigma^2)$. Let τ be a stopping time such that $\mathbb{E}_\theta[\tau] = h$, where h is a deterministic constant.
(i) Show that $\hat\theta_\tau = (X_1 + \cdots + X_\tau)/h$ is an unbiased estimator of θ. Hint: Apply Wald's first identity.
(ii) Show that

$$\mathbb{V}ar_\theta[\hat\theta_\tau] \le \frac{2\sigma^2}{h} + \frac{2\theta^2\,\mathbb{V}ar_\theta[\tau]}{h^2}.$$

Hint: Apply Wald's second identity.

Exercise 6.42. Prove Lemma 6.14.

Chapter 7

Linear Parametric Regression

7.1. Definitions and Notations

An important research area in many scientific fields is to find a functional relation between two variables, say X and Y, based on the experimental data. The variable Y is called a *response variable* (or, simply, *response*), while X is termed an *explanatory variable* or a *predictor variable.*

The relation between X and Y can be described by a *regression equation*

$$Y = f(X) + \varepsilon \tag{7.1}$$

where f is a *regression function*, and ε is a $\mathcal{N}(0, \sigma^2)$ *random error* independent of X. In this chapter we consider only *parametric regression models* for which the algebraic form of the function f is assumed to be known.

Remark 7.1. In this book we study only *simple* regressions where there is only one predictor X. □

Let f be a sum of known functions $g_0, \dots, g_k$ with unknown *regression coefficients* $\theta_0, \dots, \theta_k$,

$$f = \theta_0 \, g_0 + \theta_1 \, g_1 + \cdots + \theta_k \, g_k. \tag{7.2}$$

It is convenient to have a constant intercept θ_0 in the model, thus, without loss of generality, we assume that $g_0 = 1$. Note that the function f is linear in parameters $\theta_0, \dots, \theta_k$.

Plugging (7.2) into the regression equation (7.1), we obtain a general form of a *linear parametric regression* model

$$Y = \theta_0\, g_0(X) + \theta_1\, g_1(X) + \cdots + \theta_k\, g_k(X) + \varepsilon \tag{7.3}$$

where the random error ε has a $\mathcal{N}(0, \sigma^2)$ distribution and is independent of X.

Example 7.2. Consider a *polynomial regression*, for which $g_0(X) = 1$, $g_1(X) = X, \dots, g_k(X) = X^k$. Here the response variable Y is a polynomial function of X corrupted by a random error $\varepsilon \sim \mathcal{N}(0, \sigma^2)$,

$$Y = \theta_0 + \theta_1 X + \theta_2 X^2 + \cdots + \theta_k X^k + \varepsilon. \qquad \square$$

Suppose the observed data consist of n pairs of observations (x_i, y_i), $i = 1, \dots, n$. The collection of the observations of the explanatory variable X, denoted by $\mathcal{X} = \{x_1, \dots, x_n\}$, is called a *design*. According to (7.1), the data points satisfy the equations

$$y_i = f(x_i) + \varepsilon_i, \quad i = 1, \dots, n, \tag{7.4}$$

where the ε_i's are independent $\mathcal{N}(0, \sigma^2)$ random variables. In particular, the linear parametric regression model (7.3) for the observations takes the form

$$y_i = \theta_0\, g_0(x_i) + \theta_1\, g_1(x_i) + \cdots + \theta_k\, g_k(x_i) + \varepsilon_i, \quad i = 1, \dots, n, \tag{7.5}$$

where the ε_i's are independent $\mathcal{N}(0, \sigma^2)$.

A *scatter plot* is the collection of data points with the coordinates (x_i, y_i), for $i = 1, \dots, n$. A typical scatter plot for a polynomial regression is shown in Figure 3.

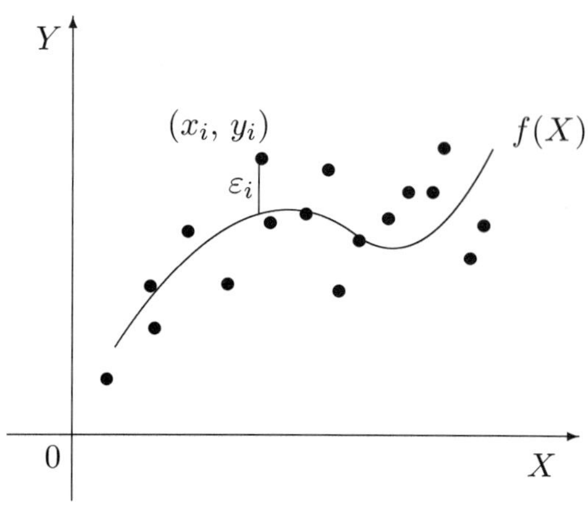

Figure 3. A scatter plot with a fitted polynomial regression function.

It is convenient to write (7.5) using vectors. To this end, introduce column vectors

$$\mathbf{y} = \big(y_1, \dots, y_n\big)', \quad \boldsymbol{\varepsilon} = \big(\varepsilon_1, \dots, \varepsilon_n\big)'$$

and

$$\mathbf{g}_j = \big(g_j(x_1), \dots, g_j(x_n)\big)', \; j = 0, \dots k.$$

Here the prime indicates the operation of vector transposition. In this notation, the equations (7.5) turn into

$$\mathbf{y} = \theta_0\,\mathbf{g}_0 + \theta_1\,\mathbf{g}_1 + \cdots + \theta_k\,\mathbf{g}_k + \boldsymbol{\varepsilon} \tag{7.6}$$

where $\boldsymbol{\varepsilon} \sim \mathcal{N}_n(\mathbf{0}, \sigma^2\,\mathbf{I}_n)$. That is, $\boldsymbol{\varepsilon}$ has an n-variate normal distribution with mean $\mathbf{0} = (0, \dots, 0)'$ and covariance matrix $\mathbb{E}\big[\,\boldsymbol{\varepsilon}\,\boldsymbol{\varepsilon}'\,\big] = \sigma^2\,\mathbf{I}_n$, where $\mathbf{I}_n$ is the $n \times n$ identity matrix.

Denote a *linear span-space* generated by the vectors $\mathbf{g}_0, \dots, \mathbf{g}_k$ by

$$\mathcal{S} = \operatorname{span}\{\,\mathbf{g}_0, \dots, \mathbf{g}_k\,\} \subseteq \mathbb{R}^n.$$

The vectors $\mathbf{g}_0, \dots, \mathbf{g}_k$ are assumed to be linearly independent, so that the dimension of the span-space $\dim(\mathcal{S})$ is equal to $k + 1$. Obviously, it may happen only if $n \geq k + 1$. Typically, n is much larger than k.

Example 7.3. For the polynomial regression, the span-space $\mathcal{S}$ is generated by the vectors $\mathbf{g}_0 = (1, \dots, 1)'$, $\mathbf{g}_1 = (x_1, \dots, x_n)', \dots, \mathbf{g}_k = (x_1^k, \dots, x_n^k)'$. For distinct values $x_1, \dots, x_n$, $n \geq k+1$, these vectors are linearly independent, and the assumption $\dim(\mathcal{S}) = k+1$ is fulfilled (see Exercise 11.79). □

Define an $n \times (k + 1)$ matrix $\mathbf{G} = \big[\,\mathbf{g}_0, \dots, \mathbf{g}_k\,\big]$, called a *design matrix*, and let $\boldsymbol{\theta} = \big(\,\theta_0, \dots, \theta_k\,\big)'$ denote the vector of the regression coefficients. The linear regression (7.6) can be written in the matrix form

$$\mathbf{y} = \mathbf{G}\,\boldsymbol{\theta} + \boldsymbol{\varepsilon}, \;\; \boldsymbol{\varepsilon} \sim \mathcal{N}_n(\mathbf{0},\, \sigma^2\,\mathbf{I}_n). \tag{7.7}$$

7.2. Least-Squares Estimator

In the system of equations (7.5) (or, equivalently, in its vector form (7.6)), the parameters $\theta_0, \dots, \theta_k$ have unknown values, which should be estimated from the observations $(x_i,\, y_i)$, $i = 1, \dots, n$.

Let $\hat{\mathbf{y}} = (\hat{y}_1, \dots, \hat{y}_n)'$ denote the orthogonal projection of $\mathbf{y}$ on the span-space $\mathcal{S}$ (see Figure 4). This vector is called a *fitted* (or *predicted*) *response vector*. As any vector in $\mathcal{S}$, this projection is a linear combination of vectors $\mathbf{g}_0$, $\mathbf{g}_1, \dots, \mathbf{g}_k$, that is, there exist some constants $\hat{\theta}_0$, $\hat{\theta}_1$, $\dots, \hat{\theta}_k$ such that

$$\hat{\mathbf{y}} = \hat{\theta}_0\,\mathbf{g}_0 + \hat{\theta}_1\,\mathbf{g}_1 + \cdots + \hat{\theta}_k\,\mathbf{g}_k. \tag{7.8}$$

These coefficients $\hat{\theta}_0, \hat{\theta}_1, \dots, \hat{\theta}_k$ may serve as estimates of the unknown parameters θ_0, $\theta_1, \dots, \theta_k$. Indeed, in the absence of the random error in (7.6), that is, when $\boldsymbol{\varepsilon} = \mathbf{0}$, we have $\hat{\mathbf{y}} = \mathbf{y}$ which implies that $\hat{\theta}_j = \theta_j$ for all $j = 0, \dots, k$.

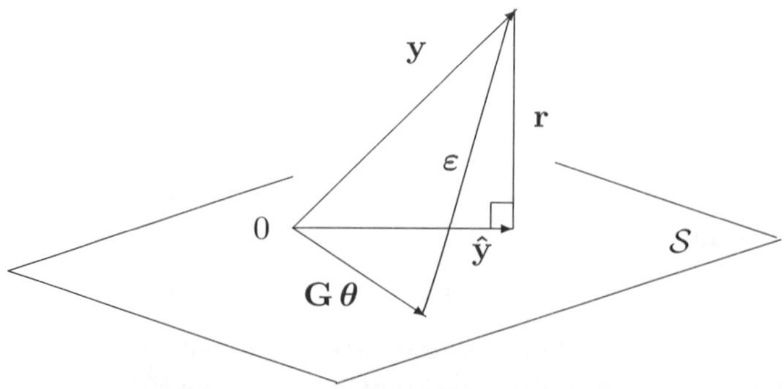

Figure 4. Geometric interpretation of the linear parametric regression.

The problem of finding the estimators $\hat{\theta}_0, \hat{\theta}_1, \dots, \hat{\theta}_k$ can be looked at as the minimization problem

$$\| \mathbf{y} - \hat{\mathbf{y}} \|^2 = \| \mathbf{y} - (\hat{\theta}_0 \mathbf{g}_0 + \cdots + \hat{\theta}_k \mathbf{g}_k) \|^2 \to \min_{\hat{\theta}_0, \dots, \hat{\theta}_k} . \tag{7.9}$$

Here $\| \cdot \|$ denotes the Euclidean norm of a vector in $\mathbb{R}^n$,

$$\| \mathbf{y} - \hat{\mathbf{y}} \|^2 = (y_1 - \hat{y}_1)^2 + \cdots + (y_n - \hat{y}_n)^2.$$

The estimation procedure consists of finding the minimum of the sum of squares of the coordinates, thus, the estimators $\hat{\theta}_0, \dots, \hat{\theta}_k$ are referred to as the *least-squares estimators.*

The easiest way to solve the minimization problem is through the geometric interpretation of linear regression. In fact, by the definition of a projection, the vector $\mathbf{y} - \hat{\mathbf{y}}$ is orthogonal to every vector in the span-space $\mathcal{S}$. In particular, its dot product with any basis vector in $\mathcal{S}$ must be equal to zero,

$$(\mathbf{y} - \hat{\mathbf{y}}, \mathbf{g}_j) = 0, \quad j = 0, \dots, k. \tag{7.10}$$

Substituting $\hat{\mathbf{y}}$ in (7.10) by its expression from (7.8), we arrive at the system of $k + 1$ linear equations with respect to the estimators $\hat{\theta}_0, \dots, \hat{\theta}_k$,

$$(\mathbf{y}, \mathbf{g}_j) - \hat{\theta}_0 (\mathbf{g}_0, \mathbf{g}_j) - \cdots - \hat{\theta}_k (\mathbf{g}_k, \mathbf{g}_j) = 0, \quad j = 0, \dots, k.$$

These equations can be rewritten in a standard form, which are known as a system of *normal equations*,

$$\hat{\theta}_0 (\mathbf{g}_0, \mathbf{g}_j) + \cdots + \hat{\theta}_k (\mathbf{g}_k, \mathbf{g}_j) = (\mathbf{y}, \mathbf{g}_j), \quad j = 0, \dots, k. \tag{7.11}$$

Let $\hat{\boldsymbol{\theta}} = (\hat{\theta}_0, \dots, \hat{\theta}_k)'$ be the vector of estimated regression coefficients. Then we can write equations (7.11) in the matrix form

$$(\mathbf{G}'\mathbf{G}) \hat{\boldsymbol{\theta}} = \mathbf{G}'\mathbf{y}. \tag{7.12}$$

By our assumption, the $(k+1) \times (k+1)$ matrix $\mathbf{G}'\mathbf{G}$, has a full rank $k+1$, and therefore is invertible. Thus, the *least-squares estimator* of $\boldsymbol{\theta}$ is the unique solution of the normal equations (7.12),

$$\hat{\boldsymbol{\theta}} = \left(\mathbf{G}'\mathbf{G}\right)^{-1}\mathbf{G}'\mathbf{y}. \tag{7.13}$$

Remark 7.4. Three Euclidean spaces are involved in the linear regression. The primary space is the (X, Y)-plane where observed as well as fitted values may be depicted. Another is the space of observations $\mathbb{R}^n$ that includes the linear subspace $\mathcal{S}$. And the third space is the space $\mathbb{R}^{k+1}$ that contains the vector of the regression coefficients $\boldsymbol{\theta}$ as well as its least-squares estimator $\hat{\boldsymbol{\theta}}$. Though the latter two spaces play an auxiliary role in practical regression analysis, they are important from the mathematical point of view. □

7.3. Properties of the Least-Squares Estimator

Consider the least-squares estimator $\hat{\boldsymbol{\theta}} = \left(\hat{\theta}_0, \dots, \hat{\theta}_k\right)'$ of the vector of the true regression coefficients $\boldsymbol{\theta} = \left(\theta_0, \dots, \theta_k\right)'$ computed by formula (7.13). In this section, we study the properties of this estimator.

Recall that we denoted by $\mathcal{X} = (x_1, \dots, x_n)$ the design in the regression model. The explanatory variable X may be assumed deterministic, or random with a certain distribution. In what follows, we use the notation $\mathbb{E}_{\boldsymbol{\theta}}[\,\cdot\,|\,\mathcal{X}\,]$ and $\mathbb{V}ar_{\boldsymbol{\theta}}[\,\cdot\,|\,\mathcal{X}\,]$ for the conditional expectation and variance with respect to the distribution of the random error ε, given the design $\mathcal{X}$. Averaging over both distributions, ε's and $\mathcal{X}$'s, will be designated by $\mathbb{E}_{\boldsymbol{\theta}}[\,\cdot\,]$. For the deterministic designs, we use the notation $\mathbb{E}_{\boldsymbol{\theta}}[\,\cdot\,|\,\mathcal{X}\,]$ only if we want to emphasize the dependence on the design $\mathcal{X}$.

Theorem 7.5. *For a fixed design $\mathcal{X}$, the least-squares estimator $\hat{\boldsymbol{\theta}}$ has a $(k+1)$-variate normal distribution with mean $\boldsymbol{\theta}$ (is unbiased) and covariance matrix $\mathbb{E}_{\boldsymbol{\theta}}\left[(\hat{\boldsymbol{\theta}} - \boldsymbol{\theta})(\hat{\boldsymbol{\theta}} - \boldsymbol{\theta})'\,|\,\mathcal{X}\right] = \sigma^2\left(\mathbf{G}'\mathbf{G}\right)^{-1}$.*

Proof. According to the matrix form of the linear regression (7.7), the conditional mean of $\mathbf{y}$, given the design $\mathcal{X}$, is $\mathbb{E}_{\boldsymbol{\theta}}\left[\mathbf{y}\,|\,\mathcal{X}\right] = \mathbf{G}\,\boldsymbol{\theta}$, and the conditional covariance matrix of $\mathbf{y}$ is equal to

$$\mathbb{E}_{\boldsymbol{\theta}}\left[\left(\mathbf{y} - \mathbf{G}\,\boldsymbol{\theta}\right)\left(\mathbf{y} - \mathbf{G}\,\boldsymbol{\theta}\right)'\,|\,\mathcal{X}\right] = \mathbb{E}_{\boldsymbol{\theta}}\left[\,\boldsymbol{\varepsilon}\,\boldsymbol{\varepsilon}'\,|\,\mathcal{X}\right] = \sigma^2\,\mathbf{I}_n.$$

Thus, the conditional mean of $\hat{\boldsymbol{\theta}}$, given the design $\mathcal{X}$, is calculated as

$$\mathbb{E}_{\boldsymbol{\theta}}\left[\hat{\boldsymbol{\theta}}\,|\,\mathcal{X}\right] = \mathbb{E}_{\boldsymbol{\theta}}\left[\left(\mathbf{G}'\mathbf{G}\right)^{-1}\mathbf{G}'\mathbf{y}\,|\,\mathcal{X}\right]$$

$$= \left(\mathbf{G}'\mathbf{G}\right)^{-1}\mathbf{G}'\,\mathbb{E}_{\boldsymbol{\theta}}\left[\mathbf{y}\,|\,\mathcal{X}\right] = \left(\mathbf{G}'\mathbf{G}\right)^{-1}\mathbf{G}'\,\mathbf{G}\,\boldsymbol{\theta} = \boldsymbol{\theta}.$$

To find an expression for the conditional covariance matrix of $\hat{\boldsymbol{\theta}}$, notice first that $\hat{\boldsymbol{\theta}} - \boldsymbol{\theta} = \left(\mathbf{G}'\mathbf{G}\right)^{-1}\mathbf{G}'(\mathbf{y} - \mathbf{G}\boldsymbol{\theta})$. Thus,

$$\mathbb{E}_{\boldsymbol{\theta}}\left[(\hat{\boldsymbol{\theta}} - \boldsymbol{\theta})(\hat{\boldsymbol{\theta}} - \boldsymbol{\theta})'\,|\,\mathcal{X}\right]$$

$$= \mathbf{E}_{\boldsymbol\theta}\Big[\Big(\big(\mathbf{G}'\mathbf{G}\big)^{-1}\mathbf{G}'(\mathbf{y}-\mathbf{G}\boldsymbol\theta)\Big)\Big(\big(\mathbf{G}'\mathbf{G}\big)^{-1}\mathbf{G}'(\mathbf{y}-\mathbf{G}\boldsymbol\theta)\Big)'\,|\,\mathcal{X}\Big]$$

$$= \big(\mathbf{G}'\mathbf{G}\big)^{-1}\mathbf{G}'\,\mathbf{E}_{\boldsymbol\theta}\Big[(\mathbf{y}-\mathbf{G}\boldsymbol\theta)(\mathbf{y}-\mathbf{G}\boldsymbol\theta)'\,|\,\mathcal{X}\Big]\,\mathbf{G}\big(\mathbf{G}'\mathbf{G}\big)^{-1}$$

$$= \big(\mathbf{G}'\mathbf{G}\big)^{-1}\mathbf{G}'\,\sigma^2\,\mathbf{I}_n\,\mathbf{G}\big(\mathbf{G}'\mathbf{G}\big)^{-1} = \sigma^2\,\big(\mathbf{G}'\mathbf{G}\big)^{-1}. \qquad \square$$

To ease the presentation, we study the regression on the interval $[0,1]$, that is, we assume that the regression function $f(x)$ and all the components in the linear regression model, $g_0(x),\dots,g_k(x)$, are defined for $x \in [0,1]$. The design points x_i, $i = 1,\dots,n$, also belong to this interval.

Define the *least-squares estimator of the regression function* $f(x)$ in (7.2), at any point $x \in [0,1]$, by

$$(7.14) \qquad \hat f_n(x) = \hat\theta_0\, g_0(x) + \cdots + \hat\theta_k\, g_k(x).$$

Here the subscript n indicates that the estimation is based on n pairs of observations $(x_i,\, y_i)$, $i = 1,\dots,n$.

A legitimate question is how close $\hat f_n(x)$ is to $f(x)$? We try to answer this question using two different loss functions. The first one is the quadratic loss function computed at a fixed point $x \in [0,1]$,

$$(7.15) \qquad w\big(\hat f_n - f\big) = \big(\hat f_n(x) - f(x)\big)^2.$$

The risk with respect to this loss is called the *mean squared risk at a point* or *mean squared error (MSE)*.

The second loss function that we consider is the mean squared difference over the design points

$$(7.16) \qquad w\big(\hat f_n - f\big) = \frac{1}{n}\sum_{i=1}^{n}\big(\hat f_n(x_i) - f(x_i)\big)^2.$$

Note that this loss function is a discrete version of the integral L_2-norm,

$$\|\,\hat f_n - f\,\|_2^2 = \int_0^1 \big(\hat f_n(x) - f(x)\big)^2\,dx.$$

The respective risk is a discrete version of the *mean integrated squared error (MISE)*.

In this section, we study the conditional risk $\mathbb{E}_{\boldsymbol\theta}\big[w(\hat f_n - f)\,|\,\mathcal{X}\big]$, given the design $\mathcal{X}$. The next two lemmas provide computational formulas for the MSE and discrete MISE, respectively.

Introduce the matrix $\mathbf{D} = \sigma^2\,(\mathbf{G}'\mathbf{G})^{-1}$ called the *covariance matrix*. Note that $\mathbf{D}$ depends on the design $\mathcal{X}$, and this dependence can be sophisticated. In particular, that if the design $\mathcal{X}$ is random, this matrix is random as well.

Lemma 7.6. *For a fixed design $\mathcal{X}$, the estimator $\hat{f}_n(x)$ is an unbiased estimator of $f(x)$ at any $x \in [0,1]$, so that its MSE equals the variance of $\hat{f}_n(x)$,*

$$\mathbb{V}ar_{\boldsymbol{\theta}}\big[\,\hat{f}_n(x)\,|\,\mathcal{X}\,\big] = \mathbb{E}_{\boldsymbol{\theta}}\Big[\,\big(\,\hat{f}_n(x) - f(x)\,\big)^2\,|\,\mathcal{X}\,\Big] = \sum_{l,\,m=0}^{k} \mathbf{D}_{l,\,m}\, g_l(x)\, g_m(x),$$

where $\mathbf{D}_{l,\,m}$ denotes the (l,m)-th entry of the covariance matrix $\mathbf{D}$.

Proof. By Theorem 7.5, the least-squares estimator $\hat{\boldsymbol{\theta}}$ is unbiased. This implies the unbiasedness of the estimator $\hat{f}_n(x)$. To see that, write

$$\begin{aligned}\mathbb{E}_{\boldsymbol{\theta}}\big[\,\hat{f}_n(x)\,|\,\mathcal{X}\,\big] &= \mathbb{E}_{\boldsymbol{\theta}}\big[\,\hat{\theta}_0\,|\,\mathcal{X}\,\big]g_0(x) + \cdots + \mathbb{E}_{\boldsymbol{\theta}}\big[\,\hat{\theta}_k\,|\,\mathcal{X}\,\big]g_k(x)\\ &= \theta_0\, g_0(x) + \cdots + \theta_k\, g_k(x) = f(x).\end{aligned}$$

Also, the covariance matrix of $\hat{\boldsymbol{\theta}}$ is $\mathbf{D}$, and therefore the variance of $\hat{f}_n(x)$ can be written as

$$\mathbb{E}_{\boldsymbol{\theta}}\Big[\,\big(\,\hat{f}_n(x) - f(x)\,\big)^2\,|\,\mathcal{X}\,\Big] = \mathbb{E}_{\boldsymbol{\theta}}\Big[\,\big(\,(\hat{\theta}_0 - \theta_0)\, g_0(x) + \cdots + (\hat{\theta}_k - \theta_k)\, g_k(x)\,\big)^2\,|\,\mathcal{X}\,\Big]$$

$$= \sum_{l,m=0}^{k} \mathbb{E}_{\boldsymbol{\theta}}\big[(\hat{\theta}_l - \theta_l)(\hat{\theta}_m - \theta_m)|\mathcal{X}\big]\, g_l(x) g_m(x) = \sum_{l,m=0}^{k} \mathbf{D}_{l,m} g_l(x) g_m(x). \quad \square$$

Lemma 7.7. *For a fixed design $\mathcal{X}$, the mean squared difference*

$$\frac{1}{n}\sum_{i=1}^{n} \big(\,\hat{f}_n(x_i) - f(x_i)\,\big)^2 = (\sigma^2/n)\,\chi^2_{k+1}$$

where χ^2_{k+1} denotes a chi-squared random variable with $k+1$ degrees of freedom. In particular, the MISE equals to

$$\mathbb{E}_{\boldsymbol{\theta}}\Big[\frac{1}{n}\sum_{i=1}^{n} \big(\,\hat{f}_n(x_i) - f(x_i)\,\big)^2\,|\,\mathcal{X}\,\Big] = \frac{\sigma^2(k+1)}{n}.$$

Proof. Applying the facts that $\sigma^2\,\mathbf{G}'\mathbf{G} = \mathbf{D}^{-1}$, and that the matrix $\mathbf{D}$ is symmetric and positive definite (therefore, $\mathbf{D}^{1/2}$ exists), we have the equations

$$\frac{1}{n}\sum_{i=1}^{n} \big(\,\hat{f}_n(x_i) - f(x_i)\,\big)^2 = \frac{1}{n}\,\|\,\mathbf{G}\,(\,\hat{\boldsymbol{\theta}} - \boldsymbol{\theta}\,)\,\|^2$$

$$= \frac{1}{n}\,(\,\mathbf{G}\,(\,\hat{\boldsymbol{\theta}} - \boldsymbol{\theta}\,)\,)'\,(\,\mathbf{G}\,(\,\hat{\boldsymbol{\theta}} - \boldsymbol{\theta}\,)\,) = \frac{1}{n}\,(\,\hat{\boldsymbol{\theta}} - \boldsymbol{\theta}\,)'\,\mathbf{G}'\mathbf{G}\,(\,\hat{\boldsymbol{\theta}} - \boldsymbol{\theta}\,)$$

$$= \sigma^2\,\frac{1}{n}\,(\,\hat{\boldsymbol{\theta}} - \boldsymbol{\theta}\,)'\,\mathbf{D}^{-1}(\,\hat{\boldsymbol{\theta}} - \boldsymbol{\theta}\,) = \frac{\sigma^2}{n}\,\|\,\mathbf{D}^{-1/2}\,(\,\hat{\boldsymbol{\theta}} - \boldsymbol{\theta}\,)\,\|^2,$$

where by $\|\cdot\|$ we mean the Euclidean norm in the $\mathbb{R}^n$ space of observations.

By Theorem 7.5, the $(k+1)$-dimensional vector $\mathbf{D}^{-1/2}\,(\hat{\boldsymbol{\theta}} - \boldsymbol{\theta})$ has independent standard normal coordinates. The result of the proposition follows from the definition of the chi-squared distribution. □

Note that the vector with the components $\hat{f}_n(x_i)$ coincides with $\hat{\mathbf{y}}$, the projection of $\mathbf{y}$ on the span-space $\mathcal{S}$, that is,

$$\hat{y}_i \;=\; \hat{f}_n(x_i), \quad i = 1, \dots, n.$$

Introduce the vector $\mathbf{r} = \mathbf{y} - \hat{\mathbf{y}}$. The coordinates of this vector, called *residuals*, are the differences

$$r_i = y_i - \hat{y}_i = y_i - \hat{f}_n(x_i), \quad i = 1, \dots, n.$$

In other words, residuals are deviations of the observed responses from the predicted ones evaluated at the design points.

Graphically, residuals can be visualized in the data space $\mathbb{R}^n$. The vector of residuals $\mathbf{r}$, plotted in Figure 4, is orthogonal to the span-space $\mathcal{S}$. Also, the residuals r_i's can be depicted on a scatter plot (see Figure 5).

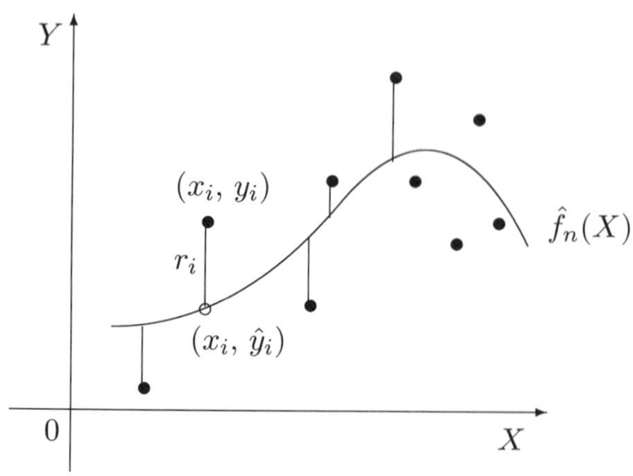

Figure 5. Residuals shown on a schematic scatter plot.

In the next lemma, we obtain the distribution of the squared norm of the residual vector $\mathbf{r}$ for a fixed design $\mathcal{X}$.

Lemma 7.8. *For a given design $\mathcal{X}$, the sum of squares of the residuals*

$$r_1^2 + \cdots + r_n^2 \;=\; \|\,\mathbf{r}\|^2 = \|\mathbf{y} \;-\; \hat{\mathbf{y}}\,\|^2 \;=\; \sigma^2\,\chi^2_{n-k-1},$$

where χ^2_{n-k-1} denotes a chi-squared random variable with $n-k-1$ degrees of freedom.

Proof. The squared Euclidean norm of the vector of random errors admits the partition

$$\|\,\boldsymbol{\varepsilon}\,\|^2 = \|\,\mathbf{y} - \mathbf{G}\,\boldsymbol{\theta}\,\|^2 = \|\,\mathbf{y} - \hat{\mathbf{y}} + \hat{\mathbf{y}} - \mathbf{G}\,\boldsymbol{\theta}\,\|^2$$

$$= \|\,\mathbf{y} - \hat{\mathbf{y}}\,\|^2 + \|\,\hat{\mathbf{y}} - \mathbf{G}\,\boldsymbol{\theta}\,\|^2 = \|\,\mathbf{r}\,\|^2 + \|\,\hat{\mathbf{y}} - \mathbf{G}\,\boldsymbol{\theta}\,\|^2.$$

Here the cross term is zero, because it is a dot product of the residual vector $\mathbf{r}$ and the vector $\hat{\mathbf{y}} - \mathbf{G}\,\boldsymbol{\theta}$ that lies in the span-space $\mathcal{S}$. Moreover, these two vectors are independent (see Exercise 7.46).

The random vector $\boldsymbol{\varepsilon}$ has $\mathcal{N}_n(\mathbf{0},\, \sigma^2\,\mathbf{I}_n)$ distribution, implying that $\|\,\boldsymbol{\varepsilon}\,\|^2 = \sigma^2\,\chi^2_n$, where χ^2_n denotes a chi-squared random variable with n degrees of freedom. Also, by Lemma 7.7,

$$\|\,\hat{\mathbf{y}} - \mathbf{G}\,\boldsymbol{\theta}\,\|^2 = \sum_{i=1}^{n} \big(\,\hat{f}_n(x_i) - f(x_i)\,\big)^2 = \sigma^2\,\chi^2_{k+1}$$

where χ^2_{k+1} has a chi-squared distribution with $k+1$ degrees of freedom.

Taking into account that vectors $\mathbf{r}$ and $\hat{\mathbf{y}} - \mathbf{G}\boldsymbol{\theta}$ are independent, it can be shown (see Exercise 7.47) that $\|\,\mathbf{r}\,\|^2$ has a chi-squared distribution with $n - (k+1)$ degrees of freedom. □

7.4. Asymptotic Analysis of the Least-Squares Estimator

In this section we focus on describing asymptotic behavior of the least-squares estimator $\hat{\boldsymbol{\theta}}$ as the sample size n goes to infinity. This task is complicated by the fact that $\hat{\boldsymbol{\theta}}$ depends on the design $\mathcal{X} = \{x_1, \dots, x_n\}$. Thus, we can expect the existence of a limiting distribution only if the design is governed by some regularity conditions.

7.4.1. Regular Deterministic Design. Take a continuous strictly positive probability density $p(x)$, $0 \le x \le 1$, and consider the cumulative distribution function $F_X(x) = \int_0^x p(t)\,dt$. Define a sequence of *regular deterministic designs* $\mathcal{X}_n = \{\,x_{n,1}, \dots, x_{n,n}\,\}$ where $x_{i,n}$ is the (i/n)-th quantile of this distribution,

$$F_X(x_{n,i}) = \frac{i}{n}, \quad i = 1, \dots, n. \tag{7.17}$$

Equivalently, the $x_{n,i}$'s satisfy the recursive equations

$$\int_{x_{n,i-1}}^{x_{n,i}} p(x)\,dx = \frac{1}{n}, \quad i = 1, \dots, n, \quad x_{n,0} = 0. \tag{7.18}$$

It is important to emphasize that the distances between consecutive points in a regular design have magnitude $O(1/n)$ as $n \to \infty$. Typical irregular designs that are avoided in asymptotic analysis have data points that are too close to each other (concentrated around one point, or even coincide), or have big gaps between each other, or both.

For simplicity we suppress the dependence on n of the regular design points, that is, we write x_i instead of $x_{n,i}$, $i = 1, \dots, n$.

Example 7.9. The data points that are spread equidistantly on the unit interval, $x_i = i/n$, $i = 1, \dots, n$, constitute a regular design, called *uniform design*, since these points are (i/n)-th quantiles of the standard uniform distribution. □

It can be shown (see Exercise 7.48) that in the case of a regular design corresponding to a probability density $p(x)$, for any continuous function $g(x)$, the Riemann sum converges to the integral

$$\frac{1}{n}\sum_{i=1}^{n} g(x_i) \to \int_0^1 g(x)\,p(x)\,dx \quad \text{as } n \to \infty. \tag{7.19}$$

If the functions $g_0, g_1, \dots, g_k$ in the linear regression model (7.5) are continuous, and the design points are regular, then the convergence in (7.19) implies the existence of the entrywise limits of the matrix $(1/n)\mathbf{D}^{-1}$ as $n \to \infty$, that is, for any l and m such that $0 \le l \le m \le k$,

$$\lim_{n\to\infty} \frac{1}{n}\mathbf{D}^{-1}_{l,m} = \lim_{n\to\infty} \frac{\sigma^2}{n}\,(\,\mathbf{G}'\mathbf{G}\,)_{l,m}$$

$$= \lim_{n\to\infty} \frac{\sigma^2}{n}\Big(\,g_l(x_1)\,g_m(x_1) + \cdots + g_l(x_n)\,g_m(x_n)\Big)$$

$$= \sigma^2 \int_0^1 g_l(x)\,g_m(x)\,p(x)\,dx. \tag{7.20}$$

Denote by $\mathbf{D}^{-1}_{\infty}$ the matrix with the elements $\sigma^2 \int_0^1 g_l(x)\,g_k(x)\,p(x)\,dx$. Assume that this matrix is positive definite. Then its inverse $\mathbf{D}_{\infty}$, called the *limiting covariance matrix*, exists, and the convergence takes place $n\,\mathbf{D} \to \mathbf{D}_{\infty}$.

Example 7.10. Consider a polynomial regression model with the uniform design on $[0, 1]$, that is, the regular design with the constant probability density $p(x) = 1$, $0 \le x \le 1$. The matrix $\mathbf{D}^{-1}_{\infty}$ has the entries

$$\sigma^2 \int_0^1 x^l\,x^m\,dx = \frac{\sigma^2}{1 + l + m}, \quad 0 \le l,\, m \le k. \tag{7.21}$$

This is a positive definite matrix, and hence the limiting covariance matrix $\mathbf{D}_{\infty}$ is well defined (see Exercise 7.49). □

We are ready to summarize our findings in the following theorem.

Theorem 7.11. *If $\mathcal{X}$ is a regular deterministic design, and $\mathbf{D}_\infty$ exists, then*

$$\sqrt{n}\,(\hat{\boldsymbol{\theta}} - \boldsymbol{\theta}) \to \mathcal{N}_{k+1}(\mathbf{0}, \mathbf{D}_\infty) \text{ as } n \to \infty.$$

Next we study the limiting behavior of the least-squares estimator $\hat{f}_n$ defined by (7.14). The lemma below shows that in the mean squared sense, $\hat{f}_n$ converges pointwise to the true regression function f at the rate $O(1/\sqrt{n})$ as $n \to \infty$. The proof of this lemma is assigned as an exercise (see Exercise 7.50).

Lemma 7.12. *Suppose $\mathcal{X}$ is a regular deterministic design such that $\mathbf{D}_\infty$ exists. Then at any fixed point $x \in [0,1]$, the estimator $\hat{f}_n$ of the regression function f is unbiased and its normalized quadratic risk satisfies the limiting equation*

$$\lim_{n\to\infty} \mathbb{E}_{\boldsymbol{\theta}}\Big[\,\big(\sqrt{n}\,(\hat{f}_n(x) - f(x))\big)^2\Big] = \sum_{l,\,m=0}^{k} (\mathbf{D}_\infty)_{l,m}\, g_l(x)\, g_m(x),$$

where $(\mathbf{D}_\infty)_{l,m}$ are the elements of the limiting covariance matrix $\mathbf{D}_\infty$.

7.4.2. Regular Random Design. We call a random design *regular*, if its points are independent with a common continuous and strictly positive probability density function $p(x)$, $x \in [0,1]$.

Suppose the functions $g_0, \dots, g_k$ are continuous on $[0,1]$. By the Law of Large Numbers, for any element of the matrix $\mathbf{D}^{-1} = \sigma^2\,\mathbf{G}'\mathbf{G}$, we have that with probability 1 (with respect to the distribution of the random design),

$$\lim_{n\to\infty} \frac{\sigma^2}{n}(\mathbf{G}'\mathbf{G})_{l,m} = \lim_{n\to\infty} \frac{\sigma^2}{n}\Big(g_l(x_1)\,g_m(x_1) + \cdots + g_l(x_n)\,g_m(x_n)\Big)$$

$$= \sigma^2 \int_0^1 g_l(x)\, g_m(x)\, p(x)\, dx. \tag{7.22}$$

Again, as in the case of a regular deterministic design, we assume that the matrix $\mathbf{D}_\infty^{-1}$ with the elements $\sigma^2 \int_0^1 g_l(x)\, g_m(x)\, p(x)\, dx$ is positive definite, so that its inverse matrix $\mathbf{D}_\infty$ exists.

The essential difference between the random and deterministic designs is that even in the case of a regular random design, for any given n, the matrix $\mathbf{G}'\mathbf{G}$ can be degenerate with a positive probability (see Exercise 7.51). If it happens, then for the sake of definiteness, we put $\hat{\boldsymbol{\theta}} = \mathbf{0}$. Fortunately, if the functions $g_0, \dots, g_k$ are continuous in $[0,1]$, then the probability of this "non-existence" is exponentially small in n as $n \to \infty$. For the proofs of the following lemma and theorem refer to Exercises 7.52 and 7.53.

Lemma 7.13. *Assume that* $|g_0|, \dots, |g_k| \le C_0$, *and that* $\mathcal{X} = \{x_1, \dots, x_n\}$ *is a regular random design. Then for any* n, *for however small* $\delta > 0$, *and for all* l *and* m *such that* $0 \le l, m \le k$, *the following inequality holds:*

$$\mathbb{P}\Big(\Big| \frac{1}{n} \sum_{i=1}^{n} g_l(x_i) g_m(x_i) - \int_0^1 g_l(x) g_m(x)\, dx \Big| > \delta \Big) \le 2 \exp\Big\{ -\frac{\delta^2 n}{2C_0^4} \Big\}.$$

Assume that for a regular random design $\mathcal{X}$, the estimator $\hat{\boldsymbol{\theta}}$ is properly defined with probability 1. Then, as the next theorem shows, the distribution of the normalized estimator $\sqrt{n}(\hat{\boldsymbol{\theta}} - \boldsymbol{\theta})$ is asymptotically normal.

Theorem 7.14. *If* $\mathcal{X}$ *is a regular random design and* $\mathbf{D}_\infty$ *exists, then as* $n \to \infty$, $\sqrt{n}\big(\hat{\boldsymbol{\theta}} - \boldsymbol{\theta}\big)$ *converges in distribution to a* $\mathcal{N}_{k+1}(\mathbf{0}, \mathbf{D}_\infty)$ *random variable.*

Remark 7.15. An important conclusion is that the parametric least-squares estimator $\hat{f}_n$ is unbiased, and its typical rate of convergence under various norms and under regular designs is equal to $O(1/\sqrt{n})$ as $n \to \infty$. □

Exercises

Exercise 7.43. Consider the observations (x_i, y_i) in a *simple linear regression model,*

$$y_i = \theta_0 + \theta_1 x_i + \varepsilon_i, \quad i = 1, \dots, n,$$

where the ε_i's are independent $\mathcal{N}(0, \sigma^2)$ random variables. Write down the system of normal equations (7.11) and solve it explicitly.

Exercise 7.44. Show that in a simple linear regression model (see Exercise 7.43), the minimum of the variance $\mathbb{V}ar_{\boldsymbol{\theta}}\big[\hat{f}_n(x) \,|\, \mathcal{X}\big]$ in Lemma 7.6 is attained at $x = \bar{x} = \sum_{i=1}^n x_i / n$.

Exercise 7.45. (i) Prove that in a simple linear regression model (see Exercise 7.43), the sum of residuals is equal to zero, that is, $\sum_{i=1}^n r_i = \sum_{i=1}^n (y_i - \hat{y}_i) = 0$.
(ii) Consider a *simple linear regression through the origin,*

$$y_i = \theta_1 x_i + \varepsilon_i, \quad i = 1, \dots, n$$

where the ε_i's are independent $\mathcal{N}(0, \sigma^2)$ random variables. Show by giving an example that the sum of residuals is not necessarily equal to zero.

Exercise 7.46. Show that (i) the vector of residuals $\mathbf{r}$ has a multivariate normal distribution with mean zero and covariance matrix $\sigma^2 (\mathbf{I}_n - \mathbf{H})$, where $\mathbf{H} = \mathbf{G}(\mathbf{G}'\mathbf{G})^{-1} \mathbf{G}'$ is called the *hat matrix* because of the identity

$\hat{\mathbf{y}} = \mathbf{H}\mathbf{y}$.
(ii) Argue that the vectors $\mathbf{r}$ and $\hat{\mathbf{y}} - \mathbf{G}\,\boldsymbol{\theta}$ are independent.

Exercise 7.47. Let $Z = X + Y$ where X and Y are independent. Suppose Z and X have chi-squared distributions with n and m degrees of freedom, respectively, where $m < n$. Prove that Y also has a chi-squared distribution with $n - m$ degrees of freedom.

Exercise 7.48. Show the convergence of the Riemann sum in (7.19).

Exercise 7.49. Show that the matrix with the elements given by (7.21) is invertible.

Exercise 7.50. Prove Lemma 7.12.

Exercise 7.51. Let $k = 1$, and let $g_0 = 1$; $g_1(x) = x$ if $0 \le x \le 1/2$, and $g_1(x) = 1/2$ if $1/2 < x \le 1$. Assume that $\mathcal{X}$ is the uniform random design governed by the density $p(x) = 1$. Show that the system of normal equations does not have a unique solution with probability $1/2^n$.

Exercise 7.52. Prove Lemma 7.13.

Exercise 7.53. Prove Theorem 7.14.

Exercise 7.54. For the regression function $f = \theta_0\, g_0 + \cdots + \theta_k\, g_k$, show that the conditional expectation of the squared L_2-norm of the difference $\hat{f}_n - f$, given the design $\mathcal{X}$, admits the upper bound

$$\mathbb{E}_{\boldsymbol{\theta}}\Big[\, \| \hat{f}_n - f \|_2^2 \,|\, \mathcal{X} \Big] \le tr(\mathbf{D})\, \| \mathbf{g} \|_2^2$$

where the trace $tr(\mathbf{D}) = \mathbb{E}_{\boldsymbol{\theta}}\Big[\, \sum_{i=0}^{k} (\hat{\theta}_i - \theta_i)^2 \,|\, \mathcal{X} \Big]$ is the sum of the diagonal elements of the covariance matrix $\mathbf{D}$, and

$$\| \mathbf{g} \|_2^2 = \sum_{i=0}^{k} \| g_i \|_2^2 = \sum_{i=0}^{k} \int_0^1 \big(g_i(x)\big)^2\, dx$$

is the squared L_2-norm of the vector $\mathbf{g} = (g_0, \dots, g_k)'$.

Part 2

Nonparametric Regression

Chapter 8

Estimation in Nonparametric Regression

8.1. Setup and Notations

In a *nonparametric regression model* the response variable Y and the explanatory variable X are related by the same regression equation (7.1) as in a parametric regression model,

$$Y = f(X) + \varepsilon \tag{8.1}$$

with the random error $\varepsilon \sim \mathcal{N}(0, \sigma^2)$. However, unlike that in the parametric regression model, here the algebraic form of the regression function f is assumed unknown and must be evaluated from the data. The goal of the nonparametric regression analysis is to estimate the function f as a curve, rather than to estimate parameters of a guessed function.

A set of n pairs of observations $(x_1, y_1), \dots, (x_n, y_n)$ satisfy the relation

$$y_i = f(x_i) + \varepsilon_i, \quad i = 1, \dots, n, \tag{8.2}$$

where the ε_i's are independent $\mathcal{N}(0, \sigma^2)$ random errors. For simplicity we assume that the design $\mathcal{X} = \{x_1, \dots, x_n\}$ is concentrated on $[0, 1]$.

In nonparametric regression analysis, some assumptions are made a priori on the smoothness of the regression function f. Let $\beta \geq 1$ be an integer. We assume that f belongs to a *Hölder class* of functions of *smoothness* β, denoted by $\Theta(\beta, L, L_1)$. That is, we assume that (i) its derivative $f^{(\beta-1)}$ of order $\beta - 1$ satisfies the *Lipschitz condition* with a given constant L,

$$|\, f^{(\beta-1)}(x_2) \,-\, f^{(\beta-1)}(x_1)\,| \,\leq\, L\,|\,x_2 \,-\, x_1\,|, \;\; x_1, x_2 \in [0,1],$$

and (ii) there exists a constant $L_1 > 0$ such that

$$\max_{0 \leq x \leq 1} |\, f(x)\,| \leq L_1.$$

Example 8.1. If $\beta = 1$, the class $\Theta(1, L, L_1)$ is a set of bounded Lipschitz functions. Recall that a *Lipschitz function* f satisfies the inequality

$$|\, f(x_2) \,-\, f(x_1)\,| \,\leq\, L\,|\,x_2 \,-\, x_1\,|$$

where L is a constant independent of x_1 and x_2. □

Sometimes we write $\Theta(\beta)$, suppressing the constants L and L_1 in the notation of the Hölder class $\Theta(\beta, L, L_1)$.

Denote by $\hat{f}_n$ the nonparametric estimator of the regression function f. Since f is a function of $x \in [0,1]$, so should be the estimator. The latter, however, also depends on the data points. This dependence is frequently omitted in the notation,

$$\hat{f}_n(x) \,=\, \hat{f}_n\big(x\,;\,(x_1,\,y_1),\,\dots,\,(x_n,\,y_n)\big), \;\; 0 \leq x \leq 1.$$

To measure how close $\hat{f}_n$ is to f, we consider the same loss functions as in Chapter 7, the quadratic loss function computed at a fixed point $x \in [0,1]$ specified in (7.15), and the mean squared difference over the design points given by (7.16). In addition, to illustrate particular effects in nonparametric estimation, we use the *sup-norm loss function*

$$w(\hat{f}_n - f) \,=\, \|\,\hat{f}_n - f\,\|_\infty \,=\, \sup_{0 \leq x \leq 1} \big|\,\hat{f}_n(x) \,-\, f(x)\,\big|.$$

Note that in the nonparametric case, the loss functions are, in fact, functionals since they depend of f. For simplicity, we will continue calling them functions. We denote the risk function by

$$R_n(f, \hat{f}_n) \,=\, \mathbb{E}_f\big[\,w(\hat{f}_n \,-\, f)\,\big]$$

where the subscript f in the expectation refers to a fixed regression function f. If the design $\mathcal{X}$ is random, we use the conditional expectation $\mathbb{E}_f\,[\,\cdot\,|\,\mathcal{X}\,]$ to emphasize averaging over the distribution of the random error ε.

When working with the difference $\hat{f}_n \,-\, f$, it is technically more convenient to consider separately the bias $b_n(x) \,=\, \mathbb{E}_f\big[\,\hat{f}_n(x)\,\big] - f(x)$, and the stochastic part $\xi_n(x) \,=\, \hat{f}_n(x) - \mathbb{E}_f\big[\,\hat{f}_n(x)\,\big]$. Then the MSE or discrete MISE is split into a sum (see Exercise 8.55),

$$R_n(\hat{f}_n, f) \,=\, \mathbb{E}_f\big[\,w(\hat{f}_n - f)\,\big] \,=\, \mathbb{E}_f\big[\,w(\xi_n)\,\big] \,+\, w(b_n). \tag{8.3}$$

For the sup-norm loss function, the triangle inequality applies

$$R_n(\hat{f}_n, f) = \mathbb{E}_f\big[\, \|\hat{f}_n - f\|_\infty \big] \le \mathbb{E}_f\big[\, \|\xi_n\|_\infty \big] + \|b_n\|_\infty.$$

To deal with random designs, we consider the conditional bias and stochastic part of an estimator $\hat{f}_n$, given the design $\mathcal{X}$,

$$b_n(x, \mathcal{X}) = \mathbb{E}_f\big[\, \hat{f}_n(x) \,|\, \mathcal{X} \big] - f(x)$$

and

$$\xi_n(x, \mathcal{X}) = \hat{f}_n(x) - \mathbb{E}_f\big[\, \hat{f}_n(x) \,|\, \mathcal{X} \big].$$

8.2. Asymptotically Minimax Rate of Convergence. Definition

We want to estimate the regression function in the most efficient way. As a criterion of optimality we choose the asymptotically minimax rate of convergence of the estimator.

Consider a deterministic sequence of positive numbers $\psi_n \to 0$ as $n \to \infty$. Introduce a *maximum normalized risk* of an estimator $\hat{f}_n$ with respect to a loss function w by

$$r_n(\hat{f}_n, w, \psi_n) = \sup_{f \in \Theta(\beta)} \mathbb{E}_f\Big[\, w\Big(\frac{\hat{f}_n - f}{\psi_n} \Big) \Big]. \tag{8.4}$$

A sequence of positive numbers ψ_n is called an *asymptotically minimax rate of convergence* if there exist two positive constants r_* and r^* such that for any estimator $\hat{f}_n$, the maximum normalized risk $r_n(\hat{f}_n, w, \psi_n)$ is bounded from above and below,

$$r_* \le \liminf_{n\to\infty} r_n(\hat{f}_n, w, \psi_n) \le \limsup_{n\to\infty} r_n(\hat{f}_n, w, \psi_n) \le r^*. \tag{8.5}$$

This very formal definition has a transparent interpretation. It implies that for any estimator $\hat{f}_n$ and for all n large enough, the maximum of the risk is bounded from below,

$$\sup_{f \in \Theta(\beta)} \mathbb{E}_f\Big[\, w\Big(\frac{\hat{f}_n - f}{\psi_n} \Big) \Big] \ge r_* - \varepsilon, \tag{8.6}$$

where ε is an arbitrarily small positive number. On the other hand, there exists an estimator f_n^*, called the *asymptotically minimax estimator*, the maximum risk of which is bounded from above,

$$\sup_{f \in \Theta(\beta)} \mathbb{E}_f\Big[\, w\Big(\frac{f_n^* - f}{\psi_n} \Big) \Big] \le r^* + \varepsilon. \tag{8.7}$$

Note that f_n^* is not a single estimator, rather a sequence of estimators defined for all sufficiently large n.

It is worth mentioning that the asymptotically minimax rate of convergence ψ_n is not uniquely defined but admits any bounded and separated away from zero multiplier. As we have shown in Chapter 7, a typical rate of convergence in parametric regression model is $O(1/\sqrt{n})$. In nonparametric regression, on the other hand, the rates depend on a particular loss function and on the smoothness parameter β of the Hölder class of regression functions. We study these rates in the next chapters.

8.3. Linear Estimator

8.3.1. Definition. An estimator $\hat{f}_n$ is called a *linear estimator* of f, if for any $x \in [0,1]$, there exist weights $\upsilon_{n,i}(x)$ that may also depend on the design points, $\upsilon_{n,i}(x) = \upsilon_{n,i}(x, \mathcal{X})$, $i = 1, \dots, n$, such that

$$\hat{f}_n(x) = \sum_{i=1}^{n} \upsilon_{n,i}(x)\, y_i. \tag{8.8}$$

Note that the linear estimator $\hat{f}_n$ is a linear function of the response values $y_1, \dots, y_n$. The weight $\upsilon_{n,i}(x)$ determines the influence of the observation y_i on the estimator $\hat{f}_n(x)$ at point x.

An advantage of the linear estimator (8.8) is that for a given design $\mathcal{X}$, the conditional bias and variance are easily computable (see Exercise 8.56),

$$b_n(x, \mathcal{X}) = \sum_{i=1}^{n} \upsilon_{n,i}(x)\, f(x_i) - f(x) \tag{8.9}$$

and

$$\mathbb{E}_f\left[\, \xi_n^2(x, \mathcal{X}) \,|\, \mathcal{X} \,\right] = \sigma^2 \sum_{i=1}^{n} \upsilon_{n,i}^2(x). \tag{8.10}$$

These formulas are useful when either the design $\mathcal{X}$ is deterministic or integration over the distribution of a random design is not too difficult. Since the weights $\upsilon_{n,i}(x)$ may depend on the design points in a very intricate way, in general, averaging over the distribution of $x_1, \dots, x_n$ is a complicated task.

The linear estimator (8.8) is not guaranteed to be unbiased. Even in the simplest case of a constant regression function $f(x) = \theta_0$, the linear estimator is unbiased if and only if the weights sum up to one,

$$b_n(x, \mathcal{X}) = \sum_{i=1}^{n} \upsilon_{n,i}(x)\, \theta_0 - \theta_0 = \theta_0 \Big(\sum_{i=1}^{n} \upsilon_{n,i}(x) - 1 \Big) = 0.$$

For a linear regression function $f(x) = \theta_0 + \theta_1 x$, the linear estimator is unbiased if and only if the following identity holds:

$$\Big(\sum_{i=1}^{n} \upsilon_{n,i}(x) - 1\Big)\theta_0 + \Big(\sum_{i=1}^{n} \upsilon_{n,i}(x)\, x_i - x\Big)\,\theta_1 = 0,$$

which under the condition that $\sum_{i=1}^{n} \upsilon_{n,i}(x) = 1$ is tantamount to the identity

$$\sum_{i=1}^{n} \upsilon_{n,i}(x)\, x_i = x, \quad \text{uniformly in } x \in [0,1].$$

If for any $x \in [0,1]$, the linear estimator (8.8) depends on all the design points $x_1, \dots, x_n$, it is called a *global linear estimator* of the regression function. We study global estimators later in this book.

An estimator (8.8) is called a *local linear estimator* of the regression function if the weights $\upsilon_{n,i}(x)$ differ from zero only for those i's for which the design points x_i's belong to a small neighborhood of x, that is, $|\, x_i - x \,| \le h_n$, where h_n is called a *bandwidth*. We always assume that

$$h_n > 0, \quad h_n \to 0, \quad \text{and} \quad n h_n \to \infty \quad \text{as } n \to \infty. \tag{8.11}$$

In what follows we consider only designs in which for any $x \in [0,1]$, the number of the design points in the h_n-neighborhood of x has the magnitude $O(nh_n)$ as $n \to \infty$.

8.3.2. The Nadaraya-Watson Kernel Estimator. Consider a smooth or piecewise smooth function $K = K(u)$, $u \in \mathbb{R}$. Assume that the support of K is the interval $[-1,1]$, that is, $K(u) = 0$ if $|u| > 1$. The function K is called a *kernel function* or simply, a *kernel.*

Example 8.2. Some classical kernel functions frequently used in practice are:

(i) *uniform*, $K(u) = (1/2)\,\mathbb{I}(\,|u| \le 1\,)$,
(ii) *triangular*, $K(u) = (\,1 - |u|\,)\,\mathbb{I}(\,|u| \le 1\,)$,
(iii) *bi-square*, $K(u) = (15/16)\,(\,1 - u^2\,)^2\,\mathbb{I}(\,|u| \le 1\,)$,
(iv) the *Epanechnikov* kernel, $K(u) = (3/4)\,(\,1 - u^2\,)\,\mathbb{I}(\,|u| \le 1\,)$. □

Remark 8.3. Typically, kernels are normalized in such a way that they integrate to one. It can be shown (see Exercise 8.57) that all the kernels introduced above are normalized in such a way. □

For a chosen kernel and a bandwidth, define the weights $\upsilon_{n,i}(x)$ by

$$\upsilon_{n,i}(x) = K\Big(\frac{x_i - x}{h_n}\Big) \Big/ \sum_{j=1}^{n} K\Big(\frac{x_j - x}{h_n}\Big). \tag{8.12}$$

The *Nadaraya-Watson kernel estimator* $\hat{f}_n$ of the regression function f at a given point $x \in [0,1]$ is the linear estimator with the weights defined by (8.12),

$$\hat{f}_n(x) = \sum_{i=1}^{n} y_i\, K\Big(\frac{x_i - x}{h_n}\Big) \Big/ \sum_{j=1}^{n} K\Big(\frac{x_j - x}{h_n}\Big). \tag{8.13}$$

Note that the Nadaraya-Watson estimator is an example of a local linear estimator, since outside of the interval $[x - h_n\,, x + h_n]$, the weights are equal to zero.

Example 8.4. Consider the uniform kernel defined in Example 8.2 (i). Let $N(x,\, h_n)$ denote the number of the design points in the h_n-neighborhood of x. Then the weights in (8.12) have the form

$$\upsilon_{n,i}(x) = \frac{1}{N(x,\, h_n)}\, \mathbb{I}\big(x - h_n < x_i < x + h_n\big).$$

Thus, in this case, the Nadaraya-Watson estimator is the average of the observed responses that correspond to the design points in the h_n-neighborhood of x,

$$\hat{f}_n(x) = \frac{1}{N(x,\, h_n)} \sum_{i=1}^{n} y_i\, \mathbb{I}\big(x - h_n < x_i < x + h_n\big). \qquad \Box$$

8.4. Smoothing Kernel Estimator

In Section 8.3, we explained the challenge to control the conditional bias of a linear estimator even in the case of a linear regression function. The linear regression function is important as the first step because, as the following lemma shows, any regression function from a Hölder class is essentially a polynomial. The proof of this auxiliary lemma is postponed until the end of this section.

Lemma 8.5. *For any function $f \in \Theta(\beta, L, L_1)$, the following Taylor expansion holds:*

$$f(x_i) = \sum_{m=0}^{\beta-1} \frac{f^{(m)}(x)}{m!}\,(x_i - x)^m + \rho(x_i, x), \quad 0 \le x, x_i \le 1, \tag{8.14}$$

where $f^{(m)}$ denotes the m-th derivative of f. Also, for any x_i and x such that $|x_i - x| \le h_n$, the remainder term $\rho(x_i, x)$ satisfies the inequality

$$|\,\rho(x_i, x)\,| \le \frac{L h_n^{\beta}}{(\beta-1)!}. \tag{8.15}$$

It turns out that for linear estimators, regular random designs have an advantage over deterministic ones. As we demonstrate in this section, when computing the risk, averaging over the distribution of a random design helps to eliminate a significant portion of the bias.

Next we introduce a linear estimator that guarantees the zero bias for any polynomial regression function up to degree $\beta - 1$ (see Exercise 8.59). To ease the presentation, we assume that a regular random design is uniform with the probability density $p(x) = 1$, $x \in [0, 1]$. The extension to a more general case is given in Remark 8.6.

A *smoothing kernel estimator* $\hat{f}_n(x)$ of degree $\beta - 1$ is given by the formula

$$\hat{f}_n(x) = \frac{1}{nh_n} \sum_{i=1}^{n} y_i \, K\Big(\frac{x_i - x}{h_n}\Big), \quad 0 < x < 1, \tag{8.16}$$

where the *smoothing kernel* $K = K(u)$, $|u| \le 1$, is bounded, piecewise continuous, and satisfies the normalization and orthogonality conditions

$$\int_{-1}^{1} K(u)\, du = 1 \ \text{ and } \int_{-1}^{1} u^m K(u)\, du = 0 \ \text{ for } m = 1, \dots, \beta - 1. \tag{8.17}$$

Note that the smoothing kernel is orthogonal to all monomials up to degree $\beta - 1$.

Remark 8.6. For a general density $p(x)$ of the design points, the smoothing kernel estimator is defined as

$$\hat{f}_n(x) = \frac{1}{nh_n} \sum_{i=1}^{n} \frac{y_i}{p(x_i)} \, K\Big(\frac{x_i - x}{h_n}\Big) \tag{8.18}$$

where the kernel $K(u)$ satisfies the same conditions as in (8.17). □

Remark 8.7. A smoothing kernel estimator (8.16) requires that x lies strictly inside the unit interval. In fact, the definition of $\hat{f}_n(x)$ is valid for any x such that $h_n \le x \le 1 - h_n$. On the other hand, a linear estimator (8.8) is defined for any $x \in [0, 1]$, including the endpoints. Why does the smoothing kernel estimator fail if x coincides with either of the endpoints? If, for instance, $x = 0$, then for any symmetric kernel $K(u)$, the expected value

$$\frac{1}{h_n} \mathbb{E}_f \Big[K\Big(\frac{x_i}{h_n}\Big) \Big] = \frac{1}{h_n} \int_0^{h_n} K\Big(\frac{x_i}{h_n}\Big)\, dx_i = \int_0^1 K(u)\, du = \frac{1}{2}.$$

For example, in the situation when the regression function is identically equal to 1, the responses are $y_i = 1 + \varepsilon_i$, where ε_i are $\mathcal{N}(0, \sigma^2)$ random variables independent of x_i's for all $i = 1, \dots, n$. The average value of the smoothing kernel estimator at zero is

$$\mathbb{E}_f\big[\,\hat{f}_n(0)\,\big] \,=\, \mathbb{E}_f\Big[\,\frac{1}{nh_n}\sum_{i=1}^{n}(1+\varepsilon_i)\,K\Big(\frac{x_i}{h_n}\Big)\Big] \,=\, \frac{1}{2},$$

which is certainly not satisfactory.

A remedy for the endpoints is to define a one-sided kernel to preserve the normalization and orthogonality conditions (8.17). In Exercises 8.61 and 8.62 we formulate some examples related to this topic. □

The next lemma gives upper bounds for the bias and variance of the smoothing kernel estimator (8.16). The proof of the lemma can be found at the end of this section.

Lemma 8.8. *For any regression function $f \in \Theta(\beta, L, L_1)$, at any point $x \in (0,1)$, the bias and variance of the smoothing kernel estimator* (8.16) *admit the upper bounds for all large enough n,*

$$|\,b_n(x)\,| \,\leq\, A_b h_n^{\beta} \quad \text{and} \quad \mathbb{V}ar_f\big[\,\hat{f}_n(x)\,\big] \,\leq\, \frac{A_v}{nh_n}$$

with the constants

$$A_b \,=\, \frac{L\,\|K\|_1}{(\beta-1)!} \quad \text{and} \quad A_v \,=\, (L_1^2+\sigma^2)\,\|K\|_2^2$$

where $\|K\|_1 = \int_{-1}^{1}|K(u)|\,du$ and $\|K\|_2^2 = \int_{-1}^{1}K^2(u)\,du$.

Remark 8.9. The above lemma clearly indicates that as h_n increases, the upper bound for the bias increases, while that for the variance decreases. □

Applying this lemma, we can bound the mean squared risk of $\hat{f}_n(x)$ at a point $x \in (0,1)$ by

$$\text{(8.19)}\quad \mathbb{E}_f\Big[\,\big(\,\hat{f}_n(x) - f(x)\,\big)^2\,\Big] \,=\, b_n^2(x) \,+\, \mathbb{V}ar_f\big[\,\hat{f}_n(x)\,\big] \,\leq\, A_b^2 h_n^{2\beta} \,+\, \frac{A_v}{nh_n}.$$

It is easily seen that the value of h_n that minimizes the right-hand side of (8.19) satisfies the equation

$$\text{(8.20)}\qquad h_n^{2\beta} \,=\, \frac{A}{nh_n}$$

with a constant factor A independent of n. This equation is called the *balance equation* since it reflects the idea of balancing the squared bias and variance terms.

Next, we neglect the constant in the balance equation (8.20), and label the respective *optimal bandwidth* by a superscript (*). It is a solution of the equation

$$h_n^{2\beta} \,=\, \frac{1}{nh_n},$$

and is equal to

$$h_n^* = n^{-1/(2\beta+1)}.$$

Denote by $f_n^*(x)$ the smoothing kernel estimator (8.16) corresponding to the optimal bandwidth h_n^*,

$$f_n^*(x) = \frac{1}{nh_n^*} \sum_{i=1}^{n} y_i \, K\Big(\frac{x_i - x}{h_n^*} \Big). \tag{8.21}$$

We call this estimator the *optimal smoothing kernel estimator.*

For the convenience of reference, we formulate the proposition below. Its proof follows directly from the expression (8.19), and the definition of the estimator $f_n^*(x)$.

Proposition 8.10. *For all large enough* n, *and any* $f \in \Theta(\beta)$, *the quadratic risk of the optimal smoothing kernel estimator* (8.21) *at a given point* x, $0 < x < 1$, *is bounded from above by*

$$\mathbb{E}_f\Big[\big(f_n^*(x) - f(x) \big)^2 \Big] \leq (A_b^2 + A_v)\, n^{-2\beta/(2\beta+1)}.$$

Remark 8.11. Suppose the loss function is the absolute difference at a given point $x \in (0,1)$. Then the supremum over $f \in \Theta(\beta)$ of the risk of the estimator $f_n^*(x)$ is bounded from above by

$$\sup_{f \in \Theta(\beta)} \mathbb{E}_f\big[\,|\, f_n^*(x) - f(x) \,|\, \big] \leq (A_b^2 + A_v)^{1/2}\, n^{-\beta/(2\beta+1)}.$$

This follows immediately from Proposition 8.10 and the Cauchy-Schwarz inequality. □

Finally, we give the proofs of two technical lemmas stated in this section.

Proof of Lemma 8.5. We need to prove that the bound (8.15) for the remainder term is valid. For $\beta = 1$, the bound follows from the definition of the Lipschitz class of functions $\Theta(1, L, L_1)$,

$$|\, \rho(x_i, x) \,| = |\, f(x_i) - f(x) \,| \leq L|x_i - x| \leq Lh_n.$$

If $\beta \geq 2$, then the Taylor expansion with the Lagrange remainder term has the form

$$f(x_i) = \sum_{m=0}^{\beta-2} \frac{f^{(m)}(x)}{m!} (x_i - x)^m + \frac{f^{(\beta-1)}(x^*)}{(\beta-1)!} (x_i - x)^{\beta-1} \tag{8.22}$$

where x^* is an intermediate point between x and x_i, so that $|\, x^* - x \,| \leq h_n$. This remainder can be transformed into

$$\frac{f^{(\beta-1)}(x^*)}{(\beta-1)!} (x_i - x)^{\beta-1} = \frac{f^{(\beta-1)}(x)}{(\beta-1)!} (x_i - x)^{\beta-1} + \rho(x_i, x)$$

where the new remainder term $\rho(x_i, x)$, satisfies the inequality for any x_i and x such that $|x_i - x| \leq h_n$,

$$| \rho(x_i, x) | = \frac{| f^{(\beta-1)}(x^*) - f^{(\beta-1)}(x) |}{(\beta-1)!} | x_i - x |^{\beta-1}$$

$$\leq \frac{L|x^* - x|}{(\beta-1)!} |x_i - x|^{\beta-1} \leq \frac{L h_n}{(\beta-1)!} h_n^{\beta-1} = \frac{L h_n^{\beta}}{(\beta-1)!}.$$

In the above, the definition of the Hölder class $\Theta(\beta, L, L_1)$ has been applied.

□

Proof of Lemma 8.8. Using the definition of the bias and the regression equation $y_i = f(x_i) + \varepsilon_i$, we write

$$b_n(x) = \frac{1}{nh_n} \mathbb{E}_f \Big[\sum_{i=1}^{n} y_i K\Big(\frac{x_i - x}{h_n} \Big) \Big] - f(x)$$

$$= \frac{1}{nh_n} \mathbb{E}_f \Big[\sum_{i=1}^{n} (f(x_i) + \varepsilon_i) K\Big(\frac{x_i - x}{h_n} \Big) \Big] - f(x). \tag{8.23}$$

Now since ε_i has mean zero and is independent of x_i,

$$\mathbb{E}_f \Big[\sum_{i=1}^{n} \varepsilon_i K\Big(\frac{x_i - x}{h_n} \Big) \Big] = 0.$$

Also, by the normalization condition,

$$\frac{1}{h_n} \mathbb{E}_f \Big[K\Big(\frac{x_i - x}{h_n} \Big) \Big] = \frac{1}{h_n} \int_{x-h_n}^{x+h_n} K\Big(\frac{x_i - x}{h_n} \Big) dx_i = \int_{-1}^{1} K(u)\, du = 1.$$

Consequently, continuing from (8.23), we can write

$$b_n(x) = \frac{1}{nh_n} \mathbb{E}_f \Big[\sum_{i=1}^{n} (f(x_i) - f(x)) K\Big(\frac{x_i - x}{h_n} \Big) \Big]. \tag{8.24}$$

Substituting Taylor's expansion (8.14) of the function $f(x_i)$ into (8.24), we get that for any $\beta > 1$,

$$|b_n(x)| = \frac{1}{nh_n} \Big| \mathbb{E}_f \Big[\sum_{i=1}^{n} \Big(\sum_{m=1}^{\beta-1} \frac{f^{(m)}(x)(x_i - x)^m}{m!} + \rho(x_i, x) \Big) K\Big(\frac{x_i - x}{h_n} \Big) \Big] \Big|$$

$$\leq \frac{1}{h_n} \Big| \sum_{m=1}^{\beta-1} \int_{x-h_n}^{x+h_n} \frac{f^{(m)}(x)\,(x_1 - x)^m}{m!} K\Big(\frac{x_1 - x}{h_n} \Big) dx_1$$

$$+ \frac{1}{h_n} \max_{z: |z-x| \leq h_n} |\rho(z, x)| \int_{x-h_n}^{x+h_n} \Big| K\Big(\frac{x_1 - x}{h_n} \Big) \Big| dx_1.$$

In the above, we replaced x_i by x_1 due to the independence of the design points. If $\beta = 1$, we agree to define the sum over m as zero. For any $\beta > 1$, this sum equals zero as well, which can be seen from the orthogonality conditions. For $m = 1, \dots, \beta - 1$,

$$\int_{x-h_n}^{x+h_n} (x_1 - x)^m \, K\Big(\frac{x_1 - x}{h_n}\Big)\, dx_1 \;=\; h_n^{m+1} \int_{-1}^{1} u^m \, K(u)\, du \;=\; 0.$$

Thus, using the inequality (8.15) for the remainder term $\rho(x_i, x)$, we obtain that for any $\beta \geq 1$, the absolute value of the bias is bounded by

$$|b_n(x)| \;\leq\; \frac{1}{h_n} \max_{z:|z-x|\leq h_n} |\rho(z,x)| \int_{x-h_n}^{x+h_n} \Big| K\Big(\frac{x_1 - x}{h_n}\Big)\Big| \, dx_1$$

$$\leq\; \frac{L h_n^{\beta}}{(\beta-1)!} \int_{-1}^{1} |\, K(u)\,|\, du \;=\; \frac{L \|K\|_1 h_n^{\beta}}{(\beta-1)!} \;=\; A_b h_n^{\beta}.$$

Further, to find a bound for the variance of $\hat{f}_n(x)$, we use the independence of the data points to write

$$\mathbb{V}ar_f\big[\,\hat{f}_n(x)\,\big] \;=\; \mathbb{V}ar_f\Big[\,\frac{1}{nh_n} \sum_{i=1}^{n} y_i \, K\Big(\frac{x_i - x}{h_n}\Big)\Big]$$

$$=\; \frac{1}{(nh_n)^2} \sum_{i=1}^{n} \mathbb{V}ar_f\Big[\, y_i \, K\Big(\frac{x_i - x}{h_n}\Big)\Big].$$

Now we bound the variance by the second moment, and plug in the regression equation $y_i = f(x_i) + \varepsilon_i$,

$$\leq\; \frac{1}{(nh_n)^2} \sum_{i=1}^{n} \mathbb{E}_f\Big[\, y_i^2 \, K^2\Big(\frac{x_i - x}{h_n}\Big)\Big]$$

$$=\; \frac{1}{(nh_n)^2} \sum_{i=1}^{n} \mathbb{E}_f\Big[\, \big(\, f(x_i) + \varepsilon_i\,\big)^2 \, K^2\Big(\frac{x_i - x}{h_n}\Big)\Big]$$

$$=\; \frac{1}{(nh_n)^2} \sum_{i=1}^{n} \mathbb{E}_f\Big[\, \big(\, f^2(x_i) \,+\, \varepsilon_i^2\big) \, K^2\Big(\frac{x_i - x}{h_n}\Big)\Big].$$

Here the cross term disappears because of independence of ε_i and x_i, and the fact that the expected value of ε_i is zero. Finally, using the facts that $|f(x_i)| \leq L_1$ and $\mathbb{E}_f\big[\,\varepsilon_i^2\,\big] = \sigma^2$, we find

$$\leq\; \frac{1}{(nh_n)^2} \, n \, \big(L_1^2 + \sigma^2\big) \int_{x-h_n}^{x+h_n} K^2\Big(\frac{x_1 - x}{h_n}\Big)\, dx_1$$

$$=\; \frac{1}{nh_n} \big(L_1^2 + \sigma^2\big) \int_{-1}^{1} K^2(u)\, du \;=\; \frac{1}{nh_n} \big(L_1^2 + \sigma^2\big)\, \|K\|_2^2 \;=\; \frac{A_v}{nh_n}. \qquad \square$$

Exercises

Exercise 8.55. Prove (8.3) for: (i) the quadratic loss at a point

$$w(\hat{f}_n - f) = \big(\hat{f}_n(x) - f(x)\big)^2,$$

and (ii) the mean squared difference

$$w(\hat{f}_n - f) = \frac{1}{n}\sum_{i=1}^{n}\big(\hat{f}_n(x_i) - f(x_i)\big)^2.$$

Exercise 8.56. Prove (8.9) and (8.10).

Exercise 8.57. Show that the kernels introduced in Example 8.2 integrate to one.

Exercise 8.58. Consider the Nadaraya-Watson estimator defined by (8.13). Show that conditional on the design $\mathcal{X}$, its bias
(i) is equal to zero, for any constant regression function $f(x) = \theta_0$,
(ii) does not exceed $L h_n$ in absolute value, for any regression function $f \in \Theta(1, L, L_1)$.

Exercise 8.59. Prove that the smoothing kernel estimator (8.16) is unbiased if the regression function f is a polynomial up to order $\beta - 1$.

Exercise 8.60. Find the normalizing constant C such that the *tri-cube kernel function*

$$K(u) = C(1 - |u|^3)^3\, \mathbb{I}(|u| \le 1)$$

integrates to one. What is its degree? Hint: Use (8.17).

Exercise 8.61. To define a smoothing kernel estimator at either endpoint of the unit interval, we can use formula (8.16), with $K(u)$ being a one-sided kernel function (see Remark 8.7).
(i) Show that to estimate the regression function at $x = 0$, the kernel

$$K(u) = 4 - 6u, \ 0 \le u \le 1,$$

may be applied, that satisfies the normalization and orthogonality conditions

$$\int_0^1 K(u)\,du = 1 \ \text{ and } \ \int_0^1 uK(u)\,du = 0.$$

(ii) Show that at $x = 1$, the kernel

$$K(u) = 4 + 6u, \ -1 \le u \le 0,$$

may be used, which satisfies the normalization and orthogonality conditions

$$\int_{-1}^{0} K(u)\,du = 1 \text{ and } \int_{-1}^{0} uK(u)\,du = 0.$$

Exercise 8.62. Refer to Exercise 8.61. We can apply a one-sided smoothing kernel to estimate the regression function f at x where $0 \le x \le h_n$. For example, we can take $K(u) = 4 - 6u,\ 0 \le u \le 1$. However, this kernel function does not use the observations located between 0 and x.

To deal with this drawback, we can introduce a family of smoothing kernels $K_\theta(u)$ that utilize all the observations to estimate the regression function for any x such that $0 \le x \le h_n$.
(i) Let $x = x_\theta = \theta h_n,\ 0 \le \theta \le 1$. Find a family of smoothing kernels $K_\theta(u)$ with the support $[-\theta, 1]$, satisfying the normalization and orthogonal conditions

$$\int_{-\theta}^{1} K_\theta(u)\,du = 1 \text{ and } \int_{-\theta}^{1} uK_\theta(u)\,du = 0\,.$$

Hint: Search for $K_\theta(u)$ in the class of linear functions.
(ii) Let $x = x_\theta = 1 - \theta h_n,\ 0 \le \theta \le 1$. Show that the family of smoothing kernels $K_\theta(-u),\ -1 \le u \le \theta$, can be applied to estimate $f(x)$ for any x such that $1 - h_n \le x \le 1$.

Chapter 9

Local Polynomial Approximation of the Regression Function

9.1. Preliminary Results and Definition

In a small neighborhood of a fixed point $x \in [0, 1]$, an unknown nonparametric regression function $f(x)$ can be approximated by a polynomial. This method, called the *local polynomial approximation*, is introduced in this section. Below we treat the case of the point x lying strictly inside the unit interval, $0 < x < 1$. The case of x being one of the endpoints is left as an exercise (see Exercise 9.64.)

Choose a bandwidth h_n that satisfies the standard conditions (8.11),

$$h_n > 0\,, \quad h_n \to 0\,, \quad \text{and} \quad nh_n \to \infty \quad \text{as} \quad n \to \infty.$$

Let n be so large that the interval $[x - h_n, x + h_n] \subseteq [0, 1]$. Denote by N the number of observations in the interval $[x - h_n, x + h_n]$,

$$N = \#\{\, i \,:\, x_i \in [x - h_n, x + h_n]\,\}.$$

Without loss of generality, we can assume that the observations (x_i, y_i) are distinct and numbered so that the first N design points belong to this interval,

$$x - h_n \leq x_1 < \cdots < x_N \leq x + h_n.$$

Consider the restriction of the original nonparametric Hölder regression function $f \in \Theta(\beta) = \Theta(\beta, L, L_1)$ to the interval $[x - h_n, x + h_n]$. That is, consider $f = f(t)$ where $x - h_n \leq t \leq x + h_n$. Recall that every function f in $\Theta(\beta)$ is essentially a polynomial of degree $\beta - 1$ with a small remainder term described in Lemma 8.5. Let us forget for a moment about the remainder term, and let us try to approximate the nonparametric regression function by a parametric polynomial regression of degree $\beta - 1$. The least-squares estimator in the parametric polynomial regression is defined via the solution of the minimization problem with respect to the estimates of the regression coefficients $\hat{\theta}_0, \dots, \hat{\theta}_{\beta-1}$,

(9.1)

$$\sum_{i=1}^{N} \Big(y_i - \Big[\hat{\theta}_0 + \hat{\theta}_1 \Big(\frac{x_i - x}{h_n} \Big) + \cdots + \hat{\theta}_{\beta-1} \Big(\frac{x_i - x}{h_n} \Big)^{\beta-1} \Big] \Big)^2 \rightarrow \min_{\hat{\theta}_0, \dots, \hat{\theta}_{\beta-1}} .$$

In each monomial, it is convenient to subtract x as the midpoint of the interval $[x - h_n, x + h_n]$, and to scale by h_n so that the monomials do not vanish as h_n shrinks.

Recall from Chapter 7 that solving the minimization problem (9.1) is equivalent to solving the system of normal equations

$$(\mathbf{G}'\mathbf{G})\, \hat{\boldsymbol{\theta}} = \mathbf{G}'\mathbf{y} \tag{9.2}$$

where $\hat{\boldsymbol{\theta}} = \big(\hat{\theta}_0, \dots, \hat{\theta}_{\beta-1}\big)'$ and $\mathbf{G} = \big(\mathbf{g}_0, \dots, \mathbf{g}_{\beta-1}\big)$ is the design matrix. Its m-th column has the form

$$\mathbf{g}_m = \Big(\Big(\frac{x_1 - x}{h_n} \Big)^m, \dots, \Big(\frac{x_N - x}{h_n} \Big)^m \Big)', \quad m = 0, \dots, \beta - 1.$$

The system of normal equations (9.2) has a unique solution if the matrix $\mathbf{G}'\mathbf{G}$ is invertible. We always make this assumption. It suffices to require that the design points are distinct and that $N \geq \beta$.

Applying Lemma 8.5, we can present each observation y_i as the sum of the three components: a polynomial of degree $\beta - 1$, a remainder term, and a random error,

$$y_i = \sum_{m=0}^{\beta-1} \frac{f^{(m)}(x)}{m!} (x_i - x)^m + \rho(x_i, x) + \varepsilon_i \tag{9.3}$$

where

$$|\, \rho(x_i, x) \,| \leq \frac{L h_n^{\beta}}{(\beta - 1)!} = O(h_n^{\beta}), \quad i = 1, \dots, N.$$

The system of normal equations (9.2) is linear in $\mathbf{y}$, hence each component of y_i in (9.3) can be treated separately. The next lemma provides the information about the first polynomial component.

Lemma 9.1. *If each entry of* $\mathbf{y} = (y_1, \dots, y_N)'$ *has only the polynomial component, that is,*

$$y_i = \sum_{m=0}^{\beta-1} \frac{f^{(m)}(x)}{m!} (x_i - x)^m = \sum_{m=0}^{\beta-1} \frac{f^{(m)}(x)}{m!} h_n^m \left(\frac{x_i - x}{h_n} \right)^m, \quad i = 1, \dots, N,$$

then the least-squares estimates in (9.1) *are equal to*

$$\hat{\theta}_m = \frac{f^{(m)}(x)}{m!} h_n^m, \quad m = 0, \dots, \beta - 1.$$

Proof. The proof follows immediately if we apply the results of Section 7.1. Indeed, the vector $\mathbf{y}$ belongs to the span-space $\mathcal{S}$, so it stays unchanged after projecting on this space. □

To establish results concerning the remainder $\rho(x_i, x)$ and the random error term ε_i in (9.3), some technical preliminaries are needed. In view of the fact that $|(x_i - x)/h_n| \leq 1$, all elements of matrix $\mathbf{G}$ have a magnitude $O(1)$ as n increases. That is why, generally speaking, the elements of the matrix $\mathbf{G}'\mathbf{G}$ have a magnitude $O(N)$, assuming that the number of points N may grow with n. These considerations shed light on the following assumption, which plays an essential role in this chapter.

Assumption 9.2. For a given design $\mathcal{X}$, the absolute values of the elements of the covariance matrix $\left(\mathbf{G}'\mathbf{G}\right)^{-1}$ are bounded from above by $\gamma_0 N^{-1}$ with a constant γ_0 independent of n. □

The next lemma presents the results on the remainder and stochastic terms in (9.3).

Lemma 9.3. *Suppose Assumption 9.2 holds. Then the following is valid.*

(i) If $y_i = \rho(x_i, x)$, *then the solution* $\hat{\boldsymbol{\theta}}$ *of the system of normal equations* (9.2) *has the elements* $\hat{\theta}_m$, $m = 0, \dots, \beta - 1$, *bounded by*

$$|\hat{\theta}_m| \leq C_b h_n^\beta \text{ where } C_b = \frac{\gamma_0 \beta L}{(\beta - 1)!}.$$

(ii) If $y_i = \varepsilon_i$, *then the solution* $\hat{\boldsymbol{\theta}}$ *of the system of normal equations* (9.2) *has the zero-mean normal elements* $\hat{\theta}_m$, $m = 0, \dots, \beta - 1$, *the variances of which are bounded by*

$$\mathbb{V}ar_f\left[\hat{\theta}_m \,|\, \mathcal{X}\right] \leq \frac{C_v}{N} \text{ where } C_v = (\sigma\gamma_0\beta)^2.$$

Proof. (i) As the solution of the normal equations (9.2), $\hat{\boldsymbol{\theta}} = \left(\mathbf{G}'\mathbf{G}\right)^{-1}\mathbf{G}'\mathbf{y}$. All the elements of the matrix $\mathbf{G}'$ are of the form $\left((x_i - x)/h_n\right)^m$, and thus are bounded by one. Therefore, using Assumption 9.2, we conclude that the entries of the $\beta \times N$ matrix $\left(\mathbf{G}'\mathbf{G}\right)^{-1}\mathbf{G}'$ are bounded by $\gamma_0\beta/N$. Also, from (8.15), the absolute values of the entries of the vector $\mathbf{y}$ are bounded by $Lh_n^\beta/(\beta-1)!$ since they are the remainder terms. After we compute the dot product, N cancels, and we obtain the answer.

(ii) The element $\hat{\theta}_m$ is the dot product of the m-th row of the matrix $\left(\mathbf{G}'\mathbf{G}\right)^{-1}\mathbf{G}'$ and the random vector $(\varepsilon_1, \dots, \varepsilon_N)'$. Therefore, $\hat{\theta}_m$ is the sum of independent $\mathcal{N}(0,\sigma^2)$ random variables with the weights that do not exceed $\gamma_0\beta/N$. This sum has mean zero and the variance bounded by $N\sigma^2(\gamma_0\beta/N)^2 = (\sigma\gamma_0\beta)^2/N$. □

Combining the results of Lemmas 8.5, 9.1, and 9.3, we arrive at the following conclusion.

Proposition 9.4. *Suppose Assumption 9.2 holds. Then the estimate $\hat{\theta}_m$, which is the m-th element of the solution of the system of normal equations (9.2), admits the expansion*

$$\hat{\theta}_m = \frac{f^{(m)}(x)}{m!}\,h_n^m + b_m + \mathcal{N}_m\,, \quad m = 0, \dots, \beta-1,$$

where the deterministic term b_m is the conditional bias satisfying

$$|\,b_m\,| \le C_b h_n^\beta,$$

and the stochastic term $\mathcal{N}_m$ has a normal distribution with mean zero and variance bounded by

$$\mathbb{V}ar_f\big[\mathcal{N}_m\,|\,\mathcal{X}\big] \le C_v/N.$$

Finally, we are ready to introduce the *local polynomial estimator* $\hat{f}_n(t)$, which is defined for all t such that $x - h_n \le t \le x + h_n$ by

$$\hat{f}_n(t) = \hat{\theta}_0 + \hat{\theta}_1\left(\frac{t-x}{h_n}\right) + \cdots + \hat{\theta}_{\beta-1}\left(\frac{t-x}{h_n}\right)^{\beta-1} \tag{9.4}$$

where the least-squares estimators $\hat{\theta}_0, \dots, \hat{\theta}_{\beta-1}$ are as described in Proposition 9.4.

The local polynomial estimator (9.4) corresponding to the bandwidth $h_n^* = n^{-1/(2\beta+1)}$ will be denoted by $f_n^*(t)$. Recall from Section 8.4 that h_n^* is called the *optimal bandwidth*, and it solves the equation $(h_n^*)^{2\beta} = (nh_n^*)^{-1}$.

The formula (9.4) is significantly simplified if $t = x$. In this case the local polynomial estimator is just the estimate of the intercept, $\hat{f}_n(x) = \hat{\theta}_0$.

Up to this point there was no connection between the number of the design points N in the h_n-neighborhood of x and the bandwidth h_n. Such a connection is necessary if we want to balance the bias and the variance terms in Proposition 9.4.

Assumption 9.5. There exists a positive constant γ_1, independent of n, such that for all large enough n the inequality $N \ge \gamma_1 n h_n$ holds. □

Now we will prove the result on the conditional quadratic risk at a point of the local polynomial estimator.

Theorem 9.6. *Suppose Assumptions 9.2 and 9.5 hold with* $h_n = h_n^* = n^{-1/(2\beta+1)}$. *Consider the local polynomial estimator* $f_n^*(x)$ *corresponding to* h_n^*. *Then for a given design* $\mathcal{X}$, *the conditional quadratic risk of* $f_n^*(x)$ *at the point* $x \in (0,1)$ *admits the upper bound*

$$\sup_{f \in \Theta(\beta)} \mathbb{E}_f\Big[\big(f_n^*(x) - f(x)\big)^2 \,|\, \mathcal{X}\Big] \leq r^* \, n^{-2\beta/(2\beta+1)}$$

where a positive constant r^* *is independent of* n.

Proof. By Proposition 9.4, for any $f \in \Theta(\beta)$, the conditional quadratic risk of the local polynomial estimator f_n^* is equal to

$$\mathbb{E}_f\Big[\big(f_n^*(x) - f(x)\big)^2 \,|\, \mathcal{X}\Big] = \mathbb{E}_f\big[\big(\hat{\theta}_0 - f(x)\big)^2 \,|\, \mathcal{X}\big]$$

$$= \mathbb{E}_f\Big[\big(f(x) + b_0 + \mathcal{N}_0 - f(x)\big)^2 \,|\, \mathcal{X}\Big] = b_0^2 + \mathbb{E}_f\big[\mathcal{N}_0^2 \,|\, \mathcal{X}\big]$$

$$= b_0^2 + \mathbb{V}ar_f\big[\mathcal{N}_0 \,|\, \mathcal{X}\big] \leq C_b^2 (h_n^*)^{2\beta} + C_v/N\,.$$

Applying Assumption 9.5 and the fact that h_n^* satisfies the identity $(h_n^*)^{2\beta} = (nh_n^*)^{-1} = n^{-2\beta/(2\beta+1)}$, we obtain that

$$\mathbb{E}_f\Big[\big(f_n^*(x) - f(x)\big)^2 \,|\, \mathcal{X}\Big] \leq C_b^2 (h_n^*)^{2\beta} + \frac{C_v}{\gamma_1 n h_n^*} = r^* \, n^{-2\beta/(2\beta+1)}$$

with $r^* = C_b^2 + C_v/\gamma_1$. □

Remark 9.7. Proposition 9.4 also opens a way to estimate the derivatives $f^{(m)}(t)$ of the regression function f. The estimator is especially elegant if $t = x$,

$$\hat{f}_n^{(m)}(x) = \frac{m!\,\hat{\theta}_m}{h_n^m}, \quad m = 1, \dots, \beta - 1. \tag{9.5}$$

The rate of convergence becomes slower as m increases. In Exercise 9.65, an analogue of Theorem 9.6 is stated with the rate $n^{-(\beta-m)/(2\beta+1)}$. □

9.2. Polynomial Approximation and Regularity of Design

In a further study of the local polynomial approximation, we introduce some regularity rules for a design to guarantee Assumptions 9.2 and 9.5. The lemmas that we state in this section will be proved in Section 9.4.

9.2.1. Regular Deterministic Design. Recall that according to (7.18), the design points are defined on the interval $[0,1]$ as the quantiles of a distribution with a continuous strictly positive probability density $p(x)$.

Lemma 9.8. *Let the regular deterministic design be defined by* (7.18), *and suppose the bandwidth h_n satisfies the conditions $h_n \to 0$ and $nh_n \to \infty$ as $n \to \infty$. Let N denote the number of the design points in the interval $[x - h_n, x + h_n]$. Then:*

(i) $x_{i+1} - x_i = (1 + \alpha_{i,n})/(np(x))$ where $\max_{1 \le i \le N} |\alpha_{i,n}| \to 0$ as $n \to \infty$.

(ii) $\lim_{n\to\infty} N/(nh_n) = 2p(x)$.

(iii) For any continuous function $\varphi_0(u),\ u \in [-1,1]$,

$$\lim_{n\to\infty} \frac{1}{nh_n} \sum_{i=1}^{N} \varphi_0\Big(\frac{x_i - x}{h_n}\Big) = p(x) \int_{-1}^{1} \varphi_0(u)\, du.$$

Define a matrix $\mathbf{D}_\infty^{-1}$ with the (l,m)-th element given by

$$(9.6)\quad (\mathbf{D}_\infty^{-1})_{l,m} = \begin{cases} \dfrac{1}{2}\displaystyle\int_{-1}^{1} u^{l+m}\, du = \dfrac{1}{2(l+m+1)}, & \text{if } l+m \text{ is even,} \\ 0, & \text{if } l+m \text{ is odd.} \end{cases}$$

The matrix $\mathbf{D}_\infty^{-1}$ has the inverse $\mathbf{D}_\infty$ (for a proof see Exercise 9.66). The matrix $\mathbf{D}_\infty$ is a limiting covariance matrix introduced in Chapter 7.

Lemma 9.9. *Suppose the assumptions of Lemma 9.8 hold. Then the following limit exists:*

$$\lim_{n\to\infty} N^{-1}\,(\mathbf{G}'\mathbf{G}) = \mathbf{D}_\infty^{-1},$$

and the limiting matrix is invertible.

Corollary 9.10. *Under the conditions of Lemma 9.8, Assumption 9.2 is fulfilled for all sufficiently large n, and Assumption 9.5 holds with any constant $\gamma_1 < 2p(x)$.*

Corollary 9.11. *For the regular deterministic design, the local polynomial estimator $f_n^*(x)$ with the bandwidth $h_n^* = n^{-1/(2\beta+1)}$ has the quadratic risk at $x \in (0,1)$ bounded by $r^*\, n^{-2\beta/(2\beta+1)}$ where a positive constant r^* is independent of n.*

9.2.2. Random Uniform Design. To understand the key difficulties with the random design, it suffices to look at the case of the uniformly distributed design points x_i on the interval $[0,1]$. For this design the regularity in the deterministic sense does not hold. That is, it cannot be guaranteed with probability 1 that the distances between two consecutive points are $O(1/n)$ as $n \to \infty$. With a positive probability there may be no design points in the interval $[x - h_n, x + h_n]$, or it may contain some points but the system of the normal equations (9.2) may be singular (see Exercise 7.51).

In what follows, we concentrate on the case of the optimal bandwidth $h_n^* = n^{-1/(2\beta+1)}$. Take a small fixed positive number $\alpha < 1$, and introduce the random event

$$\mathcal{A} = \Big\{ \Big| \frac{N}{nh_n^*} - 2 \Big| \le \alpha \Big\}.$$

As in the case of the deterministic design, introduce the same matrix $\mathbf{D}_\infty^{-1} = \lim_{n\to\infty} N^{-1} \big(\mathbf{G}'\mathbf{G}\big)$ and its inverse $\mathbf{D}_\infty$. Denote by C_* a constant that exceeds the absolute values of all elements of $\mathbf{D}_\infty$. Define another random event

$$\mathcal{B} = \Big\{ \Big| \big(\mathbf{G}'\mathbf{G}\big)^{-1}_{l,m} \Big| \le \frac{2C_*}{nh_n^*} \text{ for all } l, m = 0, \dots, \beta - 1 \Big\}.$$

Note that these random events depend on n, but this fact is suppressed in the notation.

Recall that the local polynomial estimator (9.4) at $t = x$ is the intercept $\hat{\theta}_0$. In the case of the random uniform design, we redefine the local polynomial estimator as

$$f_n^*(x) = \begin{cases} \hat{\theta}_0, & \text{if } \mathcal{A} \cap \mathcal{B} \text{ occurs}, \\ 0, & \text{otherwise}. \end{cases} \tag{9.7}$$

If the random event $\mathcal{A}$ occurs, then Assumption 9.5 holds with $\gamma_1 = 2 - \alpha$. If also the event $\mathcal{B}$ occurs, then Assumption 9.2 holds with $\gamma_0 = 2(2+\alpha)C_*$. Thus, if both events take place, we can anticipate an upper bound for the quadratic risk similar to the one in Theorem 9.6. If $f_n^*(x) = 0$, this estimator does not estimate the regression function at all. Fortunately, as follows from the two lemmas below (see Remark 9.14), the probability that either $\mathcal{A}$ or $\mathcal{B}$ fails is negligible as $n \to \infty$. Proofs of these lemmas can be found in the last section.

Lemma 9.12. *Let $\overline{\mathcal{A}}$ be the complement of the event $\mathcal{A}$. Then*

$$\mathbb{P}_f\big(\overline{\mathcal{A}}\big) \le 2\alpha^{-2} n^{-2\beta/(2\beta+1)}.$$

Lemma 9.13. *Let $\overline{\mathcal{B}}$ denote the complement of the event $\mathcal{B}$. Then there exists a positive number C, independent of n, such that*

$$\mathbb{P}_f\big(\overline{\mathcal{B}}\big) \le C n^{-2\beta/(2\beta+1)}.$$

Remark 9.14. Applying Lemmas 9.12 and 9.13, we see that

$$\mathbb{P}_f\big(f_n^*(x) = 0\big) = \mathbb{P}_f\big(\overline{\mathcal{A} \cap \mathcal{B}}\big)$$
$$= \mathbb{P}_f\big(\overline{\mathcal{A}} \cup \overline{\mathcal{B}}\big) \le \mathbb{P}_f\big(\overline{\mathcal{A}}\big) + \mathbb{P}_f\big(\overline{\mathcal{B}}\big)$$
$$\le 2\alpha^{-2} n^{-2\beta/(2\beta+1)} + C n^{-2\beta/(2\beta+1)} \to 0 \text{ as } n \to \infty. \qquad \square$$

Now, we are in the position to prove the main result for the quadratic risk under the random uniform design.

Theorem 9.15. *Take the optimal bandwidth $h_n^* = n^{-1/(2\beta+1)}$. Let the design $\mathcal{X}$ be random and uniform on $[0,1]$. Then the quadratic risk of the local polynomial estimator $f_n^*(x)$ at x defined by (9.7) satisfies the upper bound*

$$\sup_{f\in\Theta(\beta)} \mathbb{E}_f\Big[\big(f_n^*(x)-f(x)\big)^2\Big] \le r^{**}\, n^{-2\beta/(2\beta+1)}$$

where a positive constant r^{**} is independent of n.

Proof. Note that in the statement of Theorem 9.6, the constant r^* depends on the design $\mathcal{X}$ only through the constants γ_0 and γ_1 that appear in Assumptions 9.2 and 9.5. Thus, if the assumptions hold, then r^* is non-random, and averaging over the distribution of the design points does not affect the upper bound. Hence,

$$\mathbb{E}_f\Big[\big(f_n^*(x)-f(x)\big)^2\,\mathbb{I}(\mathcal{A}\cap\mathcal{B})\Big] \le r^*\, n^{-2\beta/(2\beta+1)}.$$

Applying this inequality and Lemmas 9.12 and 9.13, we have that for all sufficiently large n and for any $f\in\Theta(\beta,L,L_1)$,

$$\mathbb{E}_f\Big[\big(f_n^*(x)-f(x)\big)^2\Big] \le \mathbb{E}_f\Big[\big(f_n^*(x)-f(x)\big)^2\,\mathbb{I}(\mathcal{A}\cap\mathcal{B})\Big]$$

$$+\mathbb{E}_f\Big[\big(f_n^*(x)-f(x)\big)^2\,\mathbb{I}(\overline{\mathcal{A}})\Big] + \mathbb{E}_f\Big[\big(f_n^*(x)-f(x)\big)^2\,\mathbb{I}(\overline{\mathcal{B}})\Big]$$

$$\le r^*\, n^{-2\beta/(2\beta+1)} + L_1^2\big[\mathbb{P}_f(\overline{\mathcal{A}}) + \mathbb{P}_f(\overline{\mathcal{B}})\big]$$

$$\le \big[r^* + 2L_1^2\alpha^{-2} + CL_1^2\big]\, n^{-2\beta/(2\beta+1)}.$$

Finally, we choose $r^{**} = r^* + 2L_1^2\alpha^{-2} + CL_1^2$, and the result follows. □

9.3. Asymptotically Minimax Lower Bound

For the quadratic risk at a point, the results of the previous sections confirm the existence of estimators with the asymptotic rate of convergence $\psi_n = n^{-\beta/(2\beta+1)}$ in the sense of the definition (8.5). This rate is uniform over the Hölder class of regression functions $\Theta(\beta)$. To make sure that we do not miss any better estimator with a faster rate of convergence, we have to prove the lower bound for the minimax risk. In this section, we show that for all large n, and for any estimator $\hat{f}_n$ of the regression function f, the inequality

$$\sup_{f\in\Theta(\beta)} \mathbb{E}_f\Big[\big(\hat{f}_n(x)-f(x)\big)^2\Big] \ge r_*\, n^{-2\beta/(2\beta+1)} \tag{9.8}$$

holds with a positive constant r_* independent of n.

Clearly, the inequality (9.8) does not hold for any design $\mathcal{X}$. For example, if all the design points are concentrated at one point $x_1 = \cdots = x_n = x$, then our observations (x_i, y_i) are actually observations in the parametric model

$$y_i = f(x) + \varepsilon_i, \quad i = 1, \dots, n,$$

with a real-valued parameter $\theta = f(x)$. This parameter can be estimated $1/\sqrt{n}$-consistently by the simple averaging of the response values y_i. On the other hand, if the design points $x_1, \dots, x_n$ are regular, then the lower bound (9.8) turns out to be true.

9.3.1. Regular Deterministic Design. We start with the case of a deterministic regular design, and prove the following theorem.

Theorem 9.16. *Let the deterministic design points be defined by* (7.18) *with a continuous and strictly positive density* $p(x)$, $x \in [0,1]$. *Then for any fixed* x, *the inequality* (9.8) *holds.*

Proof. To prove the lower bound in (9.8), we use the same trick as in the parametric case (refer to the proof of Lemma 3.4). We substitute the supremum over $\Theta(\beta)$ by the Bayes prior distribution concentrated at two points. This time, however, the two points are represented by two regression functions, called the *test functions*,

$$f_0 = f_0(x) = 0 \text{ and } f_1 = f_1(x) \neq 0, \ f_1 \in \Theta(\beta), \ x \in [0,1].$$

Note that for any estimator $\hat{f}_n = \hat{f}_n(x)$, the supremum exceeds the mean value,

$$\sup_{f \in \Theta(\beta)} \mathbb{E}_f\Big[\big(\hat{f}_n(x) - f(x)\big)^2\Big]$$

$$\geq \frac{1}{2}\mathbb{E}_{f_0}\Big[\hat{f}_n^2(x)\Big] + \frac{1}{2}\mathbb{E}_{f_1}\Big[\big(\hat{f}_n(x) - f_1(x)\big)^2\Big]. \tag{9.9}$$

The expected values $\mathbb{E}_{f_0}$ and $\mathbb{E}_{f_1}$ denote the integration with respect to the distribution of y_i, given the corresponding regression function. Under the hypothesis $f = f_0 = 0$, the response $y_i = \varepsilon_i \sim \mathcal{N}(0, \sigma^2)$, while under the alternative $f = f_1, y_i \sim \mathcal{N}\big(f_1(x_i), \sigma^2\big)$. Changing the probability measure of integration, we can write the expectation $\mathbb{E}_{f_1}$ in terms of $\mathbb{E}_{f_0}$,

$$\mathbb{E}_{f_1}\Big[\big(\hat{f}_n(x) - f_1(x)\big)^2\Big]$$

$$= \mathbb{E}_{f_0}\Big[\big(\hat{f}_n(x) - f_1(x)\big)^2 \prod_{i=1}^n \frac{\exp\big\{-(y_i - f_1(x_i))^2/(2\sigma^2)\big\}}{\exp\big\{-y_i^2/(2\sigma^2)\big\}}\Big]$$

$$= \mathbb{E}_{f_0}\Big[\big(\hat{f}_n(x) - f_1(x)\big)^2 \exp\Big\{\sum_{i=1}^n \frac{y_i f_1(x_i)}{\sigma^2} - \sum_{i=1}^n \frac{f_1^2(x_i)}{2\sigma^2}\Big\}\Big]. \tag{9.10}$$

Now, for a given Hölder class $\Theta(\beta, L, L_1)$, we will explicitly introduce a function f_1 that belongs to this class. Take a continuous function $\varphi(u)$, $u \in \mathbb{R}$. We assume that it is supported on the interval $[-1, 1]$, is positive at the origin, and its β-th derivative is bounded by L. That is, we assume that $\varphi(u) = 0$ if $|\,u\,| > 1$, $\varphi(0) > 0$, and $|\,\varphi^{(\beta)}(u)\,| \leq L$. Choose the bandwidth $h_n^* = n^{-1/(2\beta+1)}$, and put

$$f_1(t) = (h_n^*)^\beta \, \varphi\Big(\frac{t-x}{h_n^*}\Big), \quad t \in [0, 1].$$

Schematic graphs of the functions φ and f_1 are given in Figures 6 and 7. These graphs reflect a natural choice of the function φ as a "bump". Notice that f_1 is a rescaling of φ. Indeed, since the support of φ is $[-1, 1]$, the function f_1 is non-zero only for t such that $|(t-x)/h_n^*| \leq 1$ or, equivalently, for $t \in [x - h_n^*, x + h_n^*]$, and the value of f_1 at x is small, $f_1(x) = (h_n^*)^\beta \, \varphi(0)$.

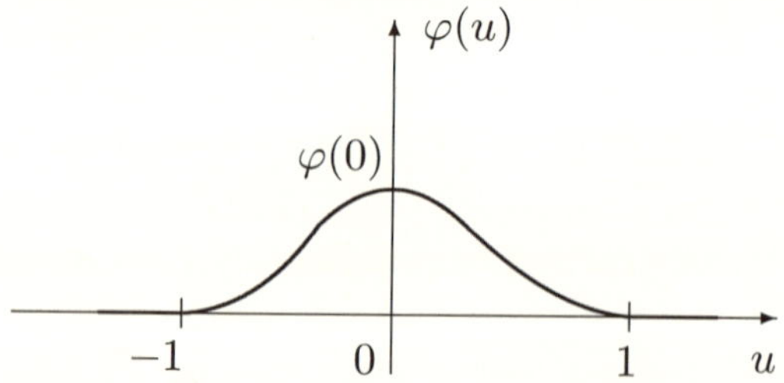

Figure 6. A graph of a "bump" function φ.

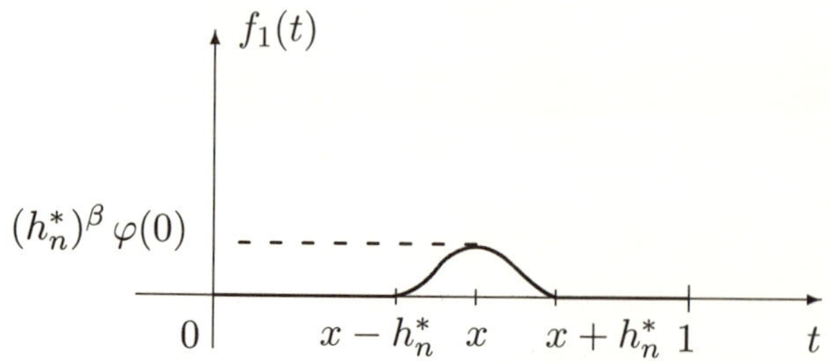

Figure 7. A graph of the function f_1.

For any n sufficiently large, the function f_1 belongs to $\Theta(\beta, L, L_1)$. Indeed, since the function φ is bounded and h_n^* is small, $|f_1|$ is bounded by L_1. Also, for any $t_1, t_2 \in [0, 1]$,

$$|\, f_1^{(\beta-1)}(t_1) - f_1^{(\beta-1)}(t_2)\,| = \Big|\, h_n^* \, \varphi^{(\beta-1)}\Big(\frac{t_1 - x}{h_n^*}\Big) - h_n^* \, \varphi^{(\beta-1)}\Big(\frac{t_2 - x}{h_n^*}\Big)\,\Big|$$

$$\leq h_n^* \max_{-1 \leq u \leq 1} |\, \varphi^{(\beta)}(u)\,| \, \Big|\frac{t_1 - x}{h_n^*} - \frac{t_2 - x}{h_n^*}\Big| \leq L\,|\, t_1 - t_2\,|.$$

Introduce a random event

$$\mathcal{E} = \Big\{ \sum_{i=1}^{n} \frac{y_i\, f_1(x_i)}{\sigma^2} - \sum_{i=1}^{n} \frac{f_1^2(x_i)}{2\sigma^2} \geq 0 \Big\}.$$

From (9.9) and (9.10), we find that

$$\sup_{f\,\in\,\Theta(\beta)} \mathbb{E}_f\Big[\,\big(\,\hat{f}_n(x) - f(x)\big)^2\,\Big]$$

$$\geq \frac{1}{2}\mathbb{E}_{f_0}\Big[\,\hat{f}_n^2(x) + \big(\,\hat{f}_n(x) - f_1(x)\,\big)^2 \exp\Big\{ \sum_{i=1}^{n} \frac{y_i\, f_1(x_i)}{\sigma^2} - \sum_{i=1}^{n} \frac{f_1^2(x_i)}{2\sigma^2}\Big\}\Big]$$

$$\geq \frac{1}{2}\mathbb{E}_{f_0}\Big[\Big(\,\hat{f}_n^2(x) + \big(\,\hat{f}_n(x) - f_1(x)\,\big)^2\Big)\,\mathbb{I}(\mathcal{E})\Big].$$

Here we bound the exponent from below by one, which is true under the event $\mathcal{E}$. Next, by the elementary inequality $a^2 + (a-b)^2 \geq b^2/2$ with $a = \hat{f}_n(x)$ and $b = f_1(x) = (h_n^*)^\beta\,\varphi(0)$, we get the following bound:

$$\sup_{f\,\in\,\Theta(\beta)} \mathbb{E}_f\Big[\,\big(\,\hat{f}_n(x) - f(x)\,\big)^2\Big] \geq \frac{1}{4}\,(h_n^*)^{2\beta}\,\varphi^2(0)\,\mathbb{P}_{f_0}(\,\mathcal{E}\,). \tag{9.11}$$

What is left to show is that the probability $\mathbb{P}_{f_0}(\,\mathcal{E}\,)$ is separated away from zero,

$$\mathbb{P}_{f_0}(\,\mathcal{E}\,) \geq p_0 \tag{9.12}$$

where p_0 is a positive constant independent of n. In this case, (9.8) holds,

$$\sup_{f\,\in\,\Theta(\beta)} \mathbb{E}_f\Big[\,\big(\,\hat{f}_n(x)-f(x)\,\big)^2\Big] \geq \frac{1}{4}\,(h_n^*)^{2\beta}\,\varphi^2(0)\,p_0 = r_*\,n^{-2\beta/(2\beta+1)} \tag{9.13}$$

with $r_* = (1/4)\,\varphi^2(0)\,p_0$. To verify (9.12), note that under the hypothesis $f = f_0 = 0$, the random variable

$$Z = \Big[\sigma^2 \sum_{i=1}^{n} f_1^2(x_i)\Big]^{-1/2} \sum_{i=1}^{n} y_i\, f_1(x_i)$$

has the standard normal distribution. Thus,

$$\lim_{n\to\infty} \mathbb{P}_{f_0}(\,\mathcal{E}\,) = \lim_{n\to\infty} \mathbb{P}_{f_0}\Big(\sum_{i=1}^{n} y_i\, f_1(x_i) \geq \frac{1}{2} \sum_{i=1}^{n} f_1^2(x_i)\Big)$$

$$= \lim_{n\to\infty} \mathbb{P}_{f_0}\Big(Z \geq \frac{1}{2\sigma}\Big[\sum_{i=1}^{n} f_1^2(x_i)\Big]^{1/2}\Big) = 1 - \lim_{n\to\infty} \Phi\Big(\frac{1}{2\sigma}\Big[\sum_{i=1}^{n} f_1^2(x_i)\Big]^{1/2}\Big).$$

Finally, we will show that

$$\lim_{n\to\infty} \sum_{i=1}^{n} f_1^2(x_i) = p(x)\,\|\varphi\|_2^2 = p(x)\int_{-1}^{1} \varphi^2(u)\,du > 0. \tag{9.14}$$

Indeed, recall that the optimal bandwidth h_n^* satisfies the identity $(h_n^*)^{2\beta} = 1/(nh_n^*)$. Using this fact and the assertion of part (iii) of Lemma 9.8, we have that

$$\sum_{i=1}^n f_1^2(x_i) = \sum_{i=1}^n (h_n^*)^{2\beta}\,\varphi^2\Big(\frac{x_i - x}{h_n^*}\Big)$$

$$(9.15) \qquad = \frac{1}{nh_n^*}\sum_{i=1}^n \varphi^2\Big(\frac{x_i - x}{h_n^*}\Big) \to p(x)\int_{-1}^1 \varphi^2(u)\,du \text{ as } n \to \infty.$$

Hence (9.14) is true, and the probability $\mathbb{P}_{f_0}(\mathcal{E})$ has a strictly positive limit,

$$\lim_{n\to\infty} \mathbb{P}_{f_0}(\mathcal{E}) = 1 - \Phi\Big(\frac{1}{2\sigma}\Big[p(x)\,\|\varphi\|_2^2\Big]^{1/2}\Big) > 0.$$

This completes the proof of the theorem. □

9.3.2. Regular Random Design. Do random designs represent all the points of the interval $[0, 1]$ "fairly" to ensure the lower bound (9.8)? It seems plausible, provided the probability density of the design is strictly positive. The following theorem supports this view.

Theorem 9.17. *Let the design points $x_1, \dots, x_n$ be independent identically distributed random variables with the common probability density $p(x)$ which is continuous and strictly positive on $[0, 1]$. Then at any fixed $x \in (0, 1)$, the inequality (9.8) holds.*

Proof. See Exercise 9.68. □

9.4. Proofs of Auxiliary Results

Proof of Lemma 9.8. (i) Consider the design points in the h_n-neighborhood of x. By the definition (7.17) of the regular deterministic design points, we have

$$\frac{1}{n} = \frac{i+1}{n} - \frac{i}{n} = F_{\mathcal{X}}(x_{i+1}) - F_{\mathcal{X}}(x_i) = p(x_i^*)(x_{i+1} - x_i)$$

where $x_i^* \in (x_i, x_{i+1})$. Hence,

$$x_{i+1} - x_i = \frac{1}{np(x_i^*)}.$$

From the continuity of the density $p(x)$, we have that $p(x_i^*)(1+\alpha_{i,n}) = p(x)$ where $\alpha_{i,n} = o(1) \to 0$ as $n \to \infty$. Therefore,

$$x_{i+1} - x_i = \frac{1 + \alpha_{i,n}}{np(x)}.$$

The quantity $|\alpha_{i,n}|$ can be bounded by a small constant uniformly over $i = 1, \dots, n$, so that $\alpha_n = \max_{1\le i\le N} |\alpha_{i,n}| \to 0$ as $n \to \infty$.

(ii) Note that by definition, the number N of observations in the interval $[x - h_n, x + h_n]$ can be bounded by

$$\frac{2h_n}{\max\limits_{1 \le i \le N}(x_{i+1} - x_i)} - 1 \le N \le \frac{2h_n}{\min\limits_{1 \le i \le N}(x_{i+1} - x_i)} + 1.$$

From part (i),

$$\frac{1 - \alpha_n}{np(x)} \le x_{i+1} - x_i \le \frac{1 + \alpha_n}{np(x)},$$

and, therefore, N is bounded by

$$\frac{2h_n np(x)}{1 + \alpha_n} - 1 \le N \le \frac{2h_n np(x)}{1 - \alpha_n} + 1. \tag{9.16}$$

Hence, $\lim_{n \to \infty} N/(nh_n) = 2p(x)$.

(iii) Put $u_i = (x_{i+1} - x)/h_n$. From part (i), we have that

$$\Delta u_i = u_{i+1} - u_i = \frac{x_{i+1} - x_i}{h_n} = \frac{1 + \alpha_{i,n}}{nh_n p(x)}, \tag{9.17}$$

or, equivalently,

$$\frac{1}{nh_n} = p(x) \frac{\Delta u_i}{1 + \alpha_{i,n}}.$$

Hence, the bounds take place:

$$p(x) \frac{\Delta u_i}{1 + \alpha_n} \le \frac{1}{nh_n} \le p(x) \frac{\Delta u_i}{1 - \alpha_n}.$$

Consequently,

$$\frac{p(x)}{1 + \alpha_n} \sum_{i=1}^{N} \varphi_0\Big(\frac{x_i - x}{h_n}\Big) \Delta u_i \le \frac{1}{nh_n} \sum_{i=1}^{N} \varphi_0\Big(\frac{x_i - x}{h_n}\Big)$$

$$\le \frac{p(x)}{1 - \alpha_n} \sum_{i=1}^{N} \varphi_0\Big(\frac{x_i - x}{h_n}\Big) \Delta u_i,$$

and the desired convergence follows,

$$\frac{1}{nh_n} \sum_{i=1}^{N} \varphi_0\Big(\frac{x_i - x}{h_n}\Big) \to p(x) \int_{-1}^{1} \varphi_0(u)\, du. \qquad \square$$

Proof of Lemma 9.9. By the definition of the matrix $\mathbf{G}$, we can write

$$\frac{1}{N} (\mathbf{G}'\mathbf{G})_{l,m} = \frac{1}{N} \sum_{i=1}^{N} \Big(\frac{x_i - x}{h_n}\Big)^l \Big(\frac{x_i - x}{h_n}\Big)^m = \frac{1}{N} \sum_{i=1}^{N} u_i^{l+m}. \tag{9.18}$$

Next, we want to find bounds for $1/N$. From (9.17), we have

$$\frac{1 - \alpha_n}{nh_n p(x)} \le \Delta u_i \le \frac{1 + \alpha_n}{nh_n p(x)}.$$

Combining this result with (9.16), we obtain

$$\Big(\frac{2h_n np(x)}{1+\alpha_n}-1\Big)\Big(\frac{1-\alpha_n}{nh_n p(x)}\Big) \leq N\Delta u_i \leq \Big(\frac{2h_n np(x)}{1-\alpha_n}+1\Big)\Big(\frac{1+\alpha_n}{nh_n p(x)}\Big).$$

Put

$$\beta_n = \frac{1-\alpha_n}{1+\alpha_n} - \frac{1-\alpha_n}{nh_n p(x)} - 1 \text{ and } \tilde{\beta}_n = \frac{1+\alpha_n}{1-\alpha_n} + \frac{1-\alpha_n}{nh_n p(x)} - 1.$$

Thus, we have shown that $2+\beta_n \leq N\Delta u_i \leq 2+\tilde{\beta}_n$, or, equivalently,

$$\frac{\Delta u_i}{2+\tilde{\beta}_n} \leq \frac{1}{N} \leq \frac{\Delta u_i}{2+\beta_n}$$

where β_n and $\tilde{\beta}_n$ vanish as n goes to infinity. Therefore, using the expression (9.18), we can bound $\frac{1}{N}\big(\mathbf{G}'\mathbf{G}\big)_{l,m}$ by

$$\frac{1}{2+\tilde{\beta}_n}\sum_{i=1}^{N} u_i^{l+m}\Delta u_i \leq \frac{1}{N}\big(\mathbf{G}'\mathbf{G}\big)_{l,m} \leq \frac{1}{2+\beta_n}\sum_{i=1}^{N} u_i^{l+m}\Delta u_i.$$

Both bounds converge as $n \to \infty$ to the integral in the definition (9.6) of $\mathbf{D}_\infty^{-1}$. The proof that this matrix is invertible is left as an exercise (see Exercise 9.66). □

Before we turn to Lemmas 9.12 and 9.13, we prove the following result. Let $g(u)$ be a continuous function such that $|g(u)| \leq 1$ for all $u \in [-1,1]$. Let $x_1,\dots,x_n$ be independent random variables with a common uniform distribution on $[0,1]$. Introduce the independent random variables η_i, $i = 1,\dots,n$, by

$$\eta_i = \begin{cases} g\Big(\frac{x_i-x}{h_n^*}\Big), & \text{if } x_i \in [x-h_n^*, x+h_n^*], \\ 0, & \text{otherwise}. \end{cases} \tag{9.19}$$

Denote by μ_n the expected value of η_i,

$$\mu_n = \mathbb{E}[\,\eta_i\,] = \int_{x-h_n^*}^{x+h_n^*} g\Big(\frac{t-x}{h_n^*}\Big)\,dt = h_n^*\int_{-1}^{1} g(u)\,du.$$

Result. For any positive number α,

$$\mathbb{P}\Big(\Big|\frac{1}{nh_n^*}(\eta_1+\dots+\eta_n) - \frac{\mu_n}{h_n^*}\Big| > \alpha\Big) \leq \frac{2}{\alpha^2 nh_n^*}. \tag{9.20}$$

Proof. Note that

$$\mathbb{V}ar[\,\eta_i\,] \leq \mathbb{E}[\,\eta_i^2\,] = h_n^*\int_{-1}^{1} g^2(u)\,du \leq 2h_n^*.$$

Thus, the Chebyshev inequality yields

$$\mathbb{P}\Big(\Big|\frac{1}{nh_n^*}(\eta_1+\dots+\eta_n) - \frac{\mu_n}{h_n^*}\Big| > \alpha\Big) \leq \frac{n\mathbb{V}ar[\,\eta_i\,]}{(\alpha nh_n^*)^2} \leq \frac{2}{\alpha^2 nh_n^*}.$$ □

Proof of Lemma 9.12. Apply the definition (9.19) of η_i with $g = 1$. In this case, $N = \eta_1 + \cdots + \eta_n$ and $\mu_n / h_n^* = 2$. Thus, from (9.20) we obtain

$$\mathbb{P}_f(\overline{\mathcal{A}}) = \mathbb{P}_f\Big(\Big|\frac{N}{nh_n^*} - 2\Big| > \alpha\Big)$$

$$= \mathbb{P}_f\Big(\Big|\frac{1}{nh_n^*}(\eta_1 + \cdots + \eta_n) - 2\Big| > \alpha\Big) \le \frac{2}{\alpha^2 nh_n^*}.$$

Finally, note that $nh_n^* = n^{2\beta/(2\beta+1)}$. □

Proof of Lemma 9.13. For an arbitrarily small $\delta > 0$, define a random event

$$\mathcal{C} = \bigcap_{l,m=0}^{\beta-1} \Big\{\Big|\frac{1}{2nh_n^*}(\mathbf{G}'\mathbf{G})_{l,m} - (\mathbf{D}_\infty^{-1})_{l,m}\Big| \le \delta\Big\}.$$

First, we want to show that the probability of the complement event

$$\overline{\mathcal{C}} = \bigcup_{l,m=0}^{\beta-1} \Big\{\Big|\frac{1}{2nh_n^*}(\mathbf{G}'\mathbf{G})_{l,m} - (\mathbf{D}_\infty^{-1})_{l,m}\Big| > \delta\Big\}$$

is bounded from above,

$$\mathbb{P}_f(\overline{\mathcal{C}}) \le 2\beta^2\delta^{-2}n^{-2\beta/(2\beta+1)}. \tag{9.21}$$

To see this, put $g(u) = (1/2)u^{l+m}$ in (9.19). Then

$$\frac{1}{2nh_n^*}(\mathbf{G}'\mathbf{G})_{l,m} = \frac{\eta_1 + \ldots + \eta_n}{nh_n^*}$$

and

$$\frac{\mu_n}{h_n^*} = \frac{1}{h_n^*}\int_{x-h_n^*}^{x+h_n^*} g\Big(\frac{t-x}{h_n^*}\Big)\,dt = \frac{1}{2}\int_{-1}^{1} u^{l+m}\,du = (\mathbf{D}_\infty^{-1})_{l,m}.$$

The inequality (9.20) provides the upper bound $2\,\delta^{-2}\,n^{-2\,\beta/(2\,\beta+1)}$ for the probability of each event in the union $\overline{\mathcal{C}}$. This proves (9.21).

Next, recall that we denoted by C_* a constant that exceeds the absolute values of all elements of the matrix $\mathbf{D}_\infty$. Due to the continuity of a matrix inversion, for any $\varepsilon \le C_*$, there exists a number $\delta = \delta(\varepsilon)$ such that

$$\mathcal{C} = \bigcap_{l,m=0}^{\beta-1} \Big\{\Big|\frac{1}{2nh_n^*}(\mathbf{G}'\mathbf{G})_{l,m} - (\mathbf{D}_\infty^{-1})_{l,m}\Big| \le \delta(\varepsilon)\Big\}$$

$$\subseteq \bigcap_{l,m=0}^{\beta-1} \Big\{\Big|(2nh_n^*)(\mathbf{G}'\mathbf{G})_{l,m}^{-1} - (\mathbf{D}_\infty)_{l,m}\Big| \le \varepsilon\Big\}$$

$$\subseteq \Big\{\Big|(\mathbf{G}'\mathbf{G})_{l,m}^{-1}\Big| \le \frac{2C_*}{nh_n^*} \text{ for all } l,m = 0,\ldots,\beta-1\Big\} = \mathcal{B}.$$

The latter inclusion follows from the fact that if $(\mathbf{G}'\mathbf{G})^{-1}_{l,m} \leq (C_* + \varepsilon)/(2nh_n^*)$ and $\varepsilon \leq C_*$, it implies that $(\mathbf{G}'\mathbf{G})^{-1}_{l,m} \leq C_*/(nh_n^*) \leq 2C_*/(nh_n^*)$. Thus, from (9.21), we obtain $\mathbb{P}_f(\overline{\mathcal{B}}) \leq \mathbb{P}_f(\overline{\mathcal{C}}) \leq Cn^{-2\beta/(2\beta+1)}$ with $C = 2\beta^2\delta^{-2}$. $\square$

Exercises

Exercise 9.63. Explain what happens to the local polynomial estimator (9.4) if one of the conditions $h_n \to 0$ or $nh_n \to \infty$ is violated.

Exercise 9.64. Take $x = h_n$, and consider the local polynomial approximation in the interval $[0, 2h_n]$. Let the estimate of the regression coefficient be defined as the solution of the respective minimization problem (9.1). Define the estimator of the regression function at the origin by $\hat{f}_n(0) = \sum_{m=0}^{\beta-1} (-1)^m \hat{\theta}_m$. Find upper bounds for the conditional bias and variance of $\hat{f}_n(0)$ for a fixed design $\mathcal{X}$.

Exercise 9.65. Prove an analogue of Theorem 9.6 for the derivative estimator (9.5),

$$\sup_{f \in \Theta(\beta)} \mathbb{E}_f\Big[\Big(\frac{m!\,\hat{\theta}_m}{(h_n^*)^m} - f^{(m)}(x)\Big)^2 \,\Big|\, \mathcal{X}\Big] \leq r^*\, n^{-2(\beta-m)/(2\beta+1)}$$

where $h_n^* = n^{-1/(2\beta+1)}$.

Exercise 9.66. Show that the matrix $\mathbf{D}_\infty^{-1}$ in Lemma 9.9 is invertible.

Exercise 9.67. Let f_1 be as defined in the proof of Theorem 9.16, and let $x_1, \dots, x_n$ be a random design with the probability density $p(x)$ on the interval $[0, 1]$.
(i) Show that the random variable

$$\sum_{i=1}^n f_1^2(x_i) = \frac{1}{n\,h_n^*} \sum_{i=1}^n \varphi^2\Big(\frac{x_i - x}{h_n^*}\Big)$$

has the expected value that converges to $p(x)\, \|\, \varphi\, \|_2^2$ as $n \to \infty$.
(ii) Prove that the variance of this random variable is $O(1/(nh_n))$ as $n \to \infty$.
(iii) Derive from parts (i) and (ii) that for all sufficiently large n,

$$\mathbb{P}_{f_0}\Big(\sum_{i=1}^n f_1^2(x_i) \leq 2p(x)\,\|\,\varphi\,\|_2^2\Big) \geq 1/2.$$

Exercise 9.68. Apply Exercise 9.67 to prove Theorem 9.17.

Chapter 10

Estimation of Regression in Global Norms

10.1. Regressogram

In Chapters 8 and 9, we gave a detailed analysis of the kernel and local polynomial estimators at a fixed point x inside the interval $(0, 1)$. The asymptotic minimax rate of convergence was found to be $\psi_n = n^{-\beta/(2\beta+1)}$, which strongly depends on the smoothness parameter β of the regression function.

What if our objective is different? What if we want to estimate the regression function $f(x)$ as a curve in the interval $[0, 1]$? The global norms serve this purpose. In this chapter, we discuss the regression estimation problems with regard to the continuous and discrete L_2-norms, and sup-norm.

In the current section, we introduce an estimator $\hat{f}_n$, called a *regressogram*. A formal definition will be given at the end of the section.

When it comes to the regression estimation in the interval $[0, 1]$, we can extend a smoothing kernel estimator (8.16) to be defined in the entire unit interval. However, the estimation at the endpoints $x = 0$ and $x = 1$ would cause difficulties. It is more convenient to introduce an estimator defined everywhere in $[0, 1]$ based on the local polynomial estimator (9.4).

Consider a partition of the interval $[0, 1]$ into small subintervals of the equal length $2h_n$. To ease the presentation assume that $Q = 1/(2h_n)$ is an integer. The number Q represents the total number of intervals in the partition. Each small interval

$$B_q = \big[2(q-1)h_n, 2qh_n\big), \quad q = 1, \dots, Q,$$

is called a *bin*. It is convenient to introduce notation for the midpoint of the bin B_q. We denote it by $c_q = (2q-1)h_n, \quad q = 1, \dots, Q$.

The local polynomial estimator (9.4) is defined separately for each bin. If we want to estimate the regression function at every point $x \in [0,1]$, we must consider a collection of the local polynomial estimators. Introduce Q minimization problems, one for each bin,

$$\sum_{i=1}^{n} \Big(y_i - \Big[\hat{\theta}_{0,\,q} + \hat{\theta}_{1,\,q} \Big(\frac{x_i - c_q}{h_n} \Big) + \dots + \hat{\theta}_{\beta-1,\,q} \Big(\frac{x_i - c_q}{h_n} \Big)^{\beta-1} \Big] \Big)^2 \mathbb{I}\big(x_i \in B_q \big)$$

$$(10.1) \qquad \to \min_{\hat{\theta}_{0,\,q},\dots,\hat{\theta}_{\beta-1,\,q}} \quad \text{for } x \in B_q, \; q = 1, \dots, Q.$$

Note that these minimization problems are totally disconnected. Each of them involves only the observations the design points of which belong to the respective bin B_q. The estimates of the regression coefficients are marked by the double subscript, representing the coefficient number and the bin number. There should also be a subscript "n", which we omit to avoid too cumbersome a notation.

As in Section 9.1, it is easier to interpret the minimization problems (10.1) if they are written in the vector notation. Denote by $N_1, \dots, N_Q$ the number of the design points in each bin, $N_1 + \dots + N_Q = n$. For a fixed $q = 1, \dots, Q$, let

$$x_{1,q} < \dots < x_{N_q,q}$$

be the design points in the bin B_q, and let the corresponding response values have matching indices $y_{1,q}, \dots, y_{N_q,q}$. Denote by

$$\hat{\boldsymbol{\theta}}_q = \big(\hat{\theta}_{0,q}, \dots, \hat{\theta}_{\beta-1,\,q} \big)'$$

the vector of the estimates of the regression coefficients in the q-th bin. The vectors $\hat{\boldsymbol{\theta}}_q$ satisfy the systems of normal equations

$$(10.2) \qquad (\mathbf{G}_q'\mathbf{G}_q)\hat{\boldsymbol{\theta}}_q = \mathbf{G}_q'\mathbf{y}_q, \quad q = 1, \dots, Q,$$

where $\mathbf{y}_q = (y_{1,\,q}, \dots, y_{N_q,\,q})'$, and the matrix $\mathbf{G}_q = \big[\mathbf{g}_{0,\,q}, \dots, \mathbf{g}_{\beta-1,\,q}\big]$ has the columns

$$\mathbf{g}_{m,\,q} = \Big(\Big(\frac{x_{1,\,q} - c_q}{h_n} \Big)^m, \dots, \Big(\frac{x_{N_q,\,q} - c_q}{h_n} \Big)^m \Big)', \quad m = 0, \dots, \beta - 1.$$

The results of Section 9.1 were based on Assumptions 9.2 and 9.5. In this section, we combine their analogues into one assumption. Provided this assumption holds, the systems of normal equations (10.2) have the unique solutions for all $q = 1, \dots, Q$.

Assumption 10.1. There exist positive constants γ_0 and γ_1, independent of n and q, such that for all $q = 1, \dots, Q$,
(i) the absolute values of the elements of the matrix $\left(\mathbf{G_q}'\mathbf{G_q}\right)^{-1}$ are bounded from above by γ_0/N_q,
(ii) the number of observations N_q in the q-th bin is bounded from below, $N_q \geq \gamma_1 n h_n$. □

Now we are ready to define the piecewise polynomial estimator $\hat{f}_n(x)$ in the entire interval $[0, 1]$. This estimator is called a *regressogram*, and is computed according to the formula

$$\hat{f}_n(x) = \hat{\theta}_{0,\,q} + \hat{\theta}_{1,\,q}\left(\frac{x - c_q}{h_n}\right) + \dots + \hat{\theta}_{\beta-1,\,q}\left(\frac{x - c_q}{h_n}\right)^{\beta-1} \quad \text{if } x \in B_q, \tag{10.3}$$

where the estimates $\hat{\theta}_{0,\,q}, \dots, \hat{\theta}_{\beta-1,\,q}$ satisfy the normal equations (10.2), $q = 1, \dots, Q$.

10.2. Integral L_2-Norm Risk for the Regressogram

Consider the regressogram $\hat{f}_n(x)$, $x \in [0, 1]$, defined by (10.3). The following statement is an adaptation of Proposition 9.4 about the components of $\hat{\boldsymbol{\theta}}_q$. We omit its proof.

Proposition 10.2. *Suppose that for a given design $\mathcal{X}$, Assumption 10.1 holds. Assume that the regression function f belongs to a Hölder class $\Theta(\beta, L, L_1)$. Then the m-th element $\hat{\theta}_{m,q}$ of the vector $\hat{\boldsymbol{\theta}}_q$, which satisfies the system of normal equations* (10.2), *admits the expansion*

$$\hat{\theta}_{m,q} = \frac{f^{(m)}(c_q)}{m!} h_n^m + b_{m,q} + \mathcal{N}_{m,q}, \quad m = 0, \dots, \beta - 1, \quad q = 1, \dots, Q,$$

where the conditional bias $b_{m,q}$ is bounded from above, $\left|b_{m,\,q}\right| \leq C_b h_n^{\beta}$, and the stochastic term $\mathcal{N}_{m,q}$ has the normal distribution with mean zero. Its variance is limited from above, $\mathbb{V}ar_f\left[\mathcal{N}_{m,q} \,\middle|\, \mathcal{X}\right] \leq C_v/N_q$. Here the constants C_b and C_v are independent of n. Conditionally, given the design $\mathcal{X}$, the random variables $\mathcal{N}_{m,q}$ are independent for different values of q.

The next theorem answers the question about the integral L_2-norm risk for the regressogram.

Theorem 10.3. *Let a design $\mathcal{X}$ be such that Assumption 10.1 holds with the bandwidth $h_n^* = n^{-1/(2\beta+1)}$. Then the mean integrated quadratic risk of* *the regressogram* $\hat{f}_n(x)$ *admits the upper bound*

$$\sup_{f \in \Theta(\beta)} \mathbb{E}_f\Big[\int_0^1 \big(\hat{f}_n(x) - f(x)\big)^2 dx \,\big|\, \mathcal{X}\Big] \le r^* n^{-2\beta/(2\beta+1)}$$

for some positive constant r^* *independent of* n.

Proof. From Lemma 8.5, for any $f \in \Theta(\beta, L, L_1)$, and for any bin B_q centered at c_q, the Taylor expansion is valid

$$f(x) = f(c_q) + f^{(1)}(c_q)(x - c_q) + \cdots + \frac{f^{(\beta-1)}(c_q)}{(\beta-1)!}(x - c_q)^{\beta-1} + \rho(x, c_q)$$

$$= \sum_{m=0}^{\beta-1} \frac{f^{(m)}(c_q)}{m!}(h_n^*)^m \Big(\frac{x - c_q}{h_n^*}\Big)^m + \rho(x, c_q)$$

where the remainder term $\rho(x, c_q)$ satisfies the inequality $\big|\rho(x, c_q)\big| \le C_\rho (h_n^*)^\beta$ with $C_\rho = L/(\beta-1)!$.

Applying Proposition 10.2, the Taylor expansion of f, and the definition of the regressogram (10.3), we get the expression for the quadratic risk

$$\mathbb{E}_f\Big[\int_0^1 \big(\hat{f}_n(x) - f(x)\big)^2 dx \,\big|\, \mathcal{X}\Big] = \sum_{q=1}^{Q} \mathbb{E}_f\Big[\int_{B_q} \big(\hat{f}_n(x) - f(x)\big)^2 dx \,\Big|\, \mathcal{X}\Big]$$

$$= \sum_{q=1}^{Q} \mathbb{E}_f\Big[\int_{B_q} \Big[\sum_{m=0}^{\beta-1} \big(b_{m,q} + \mathcal{N}_{m,q}\big)\Big(\frac{x - c_q}{h_n^*}\Big)^m - \rho(x, c_q)\Big]^2 dx \,\Big|\, \mathcal{X}\Big].$$

Using the fact that the random variables $\mathcal{N}_{m,q}$ have mean zero, we can write the latter expectation of the integral over B_q as the sum of a deterministic and stochastic terms,

$$\int_{B_q} \Big(\sum_{m=0}^{\beta-1} b_{m,q}\Big(\frac{x - c_q}{h_n^*}\Big)^m - \rho(x, c_q)\Big)^2 dx$$

$$+ \int_{B_q} \mathbb{E}_f\Big[\Big(\sum_{m=0}^{\beta-1} \mathcal{N}_{m,q}\Big(\frac{x - c_q}{h_n^*}\Big)^m\Big)^2 \Big|\mathcal{X}\Big] dx.$$

From the bounds for the bias and the remainder term, the first integrand can be estimated from above by a constant,

$$\Big(\sum_{m=0}^{\beta-1} b_{m,q}\Big(\frac{x - c_q}{h_n^*}\Big)^m - \rho(x, c_q)\Big)^2 \le \Big(\sum_{m=0}^{\beta-1} |b_{m,q}| + |\rho(x, c_q)|\Big)^2$$

$$\le \big(\beta C_b (h_n^*)^\beta + C_\rho (h_n^*)^\beta\big)^2 = C_D (h_n^*)^{2\beta} = C_D n^{-2\beta/(2\beta+1)}$$

where $C_D = (\beta C_b + C_\rho)^2$.

Note that the random variables $\mathcal{N}_{m,q}$ may be correlated for a fixed q and different m's. Using a special case of the Cauchy-Schwarz inequality $(a_0 + \cdots + a_{\beta-1})^2 \leq \beta\,(a_0^2 + \cdots + a_{\beta-1}^2)$, Proposition 10.2, and Assumption 10.1, we bound the second integrand from above by

$$\mathbb{E}_f\Big[\Big(\sum_{m=0}^{\beta-1} \mathcal{N}_{m,q}\Big(\frac{x-c_q}{h_n^*}\Big)^m\Big)^2 \,\Big|\, \mathcal{X}\Big] \leq \beta \sum_{m=0}^{\beta-1} \mathbb{V}ar_f\big[\mathcal{N}_{m,q}\,|\,\mathcal{X}\big]$$

$$\leq \frac{\beta^2 C_v}{N_q} \leq \frac{\beta^2 C_v}{\gamma_1 n h_n^*} = \frac{C_S}{n h_n^*} = C_S\, n^{-2\beta/(2\beta+1)} \text{ where } C_S = \beta^2 C_v/\gamma_1.$$

Thus, combining the deterministic and stochastic terms, we arrive at the upper bound

$$\mathbb{E}_f\Big[\int_0^1 \big(\hat{f}_n(x) - f(x)\big)^2\,dx \,\Big|\, \mathcal{X}\Big] \leq \sum_{q=1}^{Q} \int_{B_q} (C_D + C_S)\, n^{-2\beta/(2\beta+1)}\,dx$$

$$= r^* n^{-2\beta/(2\beta+1)} \text{ with } r^* = C_D + C_S. \qquad \square$$

Remark 10.4. Under Assumption 10.1, the results of Lemmas 9.8 and 9.9 stay valid uniformly over the bins B_q, $q = 1, \ldots, Q$. Therefore, we can extend the statement of Corollary 9.11 to the integral L_2-norm. For the regular deterministic design, the unconditional quadratic risk in the integral L_2-norm of the regressogram $\hat{f}_n$ with the bandwidth $h_n^* = n^{-1/(2\beta+1)}$ admits the upper bound

$$\sup_{f\in\Theta(\beta)} \mathbb{E}_f\Big[\int_0^1 \big(\hat{f}_n(x) - f(x)\big)^2\,dx\Big] \leq r^*\, n^{-2\beta/(2\beta+1)}$$

where a positive constant r^* is independent of n. A similar result is also true for the regular random design (cf. Theorem 9.15). Unfortunately, it is too technical, and we skip its proof. $\square$

Remark 10.5. The m-th derivative of the regressogram (10.3) has the form

$$\frac{d^m \hat{f}_n(x)}{dx^m} = \sum_{i=m}^{\beta-1} \hat{\theta}_{i,q}\,\frac{i!}{(i-m)!}\,\frac{1}{h_n^m}\Big(\frac{x-c_q}{h_n}\Big)^{i-m} \quad \text{if } x \in B_q,\ 0 \leq m \leq \beta-1.$$

Under the same choice of the bandwidth, $h_n^* = n^{-1/(2\beta+1)}$, this estimator admits the upper bound similar to the one in Theorem 10.3 with the rate $n^{-(\beta-m)/(2\beta+1)}$, that is,

(10.4)

$$\sup_{f\in\Theta(\beta)} \mathbb{E}_f\Big[\int_0^1 \Big(\frac{d^m \hat{f}_n(x)}{dx^m} - \frac{d^m f(x)}{dx^m}\Big)^2\,dx \,\Big|\, \mathcal{X}\Big] \leq r^*\, n^{-2(\beta-m)/(2\beta+1)}.$$

For the proof see Exercise 10.69. $\square$

10.3. Estimation in the Sup-Norm

In this section we study the asymptotic performance of the sup-norm risk of the regressogram $\hat{f}_n(x)$ defined in (10.3). The sup-norm risk for a fixed design $\mathcal{X}$ is given by

$$\mathbb{E}_f\Big[\,\|\,\hat{f}_n - f\,\|_\infty \,\Big|\, \mathcal{X}\Big] = \mathbb{E}_f\Big[\sup_{0\le x\le 1} |\,\hat{f}_n(x) - f(x)\,| \,\Big|\, \mathcal{X}\Big]. \tag{10.5}$$

Our starting point is Proposition 10.2. It is a very powerful result that allows us to control the risk under any loss function. We use this proposition to prove the following theorem.

Theorem 10.6. *Let a design $\mathcal{X}$ be such that Assumption 10.1 holds with the bandwidth $h_n = \big((\ln n)/n\big)^{1/(2\beta+1)}$. Let $\hat{f}_n$ be the regressogram that corresponds to this bandwidth. Then the conditional sup-norm risk* (10.5) *admits the upper bound*

$$\sup_{f\in\Theta(\beta)} \mathbb{E}_f\Big[\,\|\,\hat{f}_n - f\,\|_\infty \,\Big|\, \mathcal{X}\Big] \le r^*\Big(\frac{\ln n}{n}\Big)^{\beta/(2\beta+1)} \tag{10.6}$$

where r^* is a positive constant independent of n and f.

Proof. Applying Lemma 8.5, we can write the sup-norm of the difference $\hat{f}_n - f$ as

$$\|\,\hat{f}_n - f\,\|_\infty = \max_{1\le q\le Q}\sup_{x\in B_q}\Big|\sum_{m=0}^{\beta-1}\hat{\theta}_{m,q}\Big(\frac{x-c_q}{h_n}\Big)^m$$

$$- \sum_{m=0}^{\beta-1}\frac{f^{(m)}(c_q)\,(h_n)^m}{m!}\Big(\frac{x-c_q}{h_n}\Big)^m - \rho(x,c_q)\Big| \tag{10.7}$$

where $Q = 1/(2h_n)$ is the number of bins, and the q-th bin is the interval $B_q = \big[c_q - h_n\,,\, c_q + h_n\big)$ centered at $x = c_q$, $q = 1,\dots,Q$. The remainder term $\rho(x,c_q)$ satisfies the inequality $|\,\rho(x,c_q)\,| \le C_\rho h_n^\beta$ with the constant $C_\rho = L\,/\,(\beta-1)!$. Applying the formula for $\hat{\theta}_{m,q}$ from Proposition 10.2 and the fact that $|x - c_q|/h_n \le 1$, we obtain that

$$\|\,f_n - f\,\|_\infty \le \max_{1\le q\le Q}\sup_{x\in B_q}\Big|\sum_{m=0}^{\beta-1}\big(b_{m,q} + \mathcal{N}_{m,q}\big)\Big(\frac{x-c_q}{h_n}\Big)^m\Big| + C_\rho h_n^\beta$$

$$\le \Big[\beta C_b h_n^\beta + C_\rho h_n^\beta\Big] + \max_{1\le q\le Q}\sum_{m=0}^{\beta-1}|\,\mathcal{N}_{m,q}\,|. \tag{10.8}$$

Introduce the standard normal random variables

$$\mathcal{Z}_{m,q} = \big(\mathbb{V}ar_f\big[\mathcal{N}_{m,q}\,|\,\mathcal{X}\big]\big)^{-1/2}\mathcal{N}_{m,q}.$$

From the upper bound on the variance in Proposition 10.2, we find that

$$\text{(10.9)} \qquad \max_{1 \le q \le Q} \sum_{m=0}^{\beta-1} \left| \mathcal{N}_{m,q} \right| \le \max_{1 \le q \le Q} \sqrt{\frac{C_v}{N_q}} \sum_{m=0}^{\beta-1} \left| \mathcal{Z}_{m,q} \right| \le \sqrt{\frac{C_v}{\gamma_1 n h_n}} \, \mathcal{Z}^*$$

where

$$\mathcal{Z}^* = \max_{1 \le q \le Q} \sum_{m=0}^{\beta-1} \left| \mathcal{Z}_{m,q} \right|.$$

Note that the random variables $\mathcal{Z}_{m,q}$ are independent for different bins, but may be correlated for different values of m within the same bin.

Putting together (10.8) and (10.9), we get the upper bound for the sup-norm loss,

$$\text{(10.10)} \qquad \| f_n - f \|_\infty \le \left[\beta C_b + C_\rho \right] h_n^\beta + \sqrt{\frac{C_v}{\gamma_1 n h_n}} \, \mathcal{Z}^*.$$

To continue, we need the following technical result, which we ask to be proved in Exercise 10.70.

Result. *There exists a constant* $C_z > 0$ *such that*

$$\text{(10.11)} \qquad \mathbb{E}\left[\mathcal{Z}^* \,\middle|\, \mathcal{X} \right] \le C_z \sqrt{\ln n}.$$

Finally, under our choice of h_n, it is easily seen that

$$n h_n = n^{2\beta/(2\beta+1)} (\ln n)^{1/(2\beta+1)} = (\ln n) \left(\frac{n}{\ln n} \right)^{2\beta/(2\beta+1)} = (\ln n)\, h_n^{-2\beta}.$$

These results along with (10.10) yield

$$\mathbb{E}_f\Big[\| \hat{f}_n - f \|_\infty \,\big|\, \mathcal{X} \Big] \le \left[\beta C_b + C_\rho \right] h_n^\beta + C_z \sqrt{\frac{C_v \ln n}{\gamma_1 n h_n}}$$

$$\le \left[\beta C_b + C_\rho + C_z \sqrt{C_v / \gamma_1} \right] h_n^\beta = r^* h_n^\beta. \qquad \square$$

Remark 10.7. As Theorem 10.6 shows, the upper bound of the risk in the sup-norm contains an extra log-factor as compared to the case of the L_2-norm. The source of this additional factor becomes clear from (10.11). Indeed, the maximum of the random noise has the magnitude $O(\sqrt{\ln n})$ as $n \to \infty$. That is why the optimum choice of the bandwidth comes from the balance equation $h_n^\beta = \sqrt{(n h_n)^{-1} \ln n}$. $\square$

10.4. Projection on Span-Space and Discrete MISE

The objective of this section is to study the discrete mean integrated squared error (MISE) of the regressogram (10.3). The regressogram is a piecewise polynomial estimator that can be written in the form

$$\hat{f}_n(x) = \sum_{q=1}^{Q} \mathbb{I}(x \in B_q) \sum_{m=0}^{\beta-1} \hat{\theta}_{m,q} \Big(\frac{x - c_q}{h_n} \Big)^m$$

where the q-th bin $B_q = \big[2(q-1)h_n \, , 2qh_n \big)$, and $c_q = (2q-1)h_n$ is its center. Here $Q = 1/(2h_n)$ is an integer that represents the number of bins. The rates of convergence in the L_2-norm and sup-norm found in the previous sections were partially based on the fact that the bias of the regressogram has the magnitude $O(h_n^\beta)$ uniformly in $f \in \Theta(\beta)$ at any point $x \in [0, 1]$. Indeed, from Proposition 10.2, we get

$$\sup_{f \in \Theta(\beta)} \sup_{0 \le x \le 1} \Big| \mathbb{E}_f \big[\hat{f}_n(x) - f(x) \big] \Big|$$

$$\le \sup_{1 \le q \le Q} \sup_{x \in B_q} \sum_{m=0}^{\beta-1} \big| b_{m,q} \big| \Big| \frac{x - c_q}{h_n} \Big|^m \le C_b \beta h_n^\beta .$$

In turn, this upper bound for the bias is the immediate consequence of the Taylor's approximation in Lemma 8.5.

In this section, we take a different approach. Before we proceed, we need to introduce some notation. Define a set of βQ piecewise monomial functions,

(10.12)

$$\gamma_{m,q}(x) = \mathbb{I}(x \in B_q) \Big(\frac{x - c_q}{h_n} \Big)^m, \quad q = 1, \dots, Q, \quad m = 0, \dots, \beta - 1.$$

The regressogram $\hat{f}_n(x)$ is a linear combination of these monomials,

$$\hat{f}_n(x) = \sum_{q=1}^{Q} \sum_{m=0}^{\beta-1} \hat{\theta}_{m,q} \gamma_{m,q}(x) \, , \quad 0 \le x \le 1 \tag{10.13}$$

where $\hat{\theta}_{m,q}$ are the estimates of the regression coefficients in bins.

In what we used above, it was important that any function $f(x) \in \Theta(\beta)$ admits an approximation by a linear combination of $\{\gamma_{m,q}(x)\}$ with the error not exceeding $O(h_n^\beta)$. This property does not exclusively belong to the set of piecewise monomials. We will prove results in the generalized setting for which the regressogram is a special case.

What changes should be made if instead of the piecewise monomials (10.12) we use some other functions? In place of the indices m and q in the monomials (10.12) we will use a single index k for a set of functions

$\gamma_k(x),\ k = 1, \dots, K$. The number of functions $K = K(n) \to \infty$ as $n \to \infty$. In the case of the monomials (10.12), the number $K = \beta Q$.

Consider the regression observations $y_i = f(x_i) + \varepsilon_i,\ \ \varepsilon_i \sim \mathcal{N}(0, \sigma^2)$, with a Hölder regression function $f(x) \in \Theta(\beta)$. As before, we assume that the design points are distinct and ordered in the interval $[0, 1]$,

$$0 \le x_1 < x_2 < \cdots < x_n \le 1.$$

We want to estimate the regression function $f(x)$ by a linear combination

$$\hat{f}_n(x) = \hat{\theta}_1 \gamma_1(x) + \cdots + \hat{\theta}_K \gamma_K(x),\ \ x \in [0, 1], \tag{10.14}$$

in the least-squares sense over the design points. To do this, we have to solve the following minimization problem:

$$\sum_{i=1}^{n} \Big(y_i - \Big[\hat{\theta}_1 \gamma_1(x_i) + \cdots + \hat{\theta}_K \gamma_K(x_i) \Big] \Big)^2 \to \min_{\hat{\theta}_1 \dots \hat{\theta}_K}. \tag{10.15}$$

Define the design matrix $\mathbf{\Gamma}$ with columns

$$\boldsymbol{\gamma}_k = \big(\gamma_k(x_1), \dots, \gamma_k(x_n) \big)',\ \ k = 1, \dots, K. \tag{10.16}$$

From this definition, the matrix $\mathbf{\Gamma}$ has the dimensions $n \times K$. The vector $\hat{\boldsymbol{\vartheta}} = (\hat{\theta}_1, \dots, \hat{\theta}_K)'$ of estimates in (10.15) satisfies the system of normal equations

$$\mathbf{\Gamma}'\mathbf{\Gamma}\, \hat{\boldsymbol{\vartheta}} = \mathbf{\Gamma}'\mathbf{y} \tag{10.17}$$

where $\mathbf{y} = (y_1, \dots, y_n)'$.

Depending on the design $\mathcal{X}$, the normal equations (10.17) may have a unique or multiple solutions. If this system has a unique solution, then the estimate $\hat{f}_n(x)$ can be restored at any point x by (10.14). But even when (10.17) does not have a unique solution, we can still approximate the regression function $f(x)$ at the design points, relying on the geometry of the problem.

In the n-dimensional space of observations $\mathbb{R}^n$, define a linear span-space $\mathcal{S}$ generated by the columns $\boldsymbol{\gamma}_k$ of matrix $\mathbf{\Gamma}$. With a minor abuse of notation, we also denote by $\mathcal{S}$ the operator in $\mathbb{R}^n$ of the orthogonal projection on the span-space $\mathcal{S}$. Introduce a vector consisting of the values of the regression function at the design points,

$$\mathbf{f} = (f(x_1), \dots, f(x_n))',\ \ \mathbf{f} \in \mathbb{R}^n,$$

and a vector of estimates at these points,

$$\hat{\mathbf{f}}_n = \mathcal{S}\mathbf{y} = (\hat{f}(x_1), \dots, \hat{f}(x_n))'.$$

Note that this projection is correctly defined regardless of whether (10.17) has a unique solution or not.

Denote by $\boldsymbol{\varepsilon} = (\varepsilon_1, \dots, \varepsilon_n)'$ the vector of independent $\mathcal{N}(0, \sigma^2)$ - random errors. In this notation, we can interpret $\hat{\mathbf{f}}_n$ as a vector sum of two projections,

$$\hat{\mathbf{f}}_n = \mathcal{S}\mathbf{y} = \mathcal{S}\mathbf{f} + \mathcal{S}\boldsymbol{\varepsilon}. \tag{10.18}$$

Our goal is to find an upper bound on the discrete MISE. Conditionally on the design $\mathcal{X}$, the discrete MISE has the form

$$\mathbb{E}_f\Big[\frac{1}{n}\sum_{i=1}^{n}\big(\hat{f}_n(x_i) - f(x_i)\big)^2 \,\Big|\, \mathcal{X}\Big] = \frac{1}{n}\mathbb{E}_f\big[\,\|\,\mathcal{S}\mathbf{f} + \mathcal{S}\boldsymbol{\varepsilon} - \mathbf{f}\,\|^2 \,|\, \mathcal{X}\,\big]$$

$$\le \frac{2}{n}\|\,\mathcal{S}\mathbf{f} - \mathbf{f}\,\|^2 + \frac{2}{n}\mathbb{E}_f\big[\,\|\,\mathcal{S}\boldsymbol{\varepsilon}\,\|^2 \,|\, \mathcal{X}\,\big] \tag{10.19}$$

where $\|\cdot\|$ is the Euclidean norm in $\mathbb{R}^n$. Here we used the inequality $(a+b)^2 \le 2\,(a^2 + b^2)$.

Denote by $\dim(\mathcal{S})$ the dimension of the span-space $\mathcal{S}$. Note that necessarily $\dim(\mathcal{S}) \le K$. In many special cases, this inequality turns into equality, $\dim(\mathcal{S}) = K$. For example, it is true for the regressogram under Assumption 10.1 (see Exercise 10.72).

Assumption 10.8. There exists δ_n, $\delta_n \to 0$ as $n \to \infty$, such that for any $f \in \Theta(\beta)$, the inequality is fulfilled

$$\frac{1}{n}\|\,\mathcal{S}\,\mathbf{f} - \mathbf{f}\,\|^2 \le \delta_n^2. \qquad \square$$

Proposition 10.9. *Let Assumption 10.8 hold. Then the following upper bound on the discrete MISE holds:*

$$\mathbb{E}_f\Big[\frac{1}{n}\sum_{i=1}^{n}\big(\hat{f}_n(x_i) - f(x_i)\big)^2 \,\Big|\, \mathcal{X}\Big] \le 2\delta_n^2 + \frac{2\sigma^2\dim(\mathcal{S})}{n}. \tag{10.20}$$

Proof. The normal random errors ε_i, $i = 1, \dots, n$, are conditionally independent, given the design points. Therefore, the square of the Euclidean norm $\|\,\mathcal{S}\,\boldsymbol{\varepsilon}\,\|^2$ has a $\sigma^2\chi^2$-distribution with $\dim(\mathcal{S})$ degrees of freedom. Thus,

$$\mathbb{E}_f\big[\,\|\,\mathcal{S}\boldsymbol{\varepsilon}\,\|^2 \,|\, \mathcal{X}\,\big] = \sigma^2\dim(\mathcal{S}).$$

From Assumption 10.8, we find that the right-hand side of (10.19) is bounded from above by $2\delta_n^2 + 2\sigma^2\dim(\mathcal{S})/n$. $\square$

Proposition 10.10. *Assume that for any regression function $f \in \Theta(\beta)$, there exists a linear combination $a_1\gamma_1(x) + \cdots + a_K\gamma_K(x)$ such that at any design point x_i, the following inequality holds:*

$$\big|\,\big[\,a_1\gamma_1(x_i) + \cdots + a_K\gamma_K(x_i)\big] - f(x_i)\,\big| \le \delta_n, \quad i = 1, \dots, n. \tag{10.21}$$

Then the upper bound (10.20) is valid.

Proof. Recall that $\mathcal{S}$ is an operator of orthogonal projection, and, therefore, $\mathcal{S}\mathbf{f}$ is the vector in $\mathcal{S}$ closest to $\mathbf{f}$. Applying (10.21), we see that

$$\frac{1}{n}\| \mathcal{S}\mathbf{f} - \mathbf{f} \|^2 \le \frac{1}{n}\sum_{i=1}^{n} \big| \big[a_1\gamma_1(x_i) + \cdots + a_K\gamma_K(x_i) \big] - f(x_i) \big|^2 \le \delta_n^2,$$

and Assumption 10.8 holds. □

In the next theorem, we describe the asymptotical performance of the discrete MISE for the regressogram.

Theorem 10.11. *For any fixed design* $\mathcal{X}$, *the discrete MISE of the regressogram* $\hat{f}_n(x)$ *given by* (10.3) *satisfies the inequality*

$$\mathbb{E}_f\Big[\frac{1}{n}\sum_{i=1}^{n} \big(\hat{f}_n(x_i) - f(x_i) \big)^2 \,\big|\, \mathcal{X} \Big] \le 2C_\rho^2 h_n^{2\beta} + \frac{2\sigma^2\beta Q}{n} \tag{10.22}$$

where $C_\rho = L/(\beta-1)!$. *Moreover, under the optimal choice of the bandwidth* $h_n^* = n^{-1/(2\beta+1)}$, *there exists a positive constant* r^*, *independent of* n *and* $f \in \Theta(\beta)$, *such that the following upper bound holds:*

$$\mathbb{E}_f\Big[\frac{1}{n}\sum_{i=,1}^{n} \big(\hat{f}_n(x_i) - f(x_i) \big)^2 \,\big|\, \mathcal{X} \Big] \le r^* n^{-2\beta/(2\beta+2)}.$$

Proof. In the case of the regressogram, $\dim(\mathcal{S}) \le K = \beta Q$. The Taylor approximation of $f(x)$ in Lemma 8.5 within each bin guarantees the inequality (10.21) with $\delta_n = C_\rho h_n^\beta$. Hence Proposition 10.10 yields the upper bound (10.22). Now recall that $Q/n = 1/(2nh_n)$. Under the optimal choice of the bandwidth, both terms on the right-hand side of (10.22) have the same magnitude $O(n^{-2\beta/(2\beta+1)})$. □

10.5. Orthogonal Series Regression Estimator

In this section we take a different approach to estimation of the regression function f. We will be concerned with estimation of its Fourier coefficients. The functional class to which f belongs will differ from the Hölder class.

10.5.1. Preliminaries. A set of functions $\mathcal{B}$ is called an *orthonormal basis* in $L_2[0,1]$ if: (1) the L_2-norm in the interval $[0, 1]$ of any function in this set is equal to one, that is, $\int_0^1 g(x)\,dx = 1$ for any $g \in \mathcal{B}$, and (2) the dot product of any two functions in $\mathcal{B}$ is zero, that is, $\int_0^1 g_1(x)g_2(x)\,dx = 0$ for any g_1, $g_2 \in \mathcal{B}$.

Consider the following set of functions defined for all x in $[0, 1]$:

(10.23)
$$\big\{1,\ \sqrt{2}\sin(2\pi x),\ \sqrt{2}\cos(2\pi x), \dots, \sqrt{2}\sin(2\pi kx),\ \sqrt{2}\cos(2\pi kx), \dots \big\}.$$

This set is referred to as a *trigonometric basis.* The next proposition is a standard result from analysis. We omit its proof.

Proposition 10.12. *The trigonometric basis in* (10.23) *is an orthonomal basis.*

Choose the trigonometric basis as a working basis in $L_2[0,1]$ space. For any function $f(x)$, $0 \leq x \leq 1$, introduce its Fourier coefficients by

$$a_0 = \int_0^1 f(x)\,dx, \quad a_k = \int_0^1 f(x)\sqrt{2}\cos(2\pi kx)\,dx$$

and

$$b_k = \int_0^1 f(x)\sqrt{2}\cos(2\pi kx)\,dx, \quad k = 1, 2, \ldots .$$

The trigonometric basis is *complete* in the sense that if $\|f\|_2 < \infty$, and

$$f_m(x) = a_0 + \sum_{k=1}^{m} a_k\sqrt{2}\cos(2\pi kx) + \sum_{k=1}^{m} b_k\sqrt{2}\sin(2\pi kx), \quad 0 \leq x \leq 1,$$

then

$$\lim_{m\to\infty} \| f_m(\cdot) - f(\cdot) \|_2 = 0.$$

Thus, a function f with a finite L_2-norm is equivalent to its Fourier series

$$f(x) = a_0 + \sum_{k=1}^{\infty} a_k\sqrt{2}\cos(2\pi kx) + \sum_{k=1}^{\infty} b_k\sqrt{2}\sin(2\pi kx),$$

though they may differ at the points of discontinuity.

The next lemma links the decrease rate of the Fourier coefficients with β, $\beta \geq 1$, the smoothness parameter of f.

Lemma 10.13. *If* $\sum_{k=1}^{\infty}(a_k^2 + b_k^2)\,k^{2\beta} \leq L$ *for some constant* L, *then* $\| f^{(\beta)} \|_2^2 \leq (2\pi)^{2\beta} L$.

Proof. We restrict the calculations to the case $\beta = 1$. See Exercise 10.73 for the proof of the general case. For $\beta = 1$, we have

$$f'(x) = \sum_{k=1}^{\infty} \left[a_k(-2\pi k)\sqrt{2}\sin(2\pi kx) + b_k(2\pi k)\sqrt{2}\cos(2\pi kx) \right].$$

Thus,

$$\| f' \|_2^2 = (2\pi)^2 \sum_{k=1}^{\infty} k^2(a_k^2 + b_k^2) \leq (2\pi)^2 L. \qquad \square$$

10.5.2. Discrete Fourier Series and Regression. Consider the observations

$$(10.24) \qquad y_i = f(i/n) + \varepsilon_i, \ \varepsilon_i \sim \mathcal{N}(0, \sigma^2), \ i = 1, \dots, n.$$

To ease presentation, it is convenient to assume that $n = 2n_0 + 1$ is an odd number, $n_0 \geq 1$.

Our goal is to estimate the regression function f in the integral sense in $L_2[0, 1]$. Since our observations are discrete and available only at the equidistant design points $x_i = i/n$, we want to restore the regression function exclusively at these points.

Consider the values of the trigonometric basis functions in (10.23) at the design points $x_i = i/n, \ i = 1, \dots, n,$

(10.25)
$$\Big\{1, \sqrt{2}\sin\big(\frac{2\pi i}{n}\big), \sqrt{2}\cos\big(\frac{2\pi i}{n}\big), \dots, \sqrt{2}\sin\big(\frac{2\pi k i}{n}\big), \sqrt{2}\cos\big(\frac{2\pi k i}{n}\big), \dots\Big\}.$$

For any functions g, g_1 and $g_2 \in L_2[0, 1]$, define the discrete dot product and the respective squared L_2-norm by the Riemann sums

$$\big(g_1(\cdot), \, g_2(\cdot)\big)_{2,n} = \frac{1}{n}\sum_{i=1}^{n} g_1\big(i/n\big) g_2\big(i/n\big) \ \text{ and } \ \| \, g(\cdot) \, \|_{2,n}^2 = \frac{1}{n}\sum_{i=1}^{n} g^2\big(i/n\big).$$

Clearly, the values at the design points for each function in (10.25) represent a vector in $\mathbb{R}^n$. Therefore, there cannot be more than n orthonormal functions with respect to the discrete dot product. As shown in the lemma below, the functions in (10.25) corresponding to $k = 1, \dots, n_0$, form an orthonormal basis with respect to this dot product.

Lemma 10.14. *Fix $n = 2n_0 + 1$ for some $n_0 \geq 1$. For $i = 1, \dots, n$, the system of functions*

$$\Big\{1, \sqrt{2}\sin\big(\frac{2\pi i}{n}\big), \sqrt{2}\cos\big(\frac{2\pi i}{n}\big), \dots, \sqrt{2}\sin\big(\frac{2\pi n_0 i}{n}\big), \sqrt{2}\cos\big(\frac{2\pi n_0 i}{n}\big)\Big\}$$

is orthonormal with respect to the discrete dot product.

Proof. For any k and l, the elementary trigonometric identities hold:

$$(10.26) \quad \sin\big(\frac{2\pi k i}{n}\big) \sin\big(\frac{2\pi l i}{n}\big) = \frac{1}{2}\Big[\cos\big(\frac{2\pi(k-l)i}{n}\big) - \cos\big(\frac{2\pi(k+l)i}{n}\big)\Big]$$

and

$$(10.27) \quad \cos\big(\frac{2\pi k i}{n}\big) \cos\big(\frac{2\pi l i}{n}\big) = \frac{1}{2}\Big[\cos\big(\frac{2\pi(k-l)i}{n}\big) + \cos\big(\frac{2\pi(k+l)i}{n}\big)\Big].$$

Also, as shown in Exercise 10.74, for any integer $m \neq 0 (\operatorname{mod} n)$,

$$(10.28) \qquad \sum_{i=1}^{n} \cos\big(\frac{2\pi m i}{n}\big) = \sum_{i=1}^{n} \sin\big(\frac{2\pi m i}{n}\big) = 0.$$

Now fix $k \neq l$ such that $k,\, l \leq n_0$. Note that then $k \pm l \neq 0 (\mathrm{mod}\, n)$. Letting $m = k \pm l$ in (10.28), and applying (10.26)-(10.28), we obtain that

$$\sum_{i=1}^{n} \sin\Big(\frac{2\pi k i}{n}\Big) \sin\Big(\frac{2\pi l i}{n}\Big) = \frac{1}{2} \sum_{i=1}^{n} \cos\Big(\frac{2\pi (k-l) i}{n}\Big)$$

$$- \frac{1}{2} \sum_{i=1}^{n} \cos\Big(\frac{2\pi (k+l) i}{n}\Big) = 0$$

and

$$\sum_{i=1}^{n} \cos\Big(\frac{2\pi k i}{n}\Big) \cos\Big(\frac{2\pi l i}{n}\Big) = \frac{1}{2} \sum_{i=1}^{n} \cos\Big(\frac{2\pi (k-l) i}{n}\Big)$$

$$+ \frac{1}{2} \sum_{i=1}^{n} \cos\Big(\frac{2\pi (k+l) i}{n}\Big) = 0,$$

which yields that the respective dot products are zeros.

If $k = l \leq n_0$, then from (10.26)-(10.28), we have

$$\sum_{i=1}^{n} \sin^2\Big(\frac{2\pi k i}{n}\Big) = \frac{1}{2} \sum_{i=1}^{n} \cos(0) = n/2$$

and

$$\sum_{i=1}^{n} \cos^2\Big(\frac{2\pi k i}{n}\Big) = \frac{1}{2} \sum_{i=1}^{n} \cos(0) = n/2.$$

These imply the normalization condition,

$$\|\, \sqrt{2} \sin\Big(\frac{2\pi k i}{n}\Big) \|_{2,\,n}^2 = \|\, \sqrt{2} \cos\Big(\frac{2\pi k i}{n}\Big) \|_{2,\,n}^2 = 1.$$

Finally, for any $k,\, l \leq n_0$, from the identity

$$\sin\Big(\frac{2\pi k i}{n}\Big) \cos\Big(\frac{2\pi l i}{n}\Big) = \frac{1}{2}\Big[\sin\Big(\frac{2\pi (k+l) i}{n}\Big) + \sin\Big(\frac{2\pi (k-l) i}{n}\Big) \Big]$$

and (10.28), we have

$$\sum_{i=1}^{n} \sin\Big(\frac{2\pi k i}{n}\Big) \cos\Big(\frac{2\pi l i}{n}\Big) = \frac{1}{2} \sum_{i=1}^{n} \sin\Big(\frac{2\pi (k+l) i}{n}\Big)$$

$$+ \frac{1}{2} \sum_{i=1}^{n} \sin\Big(\frac{2\pi (k-l) i}{n}\Big) = 0. \qquad \square$$

For a given integer β and a positive constant L, introduce a class of functions $\Theta_{2,n} = \Theta_{2,n}(\beta, L)$ defined at the design points $x_i = i/n$, $i = 1, \dots, n$. We say that $f \in \Theta_{2,n}(\beta, L)$ if

$$f(i/n) = a_0 + \sum_{k=1}^{n_0} \Big[a_k \sqrt{2} \cos\big(\frac{2\pi k i}{n}\big) + b_k \sqrt{2} \sin\big(\frac{2\pi k i}{n}\big) \Big], \quad n = 2n_0 + 1,$$

where the Fourier coefficients a_k and b_k satisfy the condition

$$\sum_{k=1}^{n_0} \big[a_k^2 + b_k^2 \big] k^{2\beta} \le L.$$

Note that there are a total of n Fourier coefficients, $a_0, a_1, \dots, a_{n_0}$, $b_1, \dots, b_{n_0}$, that define any function f in the class $\Theta_{2,n}(\beta, L)$. It should be expected, because any such function is equivalent to a vector in $\mathbb{R}^n$.

The class $\Theta_{2,n}(\beta, L)$ replaces the Hölder class $\Theta(\beta, L, L_1)$ in our earlier studies. However, in view of Lemma 10.13, the parameter β still represents the smoothness of regression functions.

We want to estimate the regression function f in the discrete MISE. The quadratic risk, for which we preserve the notation $R_n(\hat{f}_n, f)$, has the form

$$R_n(\hat{f}_n, f) = \mathbb{E}_f\Big[\| \hat{f}_n(\cdot) - f(\cdot) \|_{2,n}^2 \Big] = \mathbb{E}_f\Big[\frac{1}{n} \sum_{i=1}^{n} \Big(\hat{f}_n(i/n) - f(i/n) \Big)^2 \Big].$$

Thus far, we have worked with the functions of sine and cosine separately. It is convenient to combine them in a single notation. Put $\varphi_0(i/n) = 1$, and $c_0 = a_0$. For $m = 1, \dots, n_0$, take

$$\varphi_{2m}(i/n) = \sqrt{2} \cos\big(\frac{2\pi m i}{n}\big), \quad c_{2m} = a_m,$$

and

$$\varphi_{2m-1}(i/n) = \sqrt{2} \sin\big(\frac{2\pi m i}{n}\big), \quad c_{2m-1} = b_m.$$

Note that altogether we have n basis functions φ_k, $k = 0, \dots, n-1$. They satisfy the orthonormality conditions

$$\big(\varphi_k(\cdot), \varphi_l(\cdot)\big)_{2,n} = \frac{1}{n} \sum_{i=1}^{n} \varphi_k(i/n)\, \varphi_l(i/n) = 0, \quad \text{for } k \ne l,$$

$$\text{and } \| \varphi_k(\cdot) \|_{2,n}^2 = \frac{1}{n} \sum_{i=1}^{n} \varphi_k^2(i/n) = 1. \tag{10.29}$$

The regression function f at the design points $x_i = i/n$ can be written as

$$f(i/n) = \sum_{k=0}^{n-1} c_k\, \varphi_k(i/n).$$

It is easier to study the estimation problem with respect to the discrete L_2-norm in the space of the Fourier coefficients c_k, called a *sequence space.* Assume that these Fourier coefficients are estimated by $\hat{c}_k$, $k = 0, \dots, n-1$. Then the estimator of the regression function in the original space can be expressed by the sum

$$\hat{f}_n(i/n) = \sum_{k=0}^{n-1} \hat{c}_k \, \varphi_k(i/n). \tag{10.30}$$

Lemma 10.15. *The discrete MISE of the estimator $\hat{f}_n$ in* (10.30) *can be presented as*

$$R_n(\hat{f}_n, f) = \mathbb{E}_f\Big[\, \| \hat{f}_n(\cdot) - f(\cdot) \|_{2,n}^2 \Big] = \mathbb{E}_f\Big[\sum_{k=0}^{n-1} (\hat{c}_k - c_k)^2 \Big].$$

Proof. By the definition of the risk function $R_n(\hat{f}_n, f)$, we have

$$R_n(\hat{f}_n, f) = \mathbb{E}_f\Big[\Big\| \sum_{k=0}^{n-1} (\hat{c}_k - c_k) \varphi_k(\cdot) \Big\|_{2,n}^2 \Big]$$

$$= \mathbb{E}_f\Big[\sum_{k,l=0}^{n-1} (\hat{c}_k - c_k)(\hat{c}_l - c_l)\big(\varphi_k(\cdot), \varphi_l(\cdot)\big)_{2,n} \Big] = \mathbb{E}_f\Big[\sum_{k=0}^{n-1} (\hat{c}_k - c_k)^2 \Big]$$

where we used the fact that the basis functions are orthonormal. □

To switch from the original observation y_i to the corresponding observation in the sequence space, consider the following transformation:

$$z_k = \big(y_i, \varphi_k(\cdot)\big)_{2,n} = \frac{1}{n} \sum_{i=1}^{n} y_i \, \varphi_k(i/n), \quad k = 0, \dots, n-1. \tag{10.31}$$

Lemma 10.16. *The random variables z_k, $k = 0, \dots, n-1$ defined by* (10.31) *satisfy the equations*

$$z_k = c_k + \sigma \, \xi_k / \sqrt{n}, \quad k = 0, \dots, n-1,$$

for some independent standard normal random variables ξ_k.

Proof. First, observe that

$$y_i = c_0 \, \varphi_0(i/n) + \dots + c_{n-1} \, \varphi_{n-1}(i/n) + \varepsilon_i$$

where the error terms ε_i's are independent $\mathcal{N}(0, \sigma^2)$-random variables, $i = 1, \dots, n$. Thus, for any $k = 0, \dots, n-1$, we can write

$$z_k = \frac{1}{n} \sum_{i=1}^{n} \big[c_0 \, \varphi_0(i/n) + \dots + c_{n-1} \, \varphi_{n-1}(i/n) \big] \, \varphi_k(i/n) + \frac{1}{n} \sum_{i=1}^{n} \varepsilon_i \, \varphi_k(i/n).$$

By the orthonormality conditions (10.29), the first sum is equal to c_k, and the second one can be written as $\sigma \xi_k / \sqrt{n}$ where

$$\xi_k = \frac{\sqrt{n}}{\sigma} \Big(\frac{1}{n} \sum_{i=1}^{n} \varepsilon_i \varphi_k(i/n) \Big)$$

$$= \Big(\frac{\sigma^2}{n} \sum_{i=1}^{n} \varphi_k^2(i/n) \Big)^{-1/2} \frac{1}{n} \sum_{i=1}^{n} \varepsilon_i \varphi_k(i/n) \sim \mathcal{N}(0,1).$$

As a result,

$$z_k = c_k + \sigma \xi_k / \sqrt{n}.$$

It remains to show that the ξ's are independent. Since they are normally distributed, it suffices to show that they are uncorrelated. This in turn follows from independence of the ε's and orthogonality of the φ's. Indeed, we have that for any $k \neq l$ such that $k, l = 0, \dots, n-1$,

$$\mathbb{C}ov(\xi_k, \xi_l) = \frac{1}{\sigma^2 n} \mathbb{E}\Big[\Big(\sum_{i=1}^{n} \varepsilon_i \varphi_k(i/n) \Big) \Big(\sum_{i=1}^{n} \varepsilon_i \varphi_l(i/n) \Big) \Big]$$

$$= \frac{1}{\sigma^2 n} \mathbb{E}\Big[\sum_{i=1}^{n} \varepsilon_i^2 \varphi_k(i/n) \varphi_l(i/n) \Big] = \frac{1}{n} \sum_{i=1}^{n} \varphi_k(i/n) \varphi_l(i/n) = 0. \quad \square$$

The *orthogonal series* (or *projection*) *estimator* of regression function f is defined by

$$\hat{f}_n(i/n) = \sum_{k=0}^{M} z_k \varphi_k(i/n), \quad i = 1, \dots, n, \tag{10.32}$$

where $M = M(n)$ is an integer parameter of the estimation procedure. Note that $\hat{f}_n(i/n)$ is indeed an estimator, because it is computable from the original observed responses $y_1, \dots, y_n$. The parameter M serves to balance the bias and variance errors of the estimation. The choice

$$M = M(n) = \lfloor (h_n^*)^{-1} \rfloor = \lfloor n^{1/(2\beta+1)} \rfloor,$$

where $\lfloor \cdot \rfloor$ denotes an integer part of a number, turns out to be optimal in the minimax sense. We will prove only the upper bound.

Theorem 10.17. *Assume that the regression function f belongs to the class $\Theta_{2,n}(\beta, L)$. Then, uniformly over this class, the quadratic risk in the discrete MISE of the orthogonal series estimator $\hat{f}_n$ given by (10.32) with $M = \lfloor n^{1/(2\beta+1)} \rfloor$ is bounded from above,*

$$R_n(\hat{f}_n, f) = \mathbb{E}_f \Big[\| \hat{f}_n - f \|_{2,n}^2 \Big] \leq \big[\sigma^2 + 4^\beta L \big] n^{-2\beta/(2\beta+1)}.$$

Proof. Consider the orthogonal series estimator $\hat{f}_n(i/n)$ specified by (10.32) with $M = \lfloor n^{1/(2\beta+1)} \rfloor$. Comparing this definition to a general form (10.30) of an estimator given by a Fourier series, we see that in this instance, the estimators of the Fourier coefficients c_k, $k = 0, \dots, n-1$, have the form

$$\hat{c}_k = \begin{cases} z_k, & \text{if } k = 0, \dots, M, \\ 0, & \text{if } k = M+1, \dots, n-1. \end{cases}$$

Now applying Lemmas 10.15 and 10.16, we get

$$\mathbb{E}_f\Big[\| \hat{f}_n - f \|_{2,n}^2 \Big] = \mathbb{E}_f\Big[\sum_{k=0}^{M} (z_k - c_k)^2 \Big] + \sum_{k=M+1}^{n-1} c_k^2$$

$$(10.33) \qquad = \frac{\sigma^2}{n} \mathbb{E}_f\Big[\sum_{k=0}^{M} \xi_k^2 \Big] + \sum_{k=M+1}^{n-1} c_k^2 = \frac{\sigma^2 M}{n} + \sum_{k=M+1}^{n-1} c_k^2 .$$

Next, let $M_0 = \lfloor (M+1)/2 \rfloor$. By the definitions of the functional space $\Theta_{2,n}$ and the basis function φ_k, the following inequalities hold:

$$\sum_{k=M+1}^{n-1} c_k^2 \leq \sum_{k=M_0}^{n_0} \big[a_k^2 + b_k^2 \big] \leq M_0^{-2\beta} \sum_{k=M_0}^{n_0} \big[a_k^2 + b_k^2 \big] k^{2\beta} \leq L M_0^{-2\beta} .$$

Substituting this estimate into (10.33), and noticing that $M_0 \leq M/2$, we obtain that

$$R_n(\hat{f}_n, f) \leq \frac{\sigma^2 M}{n} + L M_0^{-2\beta}$$

$$\leq \frac{\sigma^2 M}{n} + 2^{2\beta} L M^{-2\beta} \leq \big[\sigma^2 + 4^{\beta} L \big] n^{-2\beta/(2\beta+1)} . \qquad \square$$

Exercises

Exercise 10.69. Verify the rate of convergence in (10.4).

Exercise 10.70. Prove the inequality (10.11).

Exercise 10.71. Prove a more accurate bound in (10.11),

$$\mathbb{E}\big[\mathcal{Z}^* \,|\, \mathcal{X} \big] \leq C_z \sqrt{\ln Q} \quad \text{where } Q = 1/(2 h_n).$$

Also show that the respective balance equation for the sup-norm in Remark 10.7 is $h_n^{\beta} = \sqrt{(n h_n)^{-1} \ln h_n^{-1}}$. Solve this equation assuming that n is large.

Exercise 10.72. Prove that for the regressogram under Assumption 10.1, $\dim(\mathcal{S}) = K = \beta\, Q$.

Exercise 10.73. Prove Lemma 10.13 for any integer $\beta > 1$.

Exercise 10.74. Prove formula (10.28).

Chapter 11

Estimation by Splines

11.1. In Search of Smooth Approximation

The regressogram approach to estimation of a regression function f has several advantages. First of all, it is a piecewise polynomial approximation that requires the computation and storage of only $\beta\, Q$ coefficients. Second, the computational process for these coefficients is divided into the Q subproblems, each of dimension β which does not increase with n. Third, along with the regression function f, its derivatives $f^{(m)}$ up to the order $\beta - 1$ can be estimated (see Remark 10.5). The forth advantage is that the regressogram works in the whole interval $[0, 1]$, and the endpoints do not need special treatment.

A big disadvantage of the regressogram is that it suggests a discontinuous function as an approximation of a smooth regression function $f(x)$. An immediate idea is to smooth the regressogram, substituting it by a convolution with some kernel K,

$$\text{smoother of } \hat{f}_n(x) = \int_{x-h_n}^{x+h_n} \frac{1}{h_n} K\Big(\frac{t-x}{h_n}\Big) \hat{f}_n(t)\, dt.$$

Unfortunately, the convolution smoother has shortcomings as well. The endpoints effect exists, hence at these points the estimator should be defined separately. Besides, unless the kernel itself is a piecewise polynomial function, the smoother is no longer piecewise polynomial with ensuing computational difficulties.

A natural question arises: Is it possible to find a piecewise polynomial estimator $\hat{f}_n$ of the regression function f in the interval $[0, 1]$ that would be a smooth function up to a certain order? It turns out that the answer to this question is positive.

Suppose that we still have Q bins, and the regression function f is approximated in each bin by a polynomial. If the regression function belongs to the Hölder class $\Theta(\beta, L, L_1)$, then the justifiable order of each polynomial is $\beta - 1$. Indeed, beyond this order, we do not have any control over the deterministic remainder term in Proposition 10.2. However, it is impossible to have a smooth polynomial estimator $\hat{f}_n$ with the continuous derivatives up to the order $\beta - 1$. It would impose β constraints at each knot (breakpoint) between the bins $2h_n q$, $q = 1, \dots, Q-1$. Thus, the polynomial coefficients in each next bin would be identical to those in the previous one. It makes $\hat{f}_n(x)$ a single polynomial of order $\beta - 1$. Clearly, we cannot approximate a Hölder regression function by a single polynomial in the whole interval $[0, 1]$.

Is it possible to define a piecewise polynomial $\hat{f}_n$ that has $\beta - 2$ continuous derivatives in $[0, 1]$? This question makes sense if $\beta \geq 2$ (for $\beta = 2$ it means that the function itself is continuous). The answer to this question is affirmative. In Q bins we have βQ polynomial coefficients. At the $Q - 1$ inner knots between the bins we impose $(\beta - 1)(Q - 1) = \beta Q - (Q + \beta - 1)$ gluing conditions to guarantee the continuity of $\hat{f}_n$ with the derivatives up to the order $\beta - 2$. Still $Q + \beta - 1$ degrees of freedom are left, at least one per bin, that can be used to ensure some approximation quality of the estimator.

In the spirit of formula (10.14), we can try to define a smooth piecewise polynomial approximation by

$$\hat{f}_n(x) = \hat{\theta}_1 \gamma_1(x) + \cdots + \hat{\theta}_K \gamma_K(x), \quad x \in [0, 1],$$

where $\gamma_1(x), \dots, \gamma_K(x)$ are piecewise polynomials. We require these functions to be linearly independent in order to form a basis in $[0, 1]$. We will show that there exists theoretically and computationally convenient basis of the piecewise polynomial functions called *B-splines*. To introduce the B-splines, we need some auxiliary results presented in the next section.

11.2. Standard B-splines

Consider the linearly independent piecewise polynomial functions in Q bins, each of order $\beta - 1$ with $\beta - 2$ continuous derivatives. Let us address the question: What is the maximum number of such functions? As we argued in the previous section, the answer is $Q + \beta - 1$. We can rephrase our argument in the following way. In the first bin we can have β linearly independent polynomials (for example, all the monomials of order $m = 0, \dots, \beta - 1$). At each of the $Q - 1$ inner knots, the $\beta - 1$ constraints are imposed on the continuity of derivatives. This leaves just one degree of freedom for all the other $Q-1$ bins. Thus, the number of piecewise polynomials in the basis equals $\beta + (Q-1)$.

First, we give the definition of a standard B-spline. Here "B" is short for "basis" spline. It is defined for infinitely many bins with unit length and integer knots. A standard B-spline serves as a building block for a basis of B-splines in the interval $[0,1]$.

A *standard B-spline of order* m, denoted by $S_m(u),\ u \in \mathbb{R}$, is a function satisfying the recurrent convolution

$$S_m(u) = \int_{-\infty}^{\infty} S_{m-1}(z)\, \mathbb{I}_{[0,\,1)}(u - z)\, dz, \quad m = 2, 3, \dots, \tag{11.1}$$

with the initial standard spline $S_1(u) = \mathbb{I}_{[0,1)}(u)$. Note that, by the convolution formula, the function $S_m(u)$ is the probability density function of a sum of m independent random variables uniformly distributed on $[0,1]$.

Since the splines are piecewise continuous functions, their higher derivatives can have discontinuities at the knots. We make an agreement to define the derivatives as right-continuous functions. This is the reason to use the semi-open interval in (11.1).

It is far from being obvious that a standard B-spline meets all the requirements of a piecewise polynomial function of the certain degree of smoothness. Nevertheless, it turns out to be true. The lemmas below describe some analytical properties of standard B-splines.

Lemma 11.1. *(i) For any* $m \geq 2$,

$$S'_m(u) = S_{m-1}(u) - S_{m-1}(u-1), \quad u \in \mathbb{R}. \tag{11.2}$$

(ii) For any $m \geq 2$, $S_m(u)$ *is strictly positive in* $(0, m)$ *and is equal to zero outside of this interval.*

(iii) $S_m(u)$ *is symmetric with respect to the endpoints of the interval* $(0,m)$, *that is,*

$$S_m(u) = S_m(m-u), \quad u \in \mathbb{R}. \tag{11.3}$$

(iv) For any $m \geq 1$ *and for any* $u \in \mathbb{R}$, *the equation (called the partition of unity) holds:*

$$\sum_{j=-\infty}^{\infty} S_m(u-j) = 1. \tag{11.4}$$

Proof. (i) Differentiating formally (11.1) with respect to u, we obtain

$$S'_m(u) = \int_{-\infty}^{\infty} S_{m-1}(z) \left[\delta_{\{0\}}(u - z) - \delta_{\{1\}}(u - z) \right] dz$$

$$= S_{m-1}(u) \;-\; S_{m-1}(u-1)$$

where $\delta_{\{a\}}$ is the Dirac delta-function concentrated at a.

(ii) This part follows immediately from the definition of the standard B-spline as a probability density.

(iii) If U_j is a uniformly distributed in $[0,1]$ random variable, then $1-U_j$ has the same distribution. Hence the probability density of $U_1+\dots+U_m$ is the same as that of $m-(U_1+\dots+U_m)$.

(iv) In view of part (ii), for a fixed $u \in \mathbb{R}$, the sum $\sum_{j=-\infty}^{\infty} S_m(u-j)$ has only a finite number of non-zero terms. Using this fact and (11.2), we have

$$\Big[\sum_{j=-\infty}^{\infty} S_m(u-j)\Big]' = \sum_{j=-\infty}^{\infty} S_m'(u-j) = \sum_{j=-\infty}^{\infty} \big[S_{m-1}(u-j) - S_{m-1}(u-j-1)\big]$$

$$= \sum_{j=-\infty}^{\infty} S_{m-1}(u-j) - \sum_{j=-\infty}^{\infty} S_{m-1}(u-j-1) = 0.$$

Indeed, the last two sums are both finite, hence they have the identical values. Consequently, the sum $\sum_{j=-\infty}^{\infty} S_m(u-j)$ is a constant c, say. We write

$$c = \int_0^1 c\,du = \int_0^1 \Big[\sum_{j=-\infty}^{\infty} S_m(u-j)\Big]\,du \;=\; \sum_{j=0}^{m-1} \int_0^1 S_m(u-j)\,du.$$

Here we used part (ii) once again, and the fact that the variable of integration u belongs to the unit interval. Continuing, we obtain

$$c = \sum_{j=0}^{m-1} \int_j^{j+1} S_m(u)\,du \;=\; \int_0^m S_m(u)\,du \;=\; 1,$$

for $S_m(u)$ is the probability density for u in the interval $[0,m]$. □

Now we try to answer the question: How smooth is the standard B-spline $S_m(u)$? The answer can be found in the following lemma.

Lemma 11.2. *For any $m \geq 2$, the standard B-spline $S_m(u)$, $u \in \mathbb{R}$, is a piecewise polynomial of order $m-1$. It has continuous derivatives up to the order $m-2$, and its derivative of order $m-1$ is a piecewise constant function given by the sum*

$$S_m^{(m-1)}(u) \;=\; \sum_{j=0}^{m-1} (-1)^j \binom{m-1}{j} \mathbb{I}_{[j,\,j+1)}(u). \tag{11.5}$$

Proof. We start by stating the following result. For any $m \geq 2$, the k-th derivative of $S_m(u)$ can be written in the form

$$S_m^{(k)}(u) = \sum_{j=0}^{k} (-1)^j \binom{k}{j} S_{m-k}(u-j), \quad k \leq m-1. \tag{11.6}$$

The shortest way to verify this identity is to use induction on k starting with (11.2). We leave it as an exercise (see Exercise 11.76).

If $k \leq m-2$, then the function $S_{m-k}(u-j)$ is continuous for any j. Indeed, all the functions $S_m(u)$, $m \geq 2$, are continuous as the convolutions in (11.1). Thus, by (11.6), as a linear combination of continuous functions, $S_m^{(k)}(u)$, $k \leq m-2$, is continuous in $u \in \mathbb{R}$. Also, for $k = m-1$, the formula (11.6) yields (11.5).

It remains to show that $S_m(u)$, $u \in \mathbb{R}$, is a piecewise polynomial of order $m-1$. From (11.2), we obtain

$$S_m(u) = \int_0^u \big[S_{m-1}(z) - S_{m-1}(z-1) \big] \, dz. \tag{11.7}$$

Note that by definition, $S_1(u) = \mathbb{I}_{[0,1)}(u)$ is a piecewise polynomial of order zero. By induction, if $S_{m-1}(u)$ is a piecewise polynomial of order at most $m-2$, then so is the integrand in the above formula. Therefore, $S_m(u)$ is a piecewise polynomial of order not exceeding $m-1$. However, from (11.5), the $(m-1)$-st derivative of $S_m(u)$ is non-zero, which proves that $S_m(u)$ has order $m-1$. □

Remark 11.3. To restore a standard B-spline $S_m(u)$, it suffices to look at (11.5) as a differential equation

$$\frac{d\, S_m(u)}{d\, u^m} = \sum_{j=0}^{m-1} \lambda_j \, \mathbb{I}_{[j,j+1)}(u) \tag{11.8}$$

with the constants λ_j defined by the right-hand side of (11.5), and to solve it with the zero initial conditions,

$$S_m(0) = S_m'(0) = \cdots = S_m^{(m-2)}(0) = 0. \qquad \square$$

11.3. Shifted B-splines and Power Splines

The results of this section play a central role in our approach to spline approximation. They may look somewhat technical. From now on, we assume that the order m of all the splines under consideration is given and fixed, $m \geq 2$. We start with the definition of the shifted B-splines. Consider the *shifted B-splines* in the interval $[0, m-1)$,

$$S_m(u),\ S_m(u-1),\ \ldots,\ S_m(u-(m-2)), \quad 0 \leq u < m-1.$$

Let $\mathcal{L}_s$ be the linear space generated by the shifted B-splines,

$$\mathcal{L}_s = \{LS(u) : LS(u) = a_0 S_m(u) + a_1 S_m(u-1) + \cdots + a_{m-2} S_m(u-(m-2))\}$$

where $a_0, \ldots, a_{m-2}$ are real coefficients. Put $\mathbf{a} = (a_0, \ldots, a_{m-2})'$ for the vector of these coefficients.

We need more definitions. Consider the piecewise polynomial functions, called the *power splines*,

$$P_k(u) = \frac{1}{(m-1)!}(u-k)^{m-1}\,\mathbb{I}(u \geq k), \quad k = 0, \ldots, m-2. \tag{11.9}$$

Note that we define the power splines on the whole real axis. In what follows, however, we restrict our attention to the interval $[0, m-1)$.

Similar to $\mathcal{L}_s$, introduce the linear space $\mathcal{L}_p$ of functions generated by the power splines,

$$\mathcal{L}_p = \{\, LP(u) : LP(u) = b_0 P_0(u) + b_1 P_1(u) + \cdots + b_{m-2} P_{m-2}(u) \,\}$$

with the vector of coefficients $\mathbf{b} = (b_0, \ldots, b_{m-2})'$.

Lemma 11.4. *In the interval $[0, m-1)$, the linear spaces $\mathcal{L}_s$ and $\mathcal{L}_p$ are identical. Moreover, there exists a linear one-to-one correspondence between the coefficients* **a** *and* **b** *in the linear combinations of shifted B-splines and power splines.*

Proof. The proof is postponed until Section 11.5. □

For any particular linear combination $LS \in \mathcal{L}_s$, consider its derivatives at the right-most point $u = m-1$,

$$\nu_0 = LS^{(0)}(m-1), \ \nu_1 = LS^{(1)}(m-1), \ \ldots, \ \nu_{m-2} = LS^{(m-2)}(m-1),$$

and put $\boldsymbol{\nu} = (\nu_0, \ldots, \nu_{m-2})'$. Is it possible to restore the function $LS(u) = a_0 S_m(u) + a_1 S_m(u-1) + \cdots + a_{m-2} S_m(u-(m-2))$ from these derivatives? In other words, is it possible to restore in a unique way the vector of coefficients **a** from $\boldsymbol{\nu}$? As the following lemma shows, the answer is assertive.

Lemma 11.5. *There exists a linear one-to-one correspondence between* **a** *and* $\boldsymbol{\nu}$.

Proof. Can be found in the last section of the present chapter. □

Remark 11.6. Though our principal interest lies in the shifted B-splines, we had to involve the power splines for the following reason. The derivatives of the power splines at the right-most point provide the explicit formula (11.19), while for the shifted B-splines this relation is more complex. So, the power splines are just a technical tool for our consideration. □

Next, let us discuss the following problem. Consider the shifted B-spline from $\mathcal{L}_s$ added by another one, $S_m(u-(m-1))$. All these shifted B-splines have a non-trivial effect in the interval $[m-1, m)$. Assume that a polynomial $g(u)$ of degree $m-1$ or less is given in the interval $[m-1, m)$. Can we guarantee a representation of this polynomial in $[m-1, m)$ as a linear combination of the shifted standard B-splines? The answer to this question is revealed in the next two lemmas. The first one explains that all these splines, except the last one, can ensure the approximation of the derivatives of $g(u)$ at $u = m-1$. At this important step, we rely on Lemma 11.5. The last spline $S_m(u-(m-1))$ is used to fit the leading coefficient of $g(u)$. This is done in the second lemma. It is remarkable that the polynomial $g(u)$ not only coincides with the linear combination of B-splines in $[m-1, m)$, but it also controls the maximum of this linear combination for $u \in [0, m-1)$.

Lemma 11.7. *Denote by $\nu_0 = g^{(0)}(m-1), \ldots, \nu_{m-2} = g^{(m-2)}(m-1)$ the derivatives of the polynomial $g(u)$ at $u = m-1$. There exists a unique linear combination $LS(u) = a_0 S_m(u) + a_1 S_m(u-1) + \cdots + a_{m-2} S_m(u-(m-2))$ that solves the boundary value problem*

$$LS^{(0)}(m-1) = \nu_0, \ldots, LS^{(m-2)}(m-1) = \nu_{m-2}.$$

Moreover, there exists a constant $C(m)$ such that

$$\max_{0 \le u \le m-1} |LS(u)| \le C(m) \max\big[|\nu_0|, \ldots, |\nu_{m-2}|\big].$$

Proof. In accordance with Lemma 11.5, there exists a one-to-one linear correspondence between $\mathbf{a}$ and $\boldsymbol{\nu}$, which implies the inequality

$$\max\big[|a_0|, \ldots, |a_{m-2}|\big] \le C(m) \max\big[|\nu_0|, \ldots, |\nu_{m-2}|\big]$$

with a positive constant $C(m)$. Also, we can write

$$\begin{aligned}
\big|LS(u)\big| &= \big|a_0 S_m(u) + \cdots + a_{m-2} S_m(u-(m-2))\big| \\
&\le \max\big[|a_0|, \ldots, |a_{m-2}|\big] \sum_{j=0}^{m-2} S_m(u-j) \\
&\le \max\big[|a_0|, \ldots, |a_{m-2}|\big] \sum_{j=-\infty}^{\infty} S_m(u-j) \\
&\le \max\big[|a_0|, \ldots, |a_{m-2}|\big] \le C(m) \max\big[|\nu_0|, \ldots, |\nu_{m-2}|\big]
\end{aligned}$$

where we applied the partition of unity (11.4). □

Lemma 11.8. *For any polynomial $g(u)$ of order $m-1$, $m-1 \le u < m$, there exists a unique linear combination of the shifted standard B-splines*

$$LS^*(u) = a_0 S_m(u) + \cdots + a_{m-2} S_m(u-(m-2)) + a_{m-1} S_m(u-(m-1))$$

such that $LS^*(u) = g(u)$, if $m-1 \le u < m$.

Proof. Find $LS(u) = a_0 S_m(u) + \cdots + a_{m-2} S_m(u - (m-2))$ such that all the derivatives up to the order $m-2$ of $g(u)$ and $LS(u)$ are identical at $u = m-1$. By Lemma 11.7 such a linear combination exists and is unique. Note that $(m-1)$-st derivatives of $LS(u)$ and $g(u)$ are constants in $[m-1, m)$. If these constants are different, we can add another B-spline, $LS^*(u) = LS(u) + a_{m-1} S_m(u-(m-1))$. The newly added spline $S_m(u-(m-1))$ does not change $LS(u)$ in $[0, m-1)$. By choosing the coefficient a_{m-1} properly, we can make the $(m-1)$-st derivatives of $LS^*(u)$ and $g(u)$ identical while all the derivatives of $LS(u)$ of the smaller orders stay unchanged at $u = m-1$, because $S_m(u-(m-1))$ has all derivatives up to the order $m-2$ equal to zero at $u = m-1$. Figure 8 illustrates the statement of this lemma. □

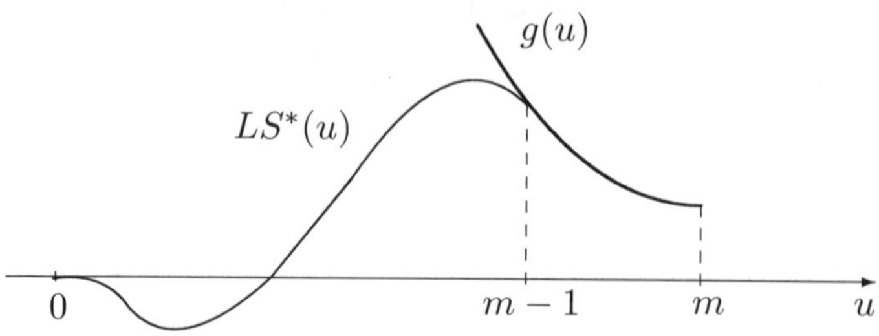

Figure 8. The linear combination $LS^*(u)$ coincides with the polynomial $g(u)$ in $[m-1,\, m)$.

11.4. Estimation of Regression by Splines

For a chosen bandwidth h_n, consider an integer number $Q = 1/(2h_n)$ of the bins

$$B_q = \big[\, 2(q-1)h_n,\ 2qh_n\, \big), \ q = 1, \dots, Q.$$

We are supposed to work with a regression function $f(x)$ that belongs to a fixed Hölder class of functions $\Theta(\beta, L, L_1)$. Let $S_\beta(u)$ be the standard B-spline of order $m = \beta$. This parameter β will determine the order of all the splines that follow. Sometimes, it will be suppressed in the notation.

The number of bins Q increases as the bandwidth $h_n \to 0$, while the order β stays a constant as $n \to \infty$. That is why it is not restrictive to assume that Q exceeds β. We make this assumption in this section. In the interval $[0,1]$, we define a set of functions

$$(11.10) \quad \gamma_k(x) = h_n^\beta S_\beta\Big(\frac{x - 2h_n k}{2h_n} \Big)\, \mathbb{I}_{[0,\,1]}(x)\,, \ k = -\beta + 1, \dots, Q-1.$$

We call these functions *scaled splines* or, simply, *splines* of order β. To visualize the behavior of the splines, notice that, if we disregard the indicator function $\mathbb{I}_{[0,1]}(x)$, then the identity $\gamma_k(x) = \gamma_{k+1}(x + 2h_n)$ holds. It means that as k ranges from $-\beta+1$ to $Q-1$, the functions $\gamma_k(x)$ move from left to right, every time shifting by $2h_n$, the size of one bin. Now we have to restrict the picture to the unit interval, truncating the functions $\gamma_{-\beta+1}, \dots, \gamma_{-1}$ below zero and $\gamma_{Q-\beta+1}, \dots, \gamma_{Q-1}$ above one. An analogy can be drawn between the performance of $\gamma_k(x)$ as k increases and taking periodic snapshots of a hunchback beast who gradually crawls into the picture from the left, crosses the space, and slowly disappears on the right still dragging its tail away. Figure 9 contains an illustration of the case $\beta = 3$.

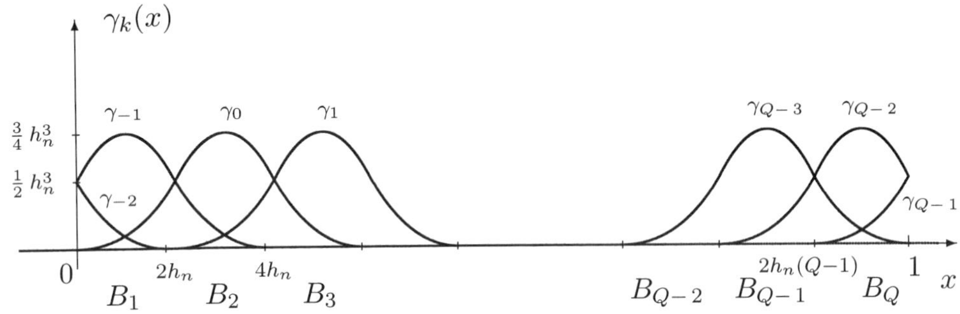

Figure 9. Graphs of functions $\gamma_k(x)$, $k = -2, \dots, Q-1$, when $\beta = 3$.

From properties of the standard B-splines, it immediately follows that within each bin B_q, the function $\gamma_k(x)$ is a polynomial of order $\beta - 1$ with continuous derivatives up to order $\beta - 2$. The knots are the endpoints of the bins. Under the assumption that Q is greater than or equal to β, there is at least one full-sized spline γ_k the support of which contains all β bins. For instance, γ_0 is such a spline.

The proof of the next lemma is postponed to the end of this chapter.

Lemma 11.9. *The set of functions $\{\gamma_k(x),\ k = -\beta+1, \dots, Q-1\}$ forms a basis in the linear sub-space of the smooth piecewise polynomials of order $\beta - 1$ that are defined in bins B_q, and have continuous derivatives up to order $\beta - 2$. That is, any $\gamma(x)$ in this space admits a unique representation*

$$\gamma(x) = \sum_{k=-\beta+1}^{Q-1} \hat{\theta}_k \, \gamma_k(x)\,, \quad x \in [0,1], \tag{11.11}$$

with some real coefficients $\hat{\theta}_{-\beta+1}, \dots, \hat{\theta}_{Q-1}$.

Now we return to the regression observations $y_i = f(x_i) + \varepsilon_i$, $i = 1, \dots, n$, where $f(x)$ is a Hölder function in $\Theta(\beta, L, L_1)$. In this section, we pursue the modest goal of the asymptotic analysis of the discrete MISE for the spline approximation of the regression function. We want to prove the result similar to Theorem 10.11, and in particular, the analogue of inequality (10.22). Note that the proof of Theorem 10.11 is heavily based on the relation between the approximation error δ_n (bias of the regressogram) in Proposition 10.10 and the bandwidth h_n. We need a similar result for approximation by splines.

In the spirit of the general approximation by functions $\gamma_k(x)$ in the space of observations $\mathbb{R}^n$, we introduce the span-space $\mathcal{S}$ as a linear sub-space generated by the vectors

$$\boldsymbol{\gamma}_k = \big(\gamma_k(x_1), \dots, \gamma_k(x_n)\big)', \;\; k = -\beta+1, \dots, Q-1.$$

Following the agreement of Section 10.4, by $\mathcal{S}$ we also denote the operator of the orthogonal projection on this span-space.

Remark 11.10. Note that $\mathcal{S}$ is a linear sub-space in $\mathbb{R}^n$ of the dimension which does not exceed $K = Q+\beta-1$. For the regular designs and sufficiently large n, this dimension should be K. But for a particular design, generally speaking, this dimension can be strictly less than K. □

The following lemma is a version of Proposition 10.10 with $\delta_n = O(h_n^\beta)$. Its proof can be found at the end of the present chapter.

Lemma 11.11. *There exists a constant C_0 independent of n such that for any regression function $f \in \Theta(\beta, L, L_1)$ and for any design $\mathcal{X} = \{x_1, \dots, x_n\}$, we can find $\mathbf{f}_n^* = \big(f_n^*(x_1), \dots, f_n^*(x_n)\big)'$ that belongs to $\mathcal{S}$, for which at any design point x_i the following inequality holds:*

$$|f_n^*(x_i) - f(x_i)| \le C_0\, h_n^\beta, \;\; i = 1, \dots, n. \tag{11.12}$$

Remark 11.12. The vector $\mathbf{f}_n^*$, as any vector in $\mathcal{S}$, admits the representation

$$\mathbf{f}_n^* = \sum_{k=-\beta+1}^{Q-1} \hat{\theta}_k \boldsymbol{\gamma}_k$$

with some real coefficients $\hat{\theta}_k$. Hence this vector can also be associated with the function defined by (11.11),

$$f_n^*(x) = \sum_{k=-\beta+1}^{Q-1} \hat{\theta}_k\, \gamma_k(x), \;\; x \in [0,1].$$

This representation, which is not necessarily unique, defines a spline approximation of the function $f(x)$. □

We are ready to extend the result stated in Theorem 10.11 for the regressogram to the approximation by splines.

Theorem 11.13. *For any design $\mathcal{X}$, the projection*

$$\hat{\mathbf{f}}_n = \mathcal{S}\,\mathbf{y} = \big(\hat{f}_n(x_1), \dots, \hat{f}_n(x_n)\big)'$$

of the regression observations $\mathbf{y} = (y_1, \dots, y_n)'$ on span-space $\mathcal{S}$ generated by the splines of order β, admits the upper bound of the discrete L_2-norm risk

$$\mathbb{E}_f\Big[\frac{1}{n}\sum_{i=1}^{n}\big(\hat{f}_n(x_i) - f(x_i)\big)^2 \,|\, \mathcal{X}\Big] \le C_1 h_n^{2\beta} + \frac{\sigma^2(Q+\beta-1)}{n}. \tag{11.13}$$

Moreover, under the optimal choice of the bandwidth $h_n = h_n^* = n^{-1/(2\beta+1)}$, the following upper bound holds:

$$\mathbb{E}_f\Big[\frac{1}{n}\sum_{i=1}^{n}\big(\hat{f}_n(x_i) - f(x_i)\big)^2 \,|\, \mathcal{X}\Big] \le r^* n^{-\beta/(2\beta+2)}. \tag{11.14}$$

In the above, the constants C_1 and r^* are positive and independent of n and $f \in \Theta(\beta, L, L_1)$.

Proof. The result follows immediately from the bound (11.12) on the approximation error by splines (cf. the proof of Theorem 10.11.) □

Remark 11.14. With the splines $\gamma_k(x)$ of this section, we could introduce the design matrix with column vectors (10.16) as well as the system of normal equations (10.17). In the case of B-splines, however, the system of normal equations does not partition into sub-systems as was the case of the regressogram. It makes the asymptotic analysis of spline approximation technically more challenging as compared to the one of the regressogram. In particular, an analogue of Proposition 10.2 with explicit control over the bias and the stochastic terms goes beyond the scope of this book. □

11.5. Proofs of Technical Lemmas

Proof of Lemma 11.4. It is easy to show (see Exercise 11.77) that, according to (11.5), the $(m-1)$-st derivative of $LS \in \mathcal{L}_s$ is a piecewise constant function

$$LS^{(m-1)}(u) = \lambda_j, \text{ if } j \le u < j+1, \tag{11.15}$$

where for $j = 0, \dots, m-2$,

$$\lambda_j = \sum_{i=0}^{j} a_i(-1)^{j-i}\binom{m-1}{j-i} = a_0(-1)^j\binom{m-1}{j} + a_1(-1)^{j-1}\binom{m-1}{j-1}$$

$$+ \cdots + a_{j-1}(-1)\binom{m-1}{1} + a_j\binom{m-1}{0}. \tag{11.16}$$

On the other hand, any power spline $LP \in \mathcal{L}_p$ also has the piecewise constant $(m-1)$-st derivative

$$LP^{(m-1)}(u) = \lambda_j, \text{ if } j \le u < j+1, \tag{11.17}$$

with

$$\lambda_j = b_0 + \cdots + b_j, \; j = 0, \ldots, m-2. \tag{11.18}$$

In (11.15) and (11.17), we have deliberately denoted the $(m-1)$-st derivative by the same λ_j's because we mean them to be identical. Introduce a vector $\boldsymbol{\lambda} = (\lambda_0, \ldots, \lambda_{m-2})'$. If we look at (11.16) and (11.18) as the systems of linear equations for $\mathbf{a}$ and $\mathbf{b}$, respectively, we find that the matrices of these systems are lower triangular with non-zero diagonal elements. Hence, these systems establish the linear one-to-one correspondence between $\mathbf{a}$ and $\boldsymbol{\lambda}$, on the one hand, and between $\boldsymbol{\lambda}$ and $\mathbf{b}$, on the other hand. Thus, there exists a linear one-to-one correspondence between $\mathbf{a}$ and $\mathbf{b}$. □

Proof of Lemma 11.5. Applying Lemma 11.4, we can find a linear combination of the power splines such that

$$LS(u) = LP(u) = b_0 P_0(u) + b_1 P_1(u) + \cdots + b_{m-2} P_{m-2}(u)$$
$$= b_0 \frac{1}{(m-1)!} u^{m-1} \mathbb{I}_{[0,\, m-1)}(u) + b_1 \frac{1}{(m-1)!} (u-1)^{m-1} \mathbb{I}_{[1,\, m-1)}(u)$$
$$+ \cdots + b_{m-2} \frac{1}{(m-1)!} (u-(m-2))^{m-1} \mathbb{I}_{[m-2,\, m-1)}(u).$$

The derivatives of the latter combination, $\nu_j = LP^{(j)}(u)$, at the right-most point $u = m-1$ are computable explicitly (see Exercise 11.78),
(11.19)

$$\nu_j = b_0 \frac{(m-1)^{m-j-1}}{(m-j-1)!} + b_1 \frac{(m-2)^{m-j-1}}{(m-j-1)!} + \cdots + b_{m-2} \frac{(1)^{m-j-1}}{(m-j-1)!}.$$

If we manage to restore the coefficients $\mathbf{b}$ from the derivatives $\boldsymbol{\nu}$, then by Lemma 11.4, we would prove the claim. Consider (11.19) as the system of linear equations. Then the matrix $\mathbf{M}$ of this system is an $(m-1) \times (m-1)$ matrix with the elements

$$\mathbf{M}_{j,k} = \frac{(m-k-1)^{m-j-1}}{(m-j-1)!}, \; j, k = 0, \ldots, m-2. \tag{11.20}$$

The matrix $\mathbf{M}$ is invertible because its determinant is non-zero (see Exercise 11.79). Thus, the lemma follows. □

Proof of Lemma 11.9. As shown above, the dimension of the space of smooth piecewise polynomials of order no greater than $\beta - 1$ equals $Q + \beta - 1$ which matches the number of functions $\gamma_k(x)$. Thus, the only question is about linear independence of functions $\gamma_k(x)$, $x \in [0,1]$. Consider the functions $\gamma_k(x)$ for $k = -\beta+1,\dots,Q-\beta$. In this set, each consecutive function has a support that contains a new bin not included in the union of all the previous supports. That is why the $\gamma_k(x)$'s are linearly independent for $k = -\beta+1,\dots,Q-\beta$. Hence, the linear combinations

$$\mathcal{L}_1 = \left\{ a_{-\beta+1}\,\gamma_{-\beta+1}(x) + \cdots + a_{Q-\beta}\,\gamma_{Q-\beta}(x) \,\middle|\, a_{-\beta+1},\dots,\, a_{Q-\beta} \in \mathbb{R} \right\}$$

form a linear space of functions of dimension Q. A similar argument shows that the linear combinations of the remaining splines

$$\mathcal{L}_2 = \left\{ a_{Q-\beta+1}\,\gamma_{Q-\beta+1}(x) + \cdots + a_{Q-1}\,\gamma_{Q-1}(x) \,\middle|\, a_{Q-\beta+1},\dots,\, a_{Q-1} \in \mathbb{R} \right\}$$

form a linear space of dimension $\beta - 1$.

Since the supports of the functions from $\mathcal{L}_1$ cover the whole semi-open interval $[0, 1)$, the "support" argument does not prove independence of $\mathcal{L}_1$ and $\mathcal{L}_2$. We have to show that these spaces intersect only at the origin. Indeed, by the definition of the standard B-splines, the first $(\beta-2)$ derivatives of any function from $\mathcal{L}_1$ at $x = 1$ are zeros. On the other hand, by Lemma 11.7, a function from $\mathcal{L}_2$ has all its first $\beta - 2$ derivatives equal to zero, if and only if all its coefficients are zeros, $a_{Q-\beta+1} = \cdots = a_{Q-1} = 0$. Thus, a zero function is the only one that simultaneously belongs to $\mathcal{L}_1$ and $\mathcal{L}_2$. □

Proof of Lemma 11.11. Put $q(l) = 1+\beta l$. Consider all the bins $B_{q(l)}$ with $l = 0,\dots,(Q-1)/\beta$. Without loss of generality, we assume that $(Q-1)/\beta$ is an integer, so that the last bin B_Q belongs to this subsequence. Note that the indices of the bins in the subsequence $B_{q(l)}$ are equal to 1 modulo β, and that any two consecutive bins $B_{q(l)}$ and $B_{q(l+1)}$ in this subsequence are separated by $(\beta - 1)$ original bins.

Let $x_l = 2(q(l) - 1)h_n$ denote the left endpoint of the bin $B_{q(l)}$, $l = 0, 1, \dots$. For any regression function $f \in \Theta(\beta, L, L_1)$, introduce the Taylor expansion of $f(x)$ around $x = x_l$,

(11.21)
$$\pi_l(x) = f(x_l) + \frac{f^{(1)}(x_l)}{1!}(x-x_l) + \cdots + \frac{f^{(\beta-1)}(x_l)}{(\beta-1)!}(x-x_l)^{\beta-1}, \quad x \in B_{q(l)}.$$

In accordance with Lemma 11.8, for any l, there exists a linear combination of the splines that coincides with $\pi_l(x)$ in $B_{q(l)}$. It implies that

$$\pi_l(x) = a_{q(l)-\beta}\,\gamma_{q(l)-\beta}(x) + \cdots + a_{q(l)-1}\,\gamma_{q(l)-1}(x), \quad x \in B_{q(l)},$$

with some uniquely defined real coefficients a_k, $k = q(l)-\beta,\dots,q(l)-1$. Note that as l runs from 1 to $(Q-1)/\beta$, each of the splines $\gamma_k(x)$, $k = -\beta+1,\dots,Q-1$, participates exactly once in these linear combinations.

Consider the sum

$$\gamma(x) = \sum_{1 \leq l \leq (Q-1)/\beta} \left[a_{q(l)-\beta}\, \gamma_{q(l)-\beta}(x) + \cdots + a_{q(l)-1}\, \gamma_{q(l)-1}(x) \right]$$

$$= \sum_{k=-\beta+1}^{Q-1} a_k\, \gamma_k(x), \quad 0 \leq x \leq 1.$$

This function $\gamma(x)$ defines a piecewise polynomial of order at most $\beta - 1$ that coincides with the Taylor polynomial (11.21) in all the bins $B_{q(l)}$ (see Figure 10). Hence, in the union of these bins $\bigcup_l B_{q(l)}$, the function $\gamma(x)$ does not deviate away from $f(x)$ by more than $O(h_n^\beta)$, this magnitude being preserved uniformly over $f \in \Theta(\beta, L, L_1)$.

Next, how close is $\gamma(x)$ to $f(x)$ in the rest of the unit interval? We want to show that the same magnitude holds for all $x \in [0, 1]$, that is,

$$\max_{0 \leq x \leq 1} \left| \gamma(x) - f(x) \right| \leq C_1\, h_n^\beta \tag{11.22}$$

with a constant C_1 independent of $f \in \Theta(\beta, L, L_1)$.

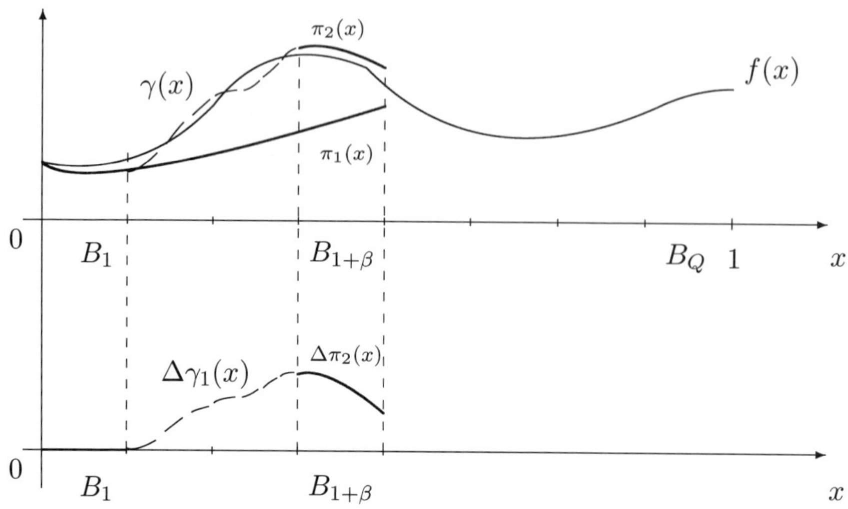

Figure 10. Schematic graphs of the functions $\gamma(x)$ and $\Delta\gamma_1(x)$ for x lying in bins B_1 through $B_{1+\beta}$.

Clearly, it is sufficient to estimate the absolute value $|\gamma(x) - f(x)|$ in the gap between two consecutive bins $B_{q(l)}$ and $B_{q(l+1)}$. Consider the interval $[x_l, x_{l+1} + 2h_n)$. It covers all the bins from $B_{q(l)}$ to $B_{q(l+1)}$, inclusively. The length of this interval is $2h_n(\beta + 1)$. Hence, the regression function $f(x)$ does not deviate away from its Taylor approximation $\pi_l(x)$ in this interval by more than $O(h_n^\beta)$ uniformly over the Hölder class. Thus, to verify (11.22), it is enough to check the magnitude of the difference $\Delta\gamma_l(x) = \gamma(x) - \pi_l(x)$, $x \in [x_l, x_{l+1} + 2h_n)$. Note that this difference is a piecewise polynomial of order at most $\beta - 1$ in the bins. In particular, it is a zero function for $x \in B_{q(l)}$, and is equal to $\Delta\pi_l(x) = \pi_{l+1}(x) - \pi_l(x)$ for $x \in B_{q(l+1)}$ (see Figure 10).

We want to rescale $\Delta\pi_l(x)$ to bring it to the scale of the integer bins of unit length. Put

$$g(u) = h_n^{-\beta}\,\Delta\pi_l(x_l + 2h_n(u+1)) \text{ with } 0 \le u \le \beta - 1,$$

so that $u = 0$ corresponds to the left endpoint of $B_{q(l)}$ and $u = \beta - 1$ corresponds the left endpoint of $B_{q(l+1)}$. Next, we compute the derivatives of $g(u)$ at $u = \beta - 1$,

$$\nu_i = \frac{d^j}{du^j} g(\beta - 1) = h_n^{-\beta}\,(2h_n)^j \frac{d^j}{dx^j} \Delta\pi_l(x_{l+1})$$

$$= 2^j h_n^{j-\beta}\Big[f^{(j)}(x_{l+1}) - \Big(f^{(j)}(x_l) + \frac{f^{(j+1)}(x_l)}{1!}(x_{l+1} - x_l) + \cdots$$

$$\cdots + \frac{f^{(\beta-1)}(x_l)}{(\beta-1-j)!}(x_{l+1} - x_l)^{\beta-1-j}\Big)\Big].$$

Note that the expression in the brackets on the right-hand side is the remainder term of the Taylor expansion of the j-th derivative $f^{(j)}(x_{l+1})$ around x_l. If $f \in \Theta(\beta, L, L_1)$, then $f^{(j)}$ belongs to the Hölder class $\Theta(\beta - j, L, L_2)$ with some positive constant L_2 (see Exercise 11.81). Similar to Lemma 10.2, this remainder term has the magnitude $O(\,|\,x_{l+1} - x_l\,|^{\beta-j}\,) = O(h_n^{\beta-j})$.

Thus, in the notation of Lemma 11.7, $\max\big[\,|\nu_0|, \dots, |\nu_{\beta-1}|\,\big] \le C_1$ where the constant C_1 does not depend on n nor l. From Lemma 11.7, the unique spline of order β with zero derivatives at $u = 0$ and the given derivatives ν_j at $u = \beta - 1$ is uniformly bounded for $0 \le u \le \beta - 1$. Since this is true for any l, we can conclude that $|g(u)| \le C_2 = C(\beta)\,C_1$ at all u where this function is defined. The constant $C(\beta)$ is introduced in Lemma 11.7, $m = \beta$. So, we proved that $\max_{0 \le x \le 1} |\,\gamma(x) - f(x)\,| = O(h_n^{\beta})$, which implies (11.12). □

Exercises

Exercise 11.75. Find explicitly the standard B-splines S_2 and S_3. Graph these functions.

Exercise 11.76. Prove (11.6).

Exercise 11.77. Prove (11.16).

Exercise 11.78. Prove (11.19).

Exercise 11.79. Show that the determinant $\det \mathbf{M}$ of the matrix $\mathbf{M}$ with the elements defined by (11.20) is non-zero. Hint: Show that this determinant is proportional to the determinant of the *generalized Vandermonde matrix*

$$\mathbf{V}_m = \begin{bmatrix} x_1 & x_1^2 & \dots & x_1^m \\ x_2 & x_2^2 & \dots & x_2^m \\ & & \dots & \\ x_m & x_m^2 & \dots & x_m^m \end{bmatrix}$$

with distinct $x_1, \dots, x_m$. Look at $\det \mathbf{V}_m$ as a function of x_m. If x_m equals either x_1, or x_2, ..., x_{m-1}, then the determinant is zero. Consequently,

$$\det \mathbf{V}_m = v(x_1, x_2, \dots, x_m)\,(x_m - x_1)(x_m - x_2)\dots(x_m - x_{n-1})$$

for some function v. Now expand along the last row to see that the determinant is a polynomial in x_m of order m with the highest coefficient equal to $\det \mathbf{V}_{m-1}$. Thus, the recursive formula holds:

$$\det \mathbf{V}_m = \det \mathbf{V}_{m-1}\, x_m(x_m - x_1)(x_m - x_2)\dots(x_m - x_{m-1}), \quad \det \mathbf{V}_1 = 1.$$

From here deduce that $\det \mathbf{V}_m = x_1\, x_2\, \dots\, x_m \prod_{i<j}(x_j - x_i) \neq 0$.

Exercise 11.80. Prove the statement similar to Lemma 11.8 for the power splines. Show that for any polynomial $g(u)$ of order $m-1$ in the interval $m-1 \leq u < m$, there exists a unique linear combination $LP^*(u)$ of power splines

$$LP^*(u) = b_0P_0(u) + b_1P_1(u) + \dots + b_{m-2}P_{m-2}(u) + b_{m-1}P_{m-1}(u)$$

such that $LS^*(u) = g(u)$ if $m-1 \leq u < m$. Apply this result to represent the function $g(u) = 2 - u^2$ in the interval $[2, 3)$ by the power splines.

Exercise 11.81. Show that if $f \in \Theta(\beta, L, L_1)$, then $f^{(j)}$ belongs to the Hölder class $\Theta(\beta - j, L, L_2)$ with some positive constant L_2.

Chapter 12

Asymptotic Optimality in Global Norms

In Chapter 10, we studied the regression estimation problem for the integral L_2-norm and the sup-norm risks. The upper bounds in Theorems 10.3 and 10.6 guarantee the rates of convergence in these norms that are $n^{-\beta/(2\beta+1)}$ and $\big(n/(\ln n)\big)^{-\beta/(2\beta+1)}$, respectively. These rates hold for any function f in the Hölder class $\Theta(\beta, L, L_1)$, and they are attained by the regressogram with the properly chosen bandwidths.

The question that we address in this chapter is whether these rates can be improved by any other estimators. The answer turns out to be negative. We will prove the lower bounds that show the minimax optimality of the regressogram.

12.1. Lower Bound in the Sup-Norm

In this section we prove that the rate

$$\psi_n = \Big(\frac{\ln n}{n}\Big)^{\beta/(2\beta+1)} \tag{12.1}$$

is the minimax rate of convergence in the Hölder class of regression functions $\Theta(\beta) = \Theta(\beta, L, L_1)$ if the losses are measured in the sup-norm. The theorem below is another example of a "lower bound" (see Theorems 9.16 and 9.17). As explained in Section 9.4, the lower bound may not hold for any design. We use the definitions of regular deterministic and random designs as in Sections 7.4 and 9.3.

Theorem 12.1. *Let the deterministic design $\mathcal{X}$ be defined by (7.17) with a continuous and strictly positive density $p(x)$ on $[0,1]$. Then for all large n, and for any estimator $\hat{f}_n$ of the regression function f, the following inequality holds:*

$$\sup_{f \in \Theta(\beta)} \mathbb{E}_f \Big[\psi_n^{-1} \, \| \hat{f}_n - f \|_\infty \Big] \geq r_* \tag{12.2}$$

where ψ_n is defined in (12.1), and a positive constant r_* is independent of n.

Proof. As in the proof of Theorem 9.16, take any "bump" function $\varphi(t)$, $t \in \mathbb{R}$, such that $\varphi(0) > 0$, $\varphi(t) = 0$ if $|t| > 1$, and $| \varphi^{(\beta)}(t) | \leq L$. Clearly, this function has a finite L_2-norm, $\|\varphi\|_2 < \infty$. We want this norm to be small, therefore, below we make an appropriate choice of $\varphi(t)$. Take a bandwidth $h_n^* = \big((\ln n)/n \big)^{1/(2\beta+1)}$, and consider the bins

$$B_q = \big[2h_n^*(q-1), \, 2h_n^* q \big), \ q = 1, \dots, Q,$$

where we assume without loss of generality that $Q = 1/(2h_n^*)$ is an integer. Introduce the *test functions* $f_0(t) = 0$, and

$$f_q(t) = (h_n^*)^\beta \varphi\Big(\frac{t - c_q}{h_n^*} \Big), \ t \in [0,1], \ q = 1, \dots, Q, \tag{12.3}$$

where c_q is the center of the bin B_q. Note that each function $f_q(t)$ takes non-zero values only within the respective bin B_q. For any small enough h_n^*, the function f_q belongs to the Hölder class $\Theta(\beta, L, L_1)$. This fact was explained in the proof of Theorem 9.16.

Recall that under the hypothesis $f = f_q$, the observations y_i in the nonparametric regression model satisfy the equation

$$y_i = f_q(x_i) + \varepsilon_i, \ i = 1, \dots, n,$$

where the x_i's are the design points, the ε_i's are independent $\mathcal{N}(0, \sigma^2)$-random variables. Put

$$d_0 = \frac{1}{2} (h_n^*)^\beta \, \| \varphi \|_\infty > 0. \tag{12.4}$$

Note that by definition,

$$\| f_l - f_q \|_\infty = 2d_0 \, , \ 1 \leq l < q \leq Q, \tag{12.5}$$

and

$$\| f_q \|_\infty = \| f_q - f_0 \|_\infty = 2d_0, \ q = 1, \dots, Q. \tag{12.6}$$

Introduce the random events

$$\mathcal{D}_q = \big\{ \| \hat{f}_n - f_q \|_\infty \geq d_0 \big\}, \ q = 0, \dots, Q.$$

Observe that for any q, $1 \leq q \leq Q$, the inclusion takes place, $\overline{\mathcal{D}}_0 \subseteq \mathcal{D}_q$. Indeed, by the triangle inequality, if $\hat{f}_n$ is closer to $f_0 = 0$ than d_0, that is if $\| \hat{f}_n \|_\infty < d_0$, then it deviates away from any f_q by no less than d_0,

$$\| \hat{f}_n - f_q \|_\infty \geq \| f_q \|_\infty - \| \hat{f}_n \|_\infty = 2\, d_0 - \| \hat{f}_n \|_\infty \geq d_0, \; q = 1, \dots, Q.$$

Further, we will need the following lemma. We postpone its proof to the end of the section.

Lemma 12.2. *Under the assumptions of Theorem 12.1, for any small $\delta > 0$, there exists a constant $c_0 > 0$ such that if $\|\varphi\|_2^2 \leq c_0$, then for all large n,*

$$\max_{0 \leq q \leq Q} \mathbb{P}_{f_q}\big(\mathcal{D}_q \big) \geq \frac{1}{2}\,(1 - \delta).$$

Now, we apply Lemma 12.2 to find that for all n large enough, the following inequalities hold:

$$\sup_{f \in \Theta(\beta)} \mathbb{E}_f\big[\| \hat{f}_n - f \|_\infty \big] \geq \max_{0 \leq q \leq Q} \mathbb{E}_{f_q}\big[\| \hat{f}_n - f_q \|_\infty \big]$$

$$\geq d_0 \max_{0 \leq q \leq Q} \mathbb{P}_{f_q}\big(\| \hat{f}_n - f_q \|_\infty \geq d_0 \big) = d_0 \max_{0 \leq q \leq Q} \mathbb{P}_{f_q}\big(\mathcal{D}_q \big)$$

$$\geq \frac{1}{2}\, d_0 (1 - \delta) = \frac{1}{4}\, (h_n^*)^\beta \, \| \varphi \|_\infty \, (1 - \delta),$$

and we can choose $r_* = (1/4) \| \varphi \|_\infty \, (1 - \delta)$. □

Remark 12.3. Contrast the proof of Theorem 12.1 with that of Theorem 9.16. The proof of the latter theorem was based on two hypotheses, $f = f_0$ or $f = f_1$, with the likelihood ratio that stayed finite as $n \to \infty$. In the sup-norm, however, the proof of the rate of convergence is complicated by the extra log-factor, which prohibits using the same idea. The likelihood ratios in the proof of Theorem 12.1 are vanishing as $n \to \infty$. To counterweigh that fact, a growing number of hypotheses is selected. Note that the number of hypotheses $Q + 1 \leq n^{1/(2\beta+1)}$ has the polynomial rate of growth as n goes to infinity. □

The next theorem handles the case of a random design. It shows that if the random design is regular, then the rate of convergence of the sup-norm risk is the same as that in the deterministic case. Since the random design can be "very bad" with a positive probability, the conditional risk for given design points does not guarantee even the consistency of estimators. That is why we study the unconditional risks. The proof of the theorem below is left as an exercise (see Exercise 12.83).

Theorem 12.4. *Let $\mathcal{X}$ be a random design such that that design points x_i are independent with a continuous and strictly positive density $p(x)$ on*

$[0,1]$. *Then for all sufficiently large* n, *and for any estimator* $\hat{f}_n(x)$ *of the regression function* $f(x)$, *the following inequality holds:*

$$\sup_{f \in \Theta(\beta)} \mathbb{E}_f \big[\psi_n^{-1} \| \hat{f}_n - f \|_\infty \big] \geq c_0 \tag{12.7}$$

with a positive constant c_0 *independent of* n.

Proof of Lemma 12.2. From the inclusion $\overline{\mathcal{D}}_0 \subseteq \mathcal{D}_q$, which holds for any $q = 1, \dots, Q$, we have that

$$\max_{0 \leq q \leq Q} \mathbb{P}_q(\mathcal{D}_q) = \max\Big(\mathbb{P}_0(\mathcal{D}_0), \max_{1 \leq q \leq Q} \mathbb{P}_q(\mathcal{D}_q) \Big)$$

$$\geq \frac{1}{2} \Big(\mathbb{P}_0(\mathcal{D}_0) + \max_{1 \leq q \leq Q} \mathbb{P}_q(\overline{\mathcal{D}}_0) \Big)$$

$$\geq \frac{1}{2} \Big(\mathbb{P}_0(\mathcal{D}_0) + \frac{1}{Q} \sum_{q=1}^{Q} \mathbb{E}_0 \big[\mathbb{I}(\overline{\mathcal{D}}_0) \exp\{L_{n,q}\} \big] \Big)$$

$$= \frac{1}{2} \Big(\mathbb{P}_0(\mathcal{D}_0) + \mathbb{E}_0 \big[\mathbb{I}(\overline{\mathcal{D}}_0) \frac{1}{Q} \sum_{q=1}^{Q} \exp\{L_{n,q}\} \big] \Big). \tag{12.8}$$

In the above, by $L_{n,q}$ we denoted the log-likelihood ratios

$$L_{n,q} = \ln \frac{d\mathbb{P}_q}{d\mathbb{P}_0}, \quad q = 1, \dots, Q.$$

They admit the asymptotic representation

$$L_{n,q} = \sigma_{n,q} \mathcal{N}_{n,q} - \frac{1}{2} \sigma_{n,q}^2 \tag{12.9}$$

where for every $q = 1, \dots, Q$,

$$\sigma_{n,q}^2 = \sigma^{-2} \sum_{i=1}^{n} f_q^2(x_i) = n\,(h_n^*)^{2\beta+1} \sigma^{-2} p(c_q) \|\varphi\|_2^2 \,(1 + o_n(1)) \tag{12.10}$$

with $o_n(1)$ vanishing as $n \to \infty$ uniformly in q. The random variables $\mathcal{N}_{n,q}$ in (12.9) are standard normal and independent for different q.

Let p^* denote the maximum of the density $p(x)$, $p^* = \max_{0 \leq x \leq 1} p(x)$. Recall that $(h_n^*)^{2\beta+1} = (\ln n)/n$. Thus, if n is large enough, then

$$\sigma_{n,q}^2 \leq 2\,\sigma^{-2} p^* \, c_0 \ln n = c_1 \ln n \tag{12.11}$$

where the constant $c_1 = 2\sigma^{-2} p^* c_0$ is small if c_0 is small. Note that the constant c_1 is independent of q.

Put $\xi_n = Q^{-1} \sum_{q=1}^{Q} \exp\{L_{n,q}\}$. The first and the second moments of ξ_n are easily computable. Indeed, since by definition, $\mathbb{E}_0\big[\exp\{L_{n,q}\}\big] = 1$, we have that $\mathbb{E}_0\big[\xi_n\big] = 1$. Applying the independence of the random variables $\mathcal{N}_{n,q}$ for different q, we find that

$$\mathbb{E}_0\big[\xi_n^2\big] = Q^{-2}\sum_{q=1}^{Q}\mathbb{E}_0\big[\exp\big\{2L_{n,q}\big\}\big]$$

$$= Q^{-2}\sum_{q=1}^{Q}\mathbb{E}_0\big[\exp\big\{2\sigma_{n,q}\mathcal{N}_{n,q} - \sigma_{n,q}\big\}\big]$$

$$= Q^{-2}\sum_{q=1}^{Q}\exp\{\sigma_{n,q}^2\} \le Q^{-1}e^{c_1\ln n} = Q^{-1}n^{c_1} = 2\,h_n^*\,n^{c_1}.$$

If we now take c_0 so small that $c_1 = 2\sigma^{-2}p^*c_0 = 1/(2\beta+1)-\epsilon$ for some small $\epsilon > 0$, then $\mathbb{E}_0\big[\xi_n^2\big] \le 2\,h_n^* n^{c_1} = 2(\ln n)^{1/(2\beta+1)}n^{-\epsilon}$ can be chosen arbitrarily small for sufficiently large n.

Next, by the Chebyshev inequality, we have

$$\mathbb{P}_0\big(\mathcal{D}_0\big) + \mathbb{E}_0\big[\mathbb{I}(\overline{\mathcal{D}}_0)\xi_n\big] \ge (1-\delta_0)\,\mathbb{P}_0\big(\xi_n \ge 1-\delta_0\big)$$

$$\ge (1-\delta_0)\,\mathbb{P}_0\big(\,|\,\xi_n - 1\,| \le \delta_0\big) = (1-\delta_0)\,\big(1 - \mathbb{P}_0\big(\,|\,\xi_n - 1\,| > \delta_0\big)\big)$$

$$\ge (1-\delta_0)\,\big(1 - \delta_0^{-2}\,\mathbb{E}_0\big[\,\xi_n^2\,\big]\big) \to 1-\delta_0 \ge 1-\delta \ \text{ if } \ \delta_0 < \delta.$$

Plugging this expression into (12.8), we obtain the result of the lemma. □

12.2. Bound in L_2-Norm. Assouad's Lemma

To prove the lower bound in the L_2-norm, a more elaborate construction is required as compared to estimation at a point (Section 9.3) or in the sup-norm (Section 12.1). The method we use here is a modified version of what is known in nonparametric statistics as *Assouad's Lemma*. This method can be relatively easily explained if we start with the definitions similar to those given for the result at a fixed point in Theorem 9.16.

We will proceed under the assumptions that the design points are deterministic, regular and controlled by a density $p(x)$ which is continuous and strictly positive in $[0, 1]$. As in Section 9.3, take a function $\varphi(u) \ge 0,\ u \in \mathbb{R}$, that satisfies all the properties mentioned in the that section. The key properties are that this function is smooth and its support is $[-1, 1]$.

In the proof of Theorem 9.16, we defined the two test functions $f_0(t)$ and $f_1(t),\ t \in [0, 1]$. To extend this definition to the L_2-norm, consider Q bins $B_1, \dots, B_Q$, centered at $c_q,\ q = 1, \dots, Q$, each of the length $2h_n^*$ where $h_n^* = n^{-1/(2\beta+1)}$. Without loss of generality, $Q = 1/(2h_n^*)$ is an integer. Denote by Ω_Q a set of Q-dimensional binary vectors

$$\Omega_Q = \big\{\boldsymbol{\omega} : \boldsymbol{\omega} = (\omega_1, \dots, \omega_Q),\ \omega_q \in \{0,\,1\}\,,\ q = 1, \dots, Q\big\}.$$

The number of elements in Ω_Q is equal to 2^Q. To study the lower bound in the L_2-norm, define 2^Q test functions by

$$f(t,\, \boldsymbol{\omega}) = \omega_1\, (h_n^*)^\beta\, \varphi\Big(\frac{t - c_1}{h_n^*}\Big) + \cdots + \omega_Q\, (h_n^*)^\beta\, \varphi\Big(\frac{t - c_Q}{h_n^*}\Big) \tag{12.12}$$

where the variable t belongs to the interval $[0,\, 1]$ and $\boldsymbol{\omega} \in \Omega_Q$. A proper choice of $\varphi(u)$ guarantees that each function $f(t,\, \boldsymbol{\omega})$ belongs to the Hölder class $\Theta(\beta, L, L_1)$.

Before continuing we introduce more notation. Define by $\mathcal{Y}_q$ the σ-algebra generated by the regression observations $y_i = f(x_i) + \varepsilon_i$ with the design points $x_i \in B_q$, $q = 1, \dots, Q$. For any estimator $\hat{f}_n$ of the regression function, we define the conditional expectation

$$\hat{f}_{n,\,q} = \mathbb{E}_f\big[\, \hat{f}_n \,|\, \mathcal{Y}_q \,\big].$$

Note that $\hat{f}_{n,\,q} = \hat{f}_{n,\,q}(t)$ depends only on the observations within the bin B_q.

For the sake of brevity, below we denote the conditional expectation $\mathbb{E}_{f(\cdot,\,\boldsymbol{\omega})}[\,\cdot \,|\, \mathcal{X}]$ by $\mathbb{E}_{\boldsymbol{\omega}}[\,\cdot\,]$, suppressing dependence on the test function and the design.

By the definition of the L_2-norm, we obtain that

$$\mathbb{E}_{\boldsymbol{\omega}}\big[\, \|\, \hat{f}_n(\cdot) - f(\cdot,\, \boldsymbol{\omega})\,\|_2^2 \,\big] = \sum_{q=1}^{Q} \mathbb{E}_{\boldsymbol{\omega}}\big[\, \|\, \hat{f}_n(\cdot) - f(\cdot,\, \boldsymbol{\omega})\,\|_{2,\,B_q}^2 \,\big].$$

Lemma 12.5. *For any estimator $\hat{f}_n$ of the regression function $f(t,\, \boldsymbol{\omega})$ the following inequality holds:*

$$\mathbb{E}_{\boldsymbol{\omega}}\big[\, \|\, \hat{f}_n(\cdot) - f(\cdot,\, \boldsymbol{\omega})\,\|_{2,\,B_q}^2 \,\big] \ge \mathbb{E}_{\boldsymbol{\omega}}\big[\, \|\, \hat{f}_{n,\,q}(\cdot) - f(\cdot,\, \boldsymbol{\omega})\,\|_{2,\,B_q}^2 \,\big]$$

$$= \mathbb{E}_{\boldsymbol{\omega}}\Big[\, \int_{B_q} \Big(\hat{f}_{n,\,q}(t) - \omega_q (h_n^*)^\beta\, \varphi\Big(\frac{t - c_q}{h_n^*}\Big)\Big)^2 dt \Big]. \tag{12.13}$$

Proof. First conditioning on the σ-algebra $\mathcal{Y}_q$, and then applying Jensen's inequality to the convex quadratic function, we obtain that for any $q = 1, \dots, Q$,

$$\mathbb{E}_{\boldsymbol{\omega}}\big[\, \|\, \hat{f}_n(\cdot) - f(\cdot,\, \boldsymbol{\omega})\,\|_{2,\,B_q}^2 \,\big] = \mathbb{E}_{\boldsymbol{\omega}}\Big[\mathbb{E}_{\boldsymbol{\omega}}\big[\, \|\, \hat{f}_n(\cdot) - f(\cdot,\, \boldsymbol{\omega})\,\|_{2,\,B_q}^2 \,|\, \mathcal{Y}_q \big] \Big]$$

$$\ge \mathbb{E}_{\boldsymbol{\omega}}\Big[\, \|\, \mathbb{E}_{\boldsymbol{\omega}}\big[\hat{f}_n(\cdot) \,|\, \mathcal{Y}_q \big] - f(\cdot,\, \boldsymbol{\omega})\,\|_{2,\,B_q}^2 \Big]$$

$$= \mathbb{E}_{\boldsymbol{\omega}}\big[\, \|\, \hat{f}_{n,\,q}(\cdot) - f(\cdot,\, \boldsymbol{\omega})\,\|_{2,\,B_q}^2 \,\big] \text{ (by definition of } \hat{f}_{n,\,q})$$

$$= \mathbb{E}_{\boldsymbol{\omega}}\Big[\, \int_{B_q} \Big(\hat{f}_{n,\,q}(t) - \omega_q (h_n^*)^\beta\, \varphi\Big(\frac{t - c_q}{h_n^*}\Big)\Big)^2 dt \Big].$$

At the last step we used the definition of the function $f(t, \boldsymbol{\omega})$ in the bin B_q. $\square$

In (12.13), the function $\hat{f}_{n,q}(t)$ depends only on the regression observations with the design points in B_q. We will denote the expectation relative to these observations by $\mathbb{E}_{\omega_q}$. We know that $\mathbb{E}_{\omega_q}$ is computed with respect to one of the two probability measures $\mathbb{P}_{\{\omega_q=0\}}$ or $\mathbb{P}_{\{\omega_q=1\}}$. These measures are controlled entirely by the performance of the test function $f(\cdot, \boldsymbol{\omega})$ in the bin B_q.

Lemma 12.6. *There exists a constant r_0, which depends only on the design density p and the chosen function φ, such that for any q, $1 \le q \le Q$, and for any $\mathcal{Y}_q$-measurable estimator $\hat{f}_{n,q}$, the following inequality holds:*

$$\max_{\omega_q \in \{0,1\}} \mathbb{E}_{\omega_q}\Big[\int_{B_q}\Big(\hat{f}_{n,q}(t) - \omega_q (h_n^*)^{\beta}\varphi\Big(\frac{t-c_q}{h_n^*}\Big)\Big)^2 dt\Big] \ge r_0/n.$$

Proof. We proceed as in the proof of Theorem 9.16. At any fixed t, $t \in B_q$, we obtain that

$$\max_{\omega_q \in \{0,1\}} \mathbb{E}_{\omega_q}\Big[\int_{B_q}\Big(\hat{f}_{n,q}(t) - \omega_q (h_n^*)^{\beta}\varphi\Big(\frac{t-c_q}{h_n^*}\Big)\Big)^2 dt\Big]$$

$$\ge \frac{1}{2}\mathbb{E}_{\{\omega_q=0\}}\Big[\int_{B_q}\hat{f}^2_{n,q}(t)dt\Big] + \frac{1}{2}\mathbb{E}_{\{\omega_q=1\}}\Big[\int_{B_q}\Big(\hat{f}_{n,q}(t) - (h_n^*)^{\beta}\varphi\Big(\frac{t-c_q}{h_n^*}\Big)\Big)^2 dt\Big]$$

$$(12.14)\qquad = \frac{1}{2}\mathbb{E}_{\{\omega_q=0\}}\Big[\int_{B_q}\hat{f}^2_{n,q}(t)\,dt + \frac{d\mathbb{P}_{\{\omega_q=1\}}}{d\mathbb{P}_{\{\omega_q=0\}}}\int_{B_q}\Big(\hat{f}_{n,q} - (h_n^*)^{\beta}\varphi\Big(\frac{t-c_q}{h_n^*}\Big)\Big)^2 dt\Big]$$

where

$$\ln\frac{d\mathbb{P}_{\{\omega_q=1\}}}{d\mathbb{P}_{\{\omega_q=0\}}} = \sigma_{n,q}\mathcal{N}_q - \frac{1}{2}\sigma^2_{n,q}$$

with a standard normal random variable $\mathcal{N}_q$ and

$$\lim_{n\to\infty}\sigma^2_{n,q} = \sigma^{-2}p(c_q)\,\|\,\varphi\,\|_2^2.$$

For all large n and any $q = 1, \dots, Q$, the standard deviation $\sigma_{n,q}$ is separated away from zero and infinity. Hence,

$$\mathbb{P}_{\{\omega_q=0\}}\big(\mathcal{N}_q > \sigma_{n,q}/2\big) \ge p_0$$

for a positive constant p_0 independent of n and q. If the random event $\{\mathcal{N}_q > \sigma_{n,q}/2\}$ holds, then we can estimate the likelihood ratio on the right-hand side of (12.14) from below by 1.

Next, note that for any functions $\hat{f}_n$ and g, the inequality is true $\|\hat{f}_n\|_2^2 + \|\hat{f}_n - g\|_2^2 \ge (1/2)\,\|g\|_2^2$. Applied to $\hat{f}_{n,q}$, it provides the lower bound

$$\int_{B_q}\Big[\hat{f}^2_{n,q}(t) + \Big(\hat{f}_{n,q} - (h_n^*)^{\beta}\varphi\Big(\frac{t-c_q}{h_n^*}\Big)\Big)^2\Big]dt$$

$$\geq \frac{1}{2}\,(h_n^*)^{2\beta} \int_{B_q} \Big[\varphi\Big(\frac{t-c_q}{h_n^*}\Big)\Big]^2\, dt \;=\; \frac{1}{2}\,(h_n^*)^{2\beta+1}\,\|\,\varphi\,\|_2^2 \;=\; \frac{1}{2n}\,\|\,\varphi\,\|_2^2.$$

Finally, combining these estimates, we obtain that

$$\max_{\omega_q \in \{0,1\}} \mathbb{E}_{\omega_q}\Big[\int_{B_q} \Big(\hat f_{n,q}(t) - \omega_q (h_n^*)^\beta \varphi\Big(\frac{t-c_q}{h_n^*}\Big)\Big)\, dt \;\geq\; p_0 \|\,\varphi\,\|_2^2/(2n) \;=\; r_0/n$$

with $r_0 \;=\; p_0 \|\,\varphi\,\|_2^2/2$. □

After these technical preparations, we are ready to formulate the minimax lower bound for estimation of the Hölder class functions in the L_2-norm.

Theorem 12.7. *Let the deterministic design $\mathcal{X}$ be defined by (7.17) with a continuous and strictly positive density $p(x)$ in $[0,1]$. There exists a positive constant r_* such that for any estimator $\hat f_n(t)$, the following asymptotic lower bound holds:*

$$\liminf_{n\to\infty} \sup_{f\in\Theta(\beta,L)} n^{2\beta/(2\beta+1)} \mathbb{E}_f\Big[\,\|\,\hat f_n - f\,\|_2^2\,\Big] \;\geq\; r_*.$$

Proof. We use the notation introduced in Lemmas 12.5 and 12.6. Applying the former lemma, we obtain the inequalities

$$\sup_{f\in\Theta(\beta,L)} \mathbb{E}_f\Big[\,\|\,\hat f_n - f\,\|_2^2\,\Big] \;\geq\; \max_{\boldsymbol{\omega}\in\Omega_Q} \mathbb{E}_{\boldsymbol{\omega}}\,\|\,\hat f_n(\cdot) - f(\cdot\,,\boldsymbol{\omega})\,\|_2^2$$

$$\geq \max_{\boldsymbol{\omega}\in\Omega_Q} \sum_{q=1}^{Q} \mathbb{E}_{\omega_q}\Big[\int_{B_q} \big[\,\hat f_{n,q}(t) \;-\; \omega_q\,(h_n^*)^\beta\,\varphi\Big(\frac{t-c_q}{h_n^*}\Big)\big]\, dt\Big].$$

Note that each term in the latter sum depends only on a single component ω_q. This is true for the expectation and the integrand. That is why the maximum over the binary vector $\boldsymbol{\omega}$ can be split into the sum of maxima. In view of Lemma 12.6, we can write

$$\sum_{q=1}^{Q} \max_{\omega_q \in \{0,1\}} \mathbb{E}_{\omega_q}\Big[\int_{B_q} \Big(\hat f_{n,q}(t) - \omega_q (h_n^*)^\beta \varphi\Big(\frac{t-c_q}{h_n^*}\Big)\Big)\, dt\Big]$$

$$\geq r_0\, Q/n \;=\; r_0\,/\,(2\,h_n^*\,n\,) \;=\; (r_0/2)\,n^{-2\beta/(2\beta+1)},$$

and the theorem follows with $r_* \;=\; r_0/2$. □

12.3. General Lower Bound

The proof of the lower bound in the previous sections explored the characteristics of the sup-norm and the L_2-norm, which do not extend very far. In particular, in the proof of the lower bound in the sup-norm, we relied on the independence of the random variables $\mathcal{N}_{q,n}$ in (12.9). A similar independence does not hold for the test functions (12.12) since their supports are overlapping. On the other hand, the idea of Assouad's lemma fails if we try to apply it to the sup-norm because the sup-norm does not split into the sum of the sup-norms over the bins.

In this section, we will suggest a more general lower bound that covers both of these norms as special cases. As above, we consider a nonparametric regression function $f(x)$, $x \in [0,1]$, of a given smoothness $\beta \geq 1$. We introduce a norm $\|f\|$ of functions in the interval $[0,1]$. This norm will be specified later in each particular case.

As in the sections above, we must care about two things: a proper set of the test functions, and the asymptotic performance of the respective likelihood ratios.

Assume that there exists a positive number d_0 and a set of $M+1$ test functions $f_0(x), \dots, f_M(x)$, $x \in [0,1]$, such that any two functions f_l and f_m are separated by at least $2d_0$, that is,

$$\| f_l - f_m \| \geq 2d_0 \text{ for any } l \neq m,\ l,m = 0,\dots,M. \tag{12.15}$$

The constant d_0 depends on n, decreases as $n \to 0$, and controls the rate of convergence. The number M typically goes to infinity as $n \to 0$. For example, in the case of the sup-norm, we had $d_0 = O\big((h_n^*)^\beta\big)$ in (12.4), and $M = Q = O\big(1/h_n^*\big)$ where $h_n^* = \big((\ln n)/n\big)^{1/(2\beta+1)}$.

In this section, we consider the regression with the regular deterministic design $\mathcal{X}$. Denote by $\mathbb{P}_m(\cdot) = \mathbb{P}_{f_m}(\cdot \,|\, \mathcal{X})$ $m = 0,\dots,M$, the probability distributions corresponding to a fixed design, and by $\mathbb{E}_m$ the respective expectations associated with the test function f_m, $m = 0,\dots,M$.

Fix one of the test functions, for instance, f_0. Consider all log-likelihood ratios for $m = 1,\dots,M$,

$$\ln \frac{d\,\mathbb{P}_0}{d\,\mathbb{P}_m} = -\frac{1}{2\sigma^2}\sum_{i=1}^n \big[\, y_i^2 - (y_i - f_m(x_i))^2 \,\big]$$

$$= \frac{1}{\sigma}\sum_{i=1}^n f_m(x_i)(-\varepsilon_i/\sigma) - \frac{1}{2\sigma^2}\sum_{i=1}^n f_m^2(x_i) = \sigma_{m,n}\mathcal{N}_{m,n} - \frac{1}{2}\sigma_{m,n}^2$$

where

$$\varepsilon_i = y_i - f(x_i) \text{ and } \sigma_{m,n}^2 = \sigma^{-2}\sum_{i=1}^n f_m^2(x_i).$$

The random variables ε_i and $\mathcal{N}_{m,n}$ are standard normal with respect to the distribution $\mathbb{P}_m$.

We need assumptions on the likelihood ratios to guarantee that they are not too small as $n \to 0$. Introduce the random events

$$\mathcal{A}_m = \{\mathcal{N}_{m,n} > 0\} \text{ with } \mathbb{P}_m\big(\mathcal{A}_m\big) = 1/2,\quad m = 1,\dots,M.$$

Assume that there exists a constant α, $0 < \alpha < 1$, such that all the variances $\sigma^2_{m,n}$ are bounded from above,

$$\max_{1 \le m \le M} \sigma^2_{m,n} \le 2\alpha \ln M. \tag{12.16}$$

If the random event $\mathcal{A}_m$ takes place and the inequality (12.16) holds, then

$$\frac{d\,\mathbb{P}_0}{d\,\mathbb{P}_m} \ge \exp\big\{ -\sigma^2_{m,n}/2 \big\} \ge \exp\{-\alpha \ln M\} = 1/M^\alpha. \tag{12.17}$$

Let $\hat{f}_n$ be an arbitrary estimator of the regression function f. Define the random events

$$\mathcal{D}_m = \{ \| \hat{f}_n - f_m \| \ge d_0 \}, \quad m = 0, \dots, M.$$

The following lemma plays the same fundamental role in the proof of the lower bound as Lemma 12.2 in the case of the sup-norm.

Lemma 12.8. *If the conditions* (12.15) *and* (12.16) *are satisfied, then the following lower bound is true:*

$$\max_{0 \le m \le M} \mathbb{P}_m(\mathcal{D}_m) \ge 1/4.$$

Proof. To start with, note that

$$\mathbb{P}_m(\overline{\mathcal{D}}_m) = \mathbb{P}_m(\overline{\mathcal{D}}_m \mathcal{A}_m) + \mathbb{P}_m(\overline{\mathcal{D}}_m \overline{\mathcal{A}}_m)$$
$$\le \mathbb{P}_m(\overline{\mathcal{D}}_m \mathcal{A}_m) + \mathbb{P}_m(\overline{\mathcal{A}}_m) = \mathbb{P}_m(\overline{\mathcal{D}}_m \mathcal{A}_m) + 1/2,$$

which implies the inequality

$$\mathbb{P}_m(\overline{\mathcal{D}}_m \mathcal{A}_m) \ge \mathbb{P}_m(\overline{\mathcal{D}}_m) - 1/2. \tag{12.18}$$

Next, the following inclusion is true:

$$\bigcup_{m=1}^{M} \overline{\mathcal{D}}_m \subseteq \mathcal{D}_0 \tag{12.19}$$

where the random events $\overline{\mathcal{D}}_m$ are mutually exclusive. Indeed, if the norm of the difference $\|\hat{f}_n - f_m\|$ is strictly less than d_0 for some m, then by the triangle inequality and (12.15), the norm $\|\hat{f}_n - f_l\|$ is not smaller than d_0 for any $l \ne m$. The inclusion (12.19) makes use of this fact for $l = 0$.

It immediately follows that

$$\mathbb{P}_0(\mathcal{D}_0) \ge \mathbb{P}_0\Big(\bigcup_{m=1}^{M} \overline{\mathcal{D}}_m\Big) = \sum_{m=1}^{M} \mathbb{P}_0(\overline{\mathcal{D}}_m) = \sum_{m=1}^{M} \mathbb{E}_m\Big[\frac{d\,\mathbb{P}_0}{d\,\mathbb{P}_m} \mathbb{I}(\overline{\mathcal{D}}_m) \Big]$$
$$\ge \sum_{m=1}^{M} \mathbb{E}_m\Big[\frac{d\,\mathbb{P}_0}{d\,\mathbb{P}_m} \mathbb{I}(\overline{\mathcal{D}}_m \mathcal{A}_m) \Big] \ge \frac{1}{M^\alpha} \sum_{m=1}^{M} \big[\mathbb{P}_m(\overline{\mathcal{D}}_m) - 1/2 \big].$$

In the latter inequality, we used (12.17).

The final step of the proof is straightforward. The maximum is estimated from below by a mean value,

$$\max_{0 \le m \le M} \mathbb{P}_m(\mathcal{D}_m) \ge \frac{1}{2}\Big[\mathbb{P}_0(\mathcal{D}_0) + \frac{1}{M}\sum_{m=1}^{M}\mathbb{P}_m(\mathcal{D}_m)\Big]$$

$$\ge \frac{1}{2}\Big[\frac{1}{M^\alpha}\sum_{m=1}^{M}\big[\mathbb{P}_m(\overline{\mathcal{D}}_m) - 1/2\big] + \frac{1}{M}\sum_{m=1}^{M}\mathbb{P}_m(\mathcal{D}_m)\Big]$$

$$\ge \frac{1}{2M}\sum_{m=1}^{M}\big[\mathbb{P}_m(\overline{\mathcal{D}}_m) + \mathbb{P}_m(\mathcal{D}_m) - 1/2\big] = 1/4. \qquad \square$$

As a consequence of Lemma 12.8, we obtain a general lower bound.

Theorem 12.9. *If the conditions* (12.15) *and* (12.16) *are satisfied, then for any estimator* $\hat{f}_n$ *and for all* n *large enough, the following lower bound holds:*

$$\sup_{f \in \Theta(\beta)} \mathbb{E}_f\Big[\,\| \hat{f}_n - f \|\,\Big] \ge d_0/4. \tag{12.20}$$

Proof. Applying Lemma 12.8, we obtain that

$$\sup_{f \in \Theta(\beta)} \mathbb{E}_f\Big[\,\| \hat{f}_n - f \|\,\Big] \ge \max_{0 \le m \le M} \mathbb{E}_m\big[\,\| \hat{f}_n - f_m \|\,\big]$$

$$\ge d_0 \max_{0 \le m \le M} \mathbb{P}_m(\mathcal{D}_m) \ge d_0/4. \qquad \square$$

12.4. Examples and Extensions

Example 12.10. The sup-norm risk. In the case of the sup-norm, the test functions are defined by (12.3) with $M = Q$. The condition (12.15) follows from (12.5) and (12.6) with $d_0 = (1/2)(h_n^*)^\beta \|\varphi\|_\infty$. Note that for all large n the following inequality holds:

$$\ln Q = \frac{1}{2\beta+1}\big(\ln n - \ln\ln n\big) - \ln 2 \ge \frac{1}{2(2\beta+1)}\ln n.$$

In view of (12.11), the variance $\sigma_{q,n}^2$ in the expansion (12.9) of $\ln\big(d\mathbb{P}_q/d\mathbb{P}_0\big)$ is bounded from above uniformly in $q = 1, \dots, Q$,

$$\sigma_{q,n}^2 \le c_1 \ln n \le 2(2\beta+1)c_1 \ln Q \le 2\alpha \ln Q = 2\alpha \ln M.$$

The latter inequality holds if the constant $c_1 = 2\sigma^{-2}p^* c_0$ is so small that $(2\beta+1)c_1 < \alpha$. Such a choice of c_1 is guaranteed because c_0 is however small. Thus, the condition (12.16) is also fulfilled. Applying Theorem 12.9, we get the lower bound

$$\sup_{f \in \Theta(\beta)} \mathbb{E}_f \Big[\psi_n^{-1} \, \| \hat{f}_n - f \, \|_\infty \Big] \ge \frac{1}{8} (h_n^*)^\beta \|\varphi\|_\infty = r_* \psi_n$$

with the constant $r_* = (1/8)\|\varphi\|_\infty$, and the rate of convergence ψ_n defined in (12.1). □

Unlike the case of the upper bounds in Chapter 9, "bad" designs do not create a problem in obtaining the lower bound in the sup-norm. Intuitively it is understandable because when we concentrate more design points in some bins, we loose them in the other bins. This process reduces the precision of the uniform estimation of the regression function. In a sense, the uniform design is optimal if we estimate the regression in the sup-norm. We will prove some results in support of these considerations.

Let a design $\mathcal{X}$ be of any kind, not necessarily regular. Assume that there exists a subset $\mathcal{M} = \mathcal{M}(\mathcal{X}) \subseteq \{1, \dots, M\}$ such that for some $\alpha \in (0, 1)$ the following inequality holds:

$$\max_{m \in \mathcal{M}} \sigma_{m,\,n}^2 \le 2\alpha \ln M. \tag{12.21}$$

Let $|\mathcal{M}|$ denote the number of elements in $\mathcal{M}$. It turns out that Lemma 12.8 remains valid in the following modification.

Lemma 12.11. *If the conditions* (12.15) *and* (12.21) *are satisfied, then the following lower bound holds:*

$$\max_{0 \le m \le M} \mathbb{P}_m(\mathcal{D}_m) \ge \frac{|\mathcal{M}|}{4M}.$$

Proof. Repeating the proof of Lemma 12.8, we find that

$$\mathbb{P}_0(\mathcal{D}_0) \ge \mathbb{P}_0\Big(\bigcup_{m=1}^{M} \overline{\mathcal{D}}_m \Big) = \sum_{m=1}^{M} \mathbb{P}_0(\overline{\mathcal{D}}_m) = \sum_{m=1}^{M} \mathbb{E}_m \Big[\frac{d\,\mathbb{P}_0}{d\,\mathbb{P}_m} \mathbb{I}(\overline{\mathcal{D}}_m) \Big]$$

$$\ge \sum_{m \in \mathcal{M}} \mathbb{E}_m \Big[\frac{d\,\mathbb{P}_0}{d\,\mathbb{P}_m} \mathbb{I}(\overline{\mathcal{D}}_m \, \mathcal{A}_m) \Big] \ge \frac{1}{M^\alpha} \sum_{m \in \mathcal{M}} \big(\mathbb{P}_m(\overline{\mathcal{D}}_m) - 1/2 \big)$$

where we have used the inequality (12.17). Under (12.21), this inequality applies only to the indices $m \in \mathcal{M}$. Continuing as in Lemma 12.8, we obtain the bound

$$\max_{0 \le m \le M} \mathbb{P}_m(\mathcal{D}_m) \ge \frac{1}{2} \Big[\mathbb{P}_0(\mathcal{D}_0) + \frac{1}{M} \sum_{m=1}^{M} \mathbb{P}_m(\mathcal{D}_m) \Big]$$

$$\ge \frac{1}{2M} \sum_{m \in \mathcal{M}} \big[\mathbb{P}_m(\overline{\mathcal{D}}_m) + \mathbb{P}_m(\mathcal{D}_m) - 1/2 \big] = \frac{|\mathcal{M}|}{4M}.$$

□

Example 12.12. The sup-norm risk (cont'd). For an arbitrary design $\mathcal{X}$, the bound (12.11) is no longer true. But it turns out (see Exercise 12.82) that for any design $\mathcal{X}$ and for any $\alpha \in (0, 1)$, there exists a "bump" function φ and a subset $\mathcal{M} = \mathcal{M}(\mathcal{X}) \subseteq \{1, \dots, Q\}$ such that

$$|\mathcal{M}| \geq Q/2 \text{ and } \max_{q \in \mathcal{M}} \sigma^2_{q,n} \leq 2\alpha \ln Q. \tag{12.22}$$

From (12.22) and Lemma 12.11, analogously to the proof of Theorem 12.9, we derive the lower bound for any design $\mathcal{X}$,

$$\sup_{f \in \Theta(\beta)} \mathbb{E}_f\Big[\psi_n^{-1} \| \hat{f}_n - f \|_\infty\Big] \geq \frac{|\mathcal{M}|}{4Q} d_0 \geq \frac{d_0}{8} = \frac{1}{16}(h_n^*)^\beta \|\varphi\|_\infty. \ \square \tag{12.23}$$

Next, we will study the case of the L_2-norm risk.

Example 12.13. The L_2-norm risk. Consider the test functions $f(t, \boldsymbol{\omega})$, $\boldsymbol{\omega} \in \Omega$, defined in (12.12). For any two functions $f(t, \boldsymbol{\omega}')$ and $f(t, \boldsymbol{\omega}'')$, the log-likelihood function has the representation

$$\ln \frac{\mathbb{P}_{f(\cdot, \boldsymbol{\omega}')}}{\mathbb{P}_{f(\cdot, \boldsymbol{\omega}'')}} = \sigma_n \mathcal{N}_n - \frac{1}{2}\sigma_n^2 \tag{12.24}$$

where $\mathcal{N}_n = \mathcal{N}_n(\boldsymbol{\omega}', \boldsymbol{\omega}'')$ is a standard normal random variable with respect to the distribution controlled by the test function $f(\cdot, \boldsymbol{\omega}'')$, and

$$\sigma_n^2 = \sigma_n^2(\boldsymbol{\omega}', \boldsymbol{\omega}'') = \sigma^{-2} \sum_{i=1}^n \big(f(x_i, \boldsymbol{\omega}') - f(x_i, \boldsymbol{\omega}'') \big)^2$$

where the x_i's are the design points (see Exercise 12.84). From the definition of the test functions, the variance σ_n^2 can be bounded from above by

$$\sigma_n^2 = \sigma^{-2} (h_n^*)^{2\beta} \sum_{q=1}^Q |\omega_q' - \omega_q''| \sum_{x_i \in B_q} \varphi^2\Big(\frac{x_i - c_q}{h_n^*}\Big)$$

$$= \sigma^{-2} \|\varphi\|_2^2 \sum_{q=1}^Q |\omega_q' - \omega_q''| p(c_q)(1 + o_{q,n}(1))$$

$$\leq \sigma^{-2} \|\varphi\|_2^2 Q\,(1 + o_n(1)) \leq 2\sigma^{-2} \|\varphi\|_2^2 Q. \tag{12.25}$$

In the above, $o_{q,n}(1) \to 0$ as $n \to \infty$ uniformly in q, $1 \leq q \leq Q$. Also, we bounded $|\omega_q' - \omega_q''|$ by 1, and used the fact that the Riemann sum of the design density approximates the integral

$$Q^{-1} \sum_{q=1}^Q p(c_q) = \int_0^1 p(x)\,dx + o_n(1) = 1 + o_n(1).$$

Next, we have to discuss the separation condition (12.15). For any test functions, the L_2-norm of the difference is easy to find,

$$\| f(x_i, \boldsymbol{\omega}') - f(x_i, \boldsymbol{\omega}'') \|_2^2 = \frac{1}{n} \|\varphi\|_2^2 \sum_{q=1}^{Q} |\omega_q' - \omega_q''|. \tag{12.26}$$

At this point, we need a result which will be proved at the end of this section.

Lemma 12.14. (Warshamov-Gilbert) *For all Q large enough, there exists a subset Ω_0, $\Omega_0 \subset \Omega$, with the number of elements no less than $1 + e^{Q/8}$ and such that for any $\boldsymbol{\omega}'$, $\boldsymbol{\omega}'' \in \Omega_0$, the following inequality holds:*

$$\sum_{q=1}^{Q} |\omega_q' - \omega_q''| \geq Q/8.$$

Continuing with the example, let $M = e^{Q/8}$. From Lemma 12.14 and (12.26), we see that there exist $M + 1$ test functions such that for any two of them,

$$\| f(x_i, \boldsymbol{\omega}') - f(x_i, \boldsymbol{\omega}'') \|_2^2 = \frac{Q}{8n} \|\varphi\|_2^2 = (2d_0)^2$$

where

$$d_0 = \frac{1}{2} \|\varphi\|_2 \sqrt{\frac{Q}{8n}} = \frac{1}{2} \|\varphi\|_2 \frac{1}{\sqrt{16 h_n^* n}} = \frac{1}{8} \|\varphi\|_2 (h_n^*)^{\beta}.$$

Hence the condition (12.15) is fulfilled with this d_0. We arbitrarily choose $f_0 = f(t, \boldsymbol{\omega}_0)$ for some $\boldsymbol{\omega}_0 \in \Omega_0$, and take $\mathcal{M}$ as a set of the rest of the functions with $\omega \in \Omega_0$. In this case, $|\mathcal{M}| = M = e^{Q/8}$.

Finally, we have to verify the condition (12.16). If we choose a "bump" function φ such that $\|\varphi\|_2^2 = \sigma^2 \alpha / 8$ where α is any number, $0 < \alpha < 1$, then it follows from (12.25) that

$$\sigma_n^2 \leq 2\sigma^{-2} \|\varphi\|_2^2 Q = 2\alpha \ln(e^{Q/8}) = 2\alpha \ln M.$$

Theorem 12.9 applies, and the lower bound of the L_2-norm risk follows for all large n,

$$\sup_{f \in \Theta(\beta)} \mathbb{E}_f \Big[\|\hat{f}_n - f\|_2 \Big] \geq \frac{1}{4}\, d_0 = \frac{1}{32} \|\varphi\|_2 (h_n^*)^{\beta} = r_* \, n^{-\beta/(2\beta+1)}. \qquad \square$$

Proof of Lemma 12.14. Define the binary vectors

$$\boldsymbol{\omega}_m = \big(\omega_{1,m}, \dots, \omega_{Q,m} \big), \quad m = 0, \dots, M,$$

with the independent Bernoulli(1/2) random components $\omega_{q,m}$. Note that for any $l \neq m$, the random variables $\xi_q = |\omega_{q,l} - \omega_{q,m}|$ are also Bernoulli(1/2), and are independent for different q.

Next, the elementary inequalities and the choice of M yield that

$$\mathbb{P}\Big(\bigcap_{0\le l<m\le M}\Big\{\sum_{q=1}^{Q}|\omega_{q,l}-\omega_{q,m}|\Big\}\ge Q/8\Big)$$

$$=1-\mathbb{P}\Big(\bigcup_{0\le l<m\le M}\Big\{\sum_{q=1}^{Q}|\omega_{q,l}-\omega_{q,m}|\Big\}<Q/8\Big)$$

$$\ge 1-\frac{M(M+1)}{2}\mathbb{P}\Big(\sum_{q=1}^{Q}\xi_q<Q/8\Big)$$

$$\ge 1-e^{Q/4}\,\mathbb{P}\Big(\sum_{q=1}^{Q}\xi_q<Q/8\Big)\ge 1-e^{Q/4}\,\mathbb{P}\Big(\sum_{q=1}^{Q}\xi_q'>(3/8)\,Q\Big)$$

where we denoted by $\xi_q'=1/2-\xi_q$ the random variables that take on values $\pm 1/2$ with equal probabilities.

Further, Chernoff's inequality $\mathbb{P}(X\ge a)\le e^{-za}\mathbb{E}\big[e^{zX}\big]$, $a,z>0$, ensures that for any positive z,

$$\mathbb{P}\Big(\sum_{q=1}^{Q}\xi_q'>(3/8)Q\Big)\le\Big(\mathbb{E}\big[\exp\{z\xi_q'\}\big]\Big)^{Q}\exp\big\{-(3/8)zQ\big\}.$$

The moment generating function of ξ_q' satisfies the inequality (see Exercise 12.85)

$$\mathbb{E}\big[\exp\{z\xi_q'\}\big]=\frac{1}{2}\big(\exp\{z/2\}+\exp\{-z/2\}\big)\le\exp\{z^2/8\}. \tag{12.27}$$

Take $z=3/2$. Then

$$\mathbb{P}\Big(\sum_{q=1}^{Q}\xi_q'>(3/8)Q\Big)\le\exp\big\{-(9/32)Q\big\}.$$

Hence,

$$\mathbb{P}\Big(\bigcap_{0\le l<m\le M}\Big\{\sum_{q=1}^{Q}|\omega_{q,l}-\omega_{q,m}|\Big\}\ge Q/8\Big)\ge 1-\exp\Big\{\Big(\frac{1}{4}-\frac{9}{32}\Big)Q\Big\}>0.$$

This proves the lemma, because what happens with a positive probability exists. □

Exercises

Exercise 12.82. Prove (12.22).

Exercise 12.83. Use (12.22) to prove Theorem 12.4.

Exercise 12.84. Verify (12.24).

Exercise 12.85. Prove (12.27).

Exercise 12.86. Let the design $\mathcal{X}$ be equidistant, that is, with the design points $x_i = i/n$. Show by giving an example that the following lower bound is false. For any large c there exists a positive p_0 independent of n such that for all large n, the following inequality holds:

$$\inf_{\hat{f}_n} \sup_{f \in \Theta(\beta)} \mathbb{P}_f\Big(\|\hat{f}_n - f\|_2 \geq c n^{-\beta/(2\beta+1)} \,\Big|\, \mathcal{X} \Big) \geq p_0.$$

Hint: Consider the case $\beta = 1$, and let f_n^* be a piecewise constant estimator in the bins. Show that the above probability goes to zero as n increases.

Part 3

Estimation in Nonparametric Models

Chapter 13

Estimation of Functionals

13.1. Linear Integral Functionals

As in the previous chapters, here we consider the observations of a regression function f in the presence of the Gaussian random noise. To ease the presentation, we concentrate on the case of the equidistant design,

$$y_i = f(i/n) + \varepsilon_i\,, \quad \varepsilon_i \sim \mathcal{N}(0,\, \sigma^2), \quad i = 1, \dots, n. \tag{13.1}$$

So far we have studied the estimation problem of the regression function. We found that the typical parametric $\sqrt{n}$-rate of convergence is not attainable in nonparametric setup. The typical minimax rate under smoothness parameter β equals $\psi_n = n^{-\beta/(2\beta+1)}$. Note that the exponent $\beta/(2\beta+1)$ approaches $1/2$ as β goes to infinity. Thus, for a very smooth nonparametric regression, the rate of convergence is close to the typical parametric rate.

In this section we focus on estimating an integral functional of the regression function, for example, $\Psi(f) = \int_0^1 f(x)dx$. We address the question: What is the minimax rate of convergence in this estimation problem? We will show that $\sqrt{n}$ is a very common rate of convergence.

We start with the easiest problem of a *linear integral functional*

$$\Psi(f) = \int_0^1 w(x) f(x)\, dx \tag{13.2}$$

where $w(x)$ is a given Lipschitz function, called the *weight function*, and $f = f(x)$ is an unknown regression observed with noise as in (13.1). Along with the integral notation, we will use the dot product notation, $\Psi(f) = (w,\, f)$ and $\|w\|_2^2 = \int_0^1 w^2(x)\,dx$.

Note that $\Psi(f)$ defined by (13.2) is a *linear functional*, that is, for any f_1 and f_2, and any constants k_1 and k_2, the following identity holds:

$$\Psi(k_1 f_1 + k_2 f_2) = \int_0^1 w(x)\big(k_1 f_1(x) + k_2 f_2(x)\big)\,dx$$

$$(13.3)\quad = k_1 \int_0^1 w(x) f_1(x)\,dx + k_2 \int_0^1 w(x) f_2(x)\,dx \;=\; k_1 \Psi(f_1) + k_2 \Psi(f_2).$$

Define an estimator of $\Psi(f)$ by

$$(13.4)\qquad \hat{\Psi}_n = \frac{1}{n}\sum_{i=1}^{n} w(i/n) y_i.$$

Example 13.1. If $w(t) = 1$, then $\Psi(f) = \int_0^1 f(x)\,dx$. Assume that $f \in \Theta(\beta, L, L_1)$ with some $\beta \geq 1$, which yields that f is a Lipschitz function. In this case, the trivial estimator, the sample mean, turns out to be $\sqrt{n}$-consistent,

$$\hat{\Psi}_n = \big(y_1 + \cdots + y_n\big)/n \;=\; \int_0^1 f(x)dx \,+\, O(n^{-1}) \,+\, \sigma Z_0/\sqrt{n}$$

where Z_0 is a standard normal random variable, and $O(n^{-1})$ represents the deterministic error of the Riemann sum approximation. Note that this deterministic error is uniform over $f \in \Theta(\beta, L, L_1)$. □

Next, we state a proposition the proof of which is straightforward (see Exercise 13.87).

Proposition 13.2. *For all* $\beta \geq 1$ *and any* $f \in \Theta(\beta, L, L_1)$*, the bias and the variance of the estimator* (13.4) *are respectively equal to*

$$b_n \;=\; \mathbb{E}_f[\hat{\Psi}_n] \;-\; \Psi(f) \;=\; O(n^{-1})$$

and

$$\mathbb{V}ar_f[\hat{\Psi}_n] \;=\; \frac{\sigma^2}{n}\int_0^1 w^2(x)\,dx \,+\, O(n^{-2}).$$

Corollary 13.3. *It immediately follows from Proposition 13.2 that for any* $f \in \Theta(\beta, L, L_1)$*, the following limit exists:*

$$\lim_{n\to\infty} \mathbb{E}_f\big[\sqrt{n}\big(\hat{\Psi}_n - \Psi(f)\big)\big]^2 \;=\; \sigma^2 \int_0^1 w^2(x)\,dx \;=\; \sigma^2 \|w\|_2^2.$$

A legitimate question is whether it is possible to improve the result of Proposition 13.2 and to find another estimator with an asymptotic variance smaller than $\sigma^2\,\|w\|_2^2$. As we could anticipate, the answer to this question is negative. To prove the lower bound, we need the following auxiliary result.

Lemma 13.4. *Let the Y_i's be independent observations of a location parameter θ in the non-homogeneous Gaussian model*

$$Y_i = \theta\,\mu_i + \varepsilon_i, \quad \varepsilon_i \sim \mathcal{N}(0,\, \sigma_i^2), \quad i = 1, \dots, n,$$

with some constant μ_i's. Assume that there exists a strictly positive limit $I_\infty = \lim_{n\to\infty} n^{-1} I_n > 0$ where $I_n = \sum_{i=1}^n \big(\mu_i/\sigma_i\big)^2$ is the Fisher information. Then for any estimator $\hat{\theta}_n$ of the location parameter θ, the following lower bound holds:

$$\liminf_{n\to\infty} \sup_{\theta\in\mathbb{R}} \mathbb{E}_\theta\big[\sqrt{nI_\infty}(\hat{\theta}_n - \theta)\big]^2 \;=\; \liminf_{n\to\infty} \sup_{\theta\in\mathbb{R}} \mathbb{E}_\theta\big[\sqrt{I_n}(\hat{\theta}_n - \theta)\big]^2 \;\geq\; 1.$$

Proof. The statement of the lemma is the Hajek-LeCam lower bound. Note that the Fisher information in the non-homogeneous Gaussian model is equal to $I_n = \sum_{i=1}^n \big(\mu_i/\sigma_i\big)^2$, and the log-likelihood ratio is normal non-asymptotically,

$$L_n(\theta) \,=\, L_n(0) \,=\, \sum_{i=1}^{n} \ln\big(f(X_i,\,\theta)/f(X_i,\,0)\big) \;=\; \sqrt{I_n}Z_{0,1}\theta \;-\; \frac{1}{2}I_n\theta^2$$

where $Z_{0,1}$ is a standard normal random variable with respect to the true distribution $\mathbb{P}_0$. Thus, as in the Hajek-LeCam case, we have the lower bound

$$\liminf_{n\to\infty} \sup_{\theta\in\mathbb{R}} \mathbb{E}_\theta\big[\sqrt{I_n}(\hat{\theta}_n - \theta)\big]^2 \;\geq\; 1. \qquad \square$$

Now we return to the functional estimation problem. Consider a one-parameter family of regression functions

$$\Theta = \big\{ f(x,\,\theta) = \theta w(x)/\|w\|_2^2, \;\; \theta\in\mathbb{R},\; \|w\|_2^2 > 0 \big\}$$

where $w(x)$ is the weight function specified in (13.2). For this family of regression functions, the functional $\Psi\big(f(\,\cdot\,,\theta)\big)$ coincides with θ identically,

$$\Psi\big(f(\,\cdot\,,\theta)\big) \;=\; \int_0^1 w(x) f(x,\theta)\,dx \;=\; \int_0^1 w(x)\,\theta\,\frac{w(x)}{\|w\|_2^2}\,dx \;=\; \theta.$$

Hence for this family of regression functions, the estimation of $\Psi(f)$ is equivalent to estimation of θ from the following observations:

$$y_i \;=\; \theta\, w(i/n)/\|w\|_2^2 \,+\, \varepsilon_i\,, \quad \varepsilon_i \sim \mathcal{N}(0\;\sigma^2), \quad i \,=\, 1, \dots, n. \tag{13.5}$$

Theorem 13.5. *For any estimator $\tilde{\Psi}_n$ from the observations* (13.5), *the following asymptotic lower bound holds:*

$$\liminf_{n\to\infty} \sup_{f(\cdot,\theta)\in\Theta} \mathbb{E}_{f(\cdot,\theta)}\Big[\sqrt{n}\Big(\tilde{\Psi}_n - \Psi\big(f(\,\cdot\,,\theta)\big)\Big)\Big]^2 \geq \sigma^2\|w\|_2^2. \tag{13.6}$$

Proof. Applying Lemma 13.4 with $\mu_i = w(i/n)/\|w\|_2^2$ and $Y_i = y_i$, we find that the Fisher information in this case is expressed as

$$I_n = \sum_{i=1}^{n} \left(\mu_i/\sigma_i\right)^2 = \sum_{i=1}^{n} \frac{w^2(i/n)}{\sigma^2 \|w\|_2^4}$$

$$= \frac{n}{\sigma^2\|w\|_2^2}\,\frac{1}{n}\sum_{i=1}^{n}\frac{w^2(i/n)}{\|w\|_2^2} = \frac{n(1+o_n(1))}{\sigma^2\|w\|_2^2}.$$

Here we used the fact that the latter sum is the Riemann sum for the integral $\int_0^1 w^2(x)/\|w\|_2^2\,dx = 1$. Thus, from Lemma 13.4, the lower bound follows

$$\liminf_{n\to\infty} \sup_{f(\cdot,\theta)\in\Theta} \mathbb{E}_{f(\cdot,\theta)}\Big[\sqrt{I_n}\Big(\tilde{\Psi}_n - \Psi(f(\,\cdot\,,\theta))\Big)\Big]^2$$

$$= \liminf_{n\to\infty} \sup_{\theta\in\mathbb{R}} \mathbb{E}_\theta\Big[\sqrt{\frac{n(1+o_n(1))}{\sigma^2\|w\|_2^2}}\,\big(\tilde{\Psi}_n - \theta\big)\Big]^2 \geq 1,$$

which is equivalent to (13.6). □

13.2. Non-Linear Functionals

As an example, suppose we want to estimate the square of the L_2-norm of the regression function f, that is, we want to estimate the *integral quadratic functional*

$$\Psi(f) = \|f\|_2^2 = \int_0^1 f^2(x)\,dx\,. \tag{13.7}$$

Clearly, this is a *non-linear functional* of f since it does not satisfy the linearity property (13.3), though it is very smooth. Can we estimate it $\sqrt{n}$-consistently? The answer is positive. The efficient estimator of the functional (13.7) is discussed in Example 13.6 below. Now we turn to general smooth functionals.

A functional $\Psi(f)$ is called *differentiable* on a set of functions Θ, if for any fixed function $f_0 \in \Theta$ and for any other function f in Θ, the following approximation holds:

$$\Psi(f) = \Psi(f_0) + \Psi'_{f_0}(f - f_0) + \rho(f, f_0) \tag{13.8}$$

where $\Psi'_{f_0}(f-f_0)$ is the first derivative of Ψ applied to the difference $f - f_0$. The functional Ψ'_{f_0} is a linear functional that depends on f_0. Moreover, we assume that for any function $g = g(x)$,

$$\Psi'_{f_0}(g) = \int_0^1 w(x, f_0)\,g(x)\,dx,$$

where $w(x,\, f_0)$ is a Lipschitz function of x and a continuous functional of f_0, that is,

$$|w(x_1,\, f_0) - w(x_2,\, f_0)| \leq L|x_1 - x_2|$$

with a Lipschitz constant L independent of f_0, and

$$\|w(\,\cdot\,,\, f) - w(\,\cdot\,,\, f_0)\|_2 \to 0 \ \text{ as } \ \|f - f_0\|_2 \to 0.$$

The remainder term $\rho(f,\, f_0)$ in (13.8) satisfies the inequality

$$\rho(f,\, f_0) \leq C_\rho \|f - f_0\|_2^2 \tag{13.9}$$

with some positive constant C_ρ independent of f and f_0. Since the functional $\Psi(f)$ is known, the weight function of its derivative $w(\,\cdot\,,\, f_0)$ is also known for all f_0.

Example 13.6. Consider the quadratic functional (13.7). From the identity

$$\|f\|_2^2 = \|f_0\|_2^2 + 2(f_0,\, f - f_0) + \|f - f_0\|_2^2,$$

we have the explicit formula for the derivative

$$\Psi'_{f_0}(g) = 2(f_0, g) = 2\int_0^1 f_0(x) g(x)\, dx.$$

This formula implies that the weight function $w(x,\, f_0) = 2f_0(x)$. The weight function is a Lipschitz function if f_0 belongs to a class of Lipschitz functions. The remainder term in this example, $\rho(f_0,\, f) = \|f - f_0\|_2^2$, meets the condition (13.9) with $C_\rho = 1$. □

The next proposition claims that a differentiable functional can be estimated $\sqrt{n}$-consistently and describes the asymptotic distribution.

Theorem 13.7. *Assume that the regression function $f \in \Theta(\beta, L, L_1)$ with some $\beta \geq 1$. Let $\Psi(f)$ be a differentiable functional on $\Theta(\beta, L, L_1)$. There exists an estimator Ψ_n^* such that the $\mathbb{P}_f$-distribution of its normalized error is asymptotically normal as $n \to \infty$,*

$$\sqrt{n}\,\big(\Psi_n^* - \Psi(f)\big) \to \mathcal{N}\Big(0,\, \sigma^2 \|w(\,\cdot\,,\, f)\|_2^2\Big).$$

Proof. Split n observations into the two sub-samples of size m and $n - m$, respectively, with $m = n^\alpha$, $5/6 \leq \alpha < 1$. Assume that n/m is an integer. Define the first sub-sample by the equidistant design points, $J_1 = \{1/m, 2/m, \dots, m/m\}$. Let the second sub-sample J_2 be composed of the rest of the design points. Note that J_2 is not necessarily regular and it contains almost all the points, $|J_2| = n(1 - n^{\alpha-1}) = n\big(1 + o_n(1)\big)$. From Theorem 10.3, even in the case of the smallest smoothing parameter, $\beta = 1$, we can choose an estimator f_n^* of f so that uniformly in f,

$$\mathbb{E}_f\big[\,\|f_n^* - f\|_2^2\,\big] \leq r^* m^{-2/3} = r^* n^{-2\alpha/3} \leq r^* n^{-5/9} = o\big(1/\sqrt{n}\big).$$

If f_n^* is not a Lipschitz function, we replace it by the projection onto the set $\Theta(\beta, L, L_1)$, which is a convex set. So that we may assume that f_n^* is a Lipschitz function and

$$\lim_{n\to\infty} \sqrt{n}\,\mathbb{E}_f\big[\, \|f_n^* - f\|_2^2 \,\big] = 0. \tag{13.10}$$

Introduce an estimator of the functional Ψ by

$$\Psi_n^* = \Psi(f_n^*) + \frac{1}{n}\sum_{i\in J_2} w(i/n,\, f_n^*)\, y_i - \int_0^1 w(x,\, f_n^*) f_n^*(x)\, dx. \tag{13.11}$$

Note that Ψ_n^* is computable from the data. The smaller portion consisting of the m observations is used in the preliminary approximation of f by f_n^*, while the larger portion of the $n - m$ observations is used in estimation of the derivative $\Psi'_{f_n^*}$, which is a linear functional. This linear functional is estimated similarly to (13.4) by $n^{-1}\sum_{i\in J_2} w(i/n,\, f_n^*)\, y_i$. From (13.11), by the definition of a differentiable functional, we obtain that

$$\sqrt{n}\,\big[\Psi_n^* - \Psi(f)\big] = \sqrt{n}\Big[\,\Psi(f_n^*) + \frac{1}{n}\sum_{i\in J_2} w(i/n,\, f_n^*)\, y_i - \int_0^1 w(x,\, f_n^*) f_n^*(x)\, dx$$

$$- \Big(\, \Psi(f_n^*) + \int_0^1 w(x,\, f_n^*)\big(f(x) - f_n^*(x)\big)\, dx + \rho(f,\, f_n^*) \Big)\Big]$$

$$= \sqrt{n}\,\Big[\frac{1}{n}\sum_{i\in J_2} w(i/n,\, f_n^*)\, y_i - \int_0^1 w(x,\, f_n^*) f(x)\, dx\Big] + \sqrt{n}\,\rho(f,\, f_n^*). \tag{13.12}$$

In view of (13.10), the remainder term in (13.12) is vanishing as $n \to \infty$,

$$\sqrt{n}\,\mathbb{E}_f\big[\, |\rho(f,\, f_n^*)| \,\big] \le C_\rho \sqrt{n}\,\mathbb{E}_f\big[\, \|f_n^* - f\|_2^2 \,\big] \to 0.$$

The normalized difference of the sum and the integral in (13.12) is normal with the expectation going to zero as $n \to \infty$, and the variance that, conditionally on f_n^*, is equal to

$$\frac{\sigma^2}{n}\sum_{i\in J_2} w^2(i/n,\, f_n^*) = \frac{\sigma^2}{n}\sum_{i=1}^{n} w^2(i/n,\, f_n^*) + O(m/n)$$

$$= \sigma^2 \int_0^1 w^2(x,\, f_n^*)\, dx + o_n(1).$$

Here we used the fact that $m/n = n^{\alpha - 1} = o_n(1) \to 0$ as $n \to 0$. By the assumption, the weight function $w(\,\cdot\,,\, f_0)$ is continuous in f_0, and

$$\int_0^1 w^2(x,\, f_n^*)\, dx \to \int_0^1 w^2(x,\, f)\, dx = \|w(\,\cdot\,,\, f)\|_2^2$$

as $\|f_n^* - f\|_2 \to 0$. Hence, the result of the theorem follows. □

Example 13.8. Consider again the integral quadratic functional Ψ defined by (13.7). We apply (13.11) to get an explicit expression for the estimator Ψ_n^* of this functional. From Example 13.6, the weight function $w(i/n,\, f_n^*) = 2f_n^*(i/n)$, therefore,

$$\Psi_n^* = \|f_n^*\|_2^2 + \frac{2}{n}\sum_{i\in J_2} f_n^*(i/n)\, y_i - 2\int_0^1 \big(f_n^*(x)\big)^2\, dx$$

$$= \frac{2}{n}\sum_{i\in J_2} f_n^*(i/n)\, y_i - \|f_n^*\|_2^2. \qquad \square$$

Remark 13.9. If the preliminary estimator f_n^* in (13.11) satisfies the condition

$$n\,\mathbb{E}_f\big[\,\|f_n^* - f\|_4^4\big] = n\,\mathbb{E}_f\Big[\int_0^1 (f_n^*(x) - f(x))^4\, dx\Big] \to 0$$

as $n \to \infty$, then the estimator Ψ_n^* converges in the mean square sense as well (see Exercise 13.90),

$$\lim_{n\to\infty} \mathbb{E}_f\Big[\Big(\sqrt{n}(\Psi_n^* - \Psi(f))\Big)^2\Big] = \sigma^2\|w(\,\cdot\,,\, f)\|_2^2. \tag{13.13}$$ $\square$

Exercises

Exercise 13.87. Verify the statement of Proposition 13.2.

Exercise 13.88. Let a regression function $f \in \Theta(\beta, L, L_1)$, $\beta \geq 1$, serve as the right-hand side of the differential equation

$$\Psi'(x) + \Psi(x) = f(x), \quad x \geq 0,$$

with the initial condition $\Psi(0) = 0$. Assume that the observations y_i, $i = 1, \dots, n$, of the regression function f satisfy (13.1). Estimate the solution $\Psi(x)$ at $x = 1$. Find the asymptotics of the estimate's bias and variance as $n \to \infty$. Hint: Check that $\Psi(x) = e^{-x}\int_0^x e^t f(t)\, dt$.

Exercise 13.89. Show that $\Psi(f) = \int_0^1 f^4(x)\, dx$ is a differentiable functional of the regression function $f \in \Theta(\beta, L, L_1)$, $\beta \geq 1$.

Exercise 13.90. Prove (13.13).

Exercise 13.91. Consider the observations $y_i = f(x_i) + \varepsilon_i$, $i = 1, \dots, n$, with a regular design governed by a density $p(x)$. Show that the sample mean $\bar{y} = (y_1 + \cdots + y_n)/n$ is the minimax efficient estimator of the functional $\Psi(f) = \int_0^1 f(x)p(x)\, dx$.

Chapter 14

Dimension and Structure in Nonparametric Regression

14.1. Multiple Regression Model

In this chapter, we revise the material of Chapters 8 and 9, and extend it to the *multiple regression model.* As in (8.1), our starting point is the regression equation $Y = f(\mathbf{X}) + \varepsilon$, $\varepsilon \sim \mathcal{N}(0, \sigma^2)$. The difference is that this time the explanatory variable is a d-dimensional vector, $\mathbf{X} = \left(X^{(1)}, \dots, X^{(d)}\right)' \in \mathbb{R}^d$. We use the upper index to label the components of this vector. A set of n observations has the form $\{ (\mathbf{x}_1, y_1), \dots, (\mathbf{x}_n, y_n) \}$ where each regressor $\mathbf{x}_i = \left(x_i^{(1)}, \dots, x_i^{(d)}\right)'$ is d-dimensional. The regression equation looks similar to (8.2),

$$y_i = f(\mathbf{x}_i) + \varepsilon_i, \quad i = 1, \dots, n, \tag{14.1}$$

where $\varepsilon_i \sim \mathcal{N}(0, \sigma^2)$ are independent normal random errors. The regression function $f : \mathbb{R}^d \to \mathbb{R}$ is a real-valued function of d variables. This function is unknown and has to be estimated from the observations (14.1). Assume that the regressors belong to the unit cube in $\mathbb{R}^d$, $\mathbf{x}_i \in [0, 1]^d$, $i = 1, \dots, n$. The design $\mathcal{X} = \{ \mathbf{x}_1, \dots, \mathbf{x}_n \}$ can be deterministic or stochastic, though in this chapter we prefer to deal with the regular deterministic designs.

Our principal objective is to explain the influence of the dimension d on the asymptotically minimax rate of convergence. We restrict ourselves to the estimation problem of $f(\mathbf{x}_0)$ at a fixed point $\mathbf{x}_0 = (x_0^{(1)}, \dots, x_0^{(d)})'$ located strictly inside the unit cube $[0,1]^d$. The asymptotically minimax rate of convergence ψ_n, defined by (8.4) and (8.5), is attached to a Hölder class $\Theta(\beta, L, L_1)$ of regression functions. Thus, we have to extend the definition of the Hölder class to the multivariate case. The direct extension via the derivatives as in the one-dimensional case is less convenient since we would have to deal with all mixed derivatives of f up to a certain order. A more fruitful approach is to use the formula (8.14) from Lemma 8.5 as a guideline.

Let β be an integer, $\beta \geq 1$, and let $\|\cdot\|$ denote the Euclidean norm in $\mathbb{R}^d$. A function $f(\mathbf{x})$, $\mathbf{x} \in [0,1]^d$, is said to belong to a *Hölder class* of functions $\Theta(\beta) = \Theta(\beta, L, L_1)$ if: (i) there exists a constant $L_1 > 0$ such that $\max_{\mathbf{x}\in[0,1]^d} |f(\mathbf{x})| \leq L_1$, and (ii) for any $\mathbf{x}_0 \in [0,1]^d$ there exists a polynomial $p(\mathbf{x}) = p(\mathbf{x}, \mathbf{x}_0, f)$ of degree $\beta - 1$ such that

$$\left| f(\mathbf{x}) - p(\mathbf{x}, \mathbf{x}_0, f) \right| \leq L\|\mathbf{x} - \mathbf{x}_0\|^\beta$$

for any $\mathbf{x} \in [0,1]^d$, with a constant L independent of f and $\mathbf{x}_0$.

To estimate the regression function at a given point $\mathbf{x}_0$ we can apply any of the methods developed in the previous chapters. Let us consider the local polynomial approximation first. Take a hypercube H_n centered at $\mathbf{x}_0$,

$$H_n = \left\{ \mathbf{x} = (x^{(1)}, \dots, x^{(d)})' : |x^{(1)} - x_0^{(1)}| \leq h_n, \dots, |x^{(d)} - x_0^{(d)}| \leq h_n \right\}$$

where the bandwidth $h_n \to 0$ as $n \to \infty$. If $\mathbf{x}_0$ belongs to an open cube $(0,1)^d$, then for any large n, the hypercube H_n is a subset of $[0,1]^d$.

Consider a polynomial $\pi(\mathbf{x})$, $\mathbf{x} \in [0,1]^d$, of degree $\beta - 1$. Note that in the case of d predictor variables, there are exactly $\binom{i+d-1}{i}$ monomials of degree $i \geq 0$ (see Exercise 14.92). That is why there are $k = k(\beta, d)$ coefficients that define the polynomial $\pi(\mathbf{x})$ where

$$k = k(\beta, d) = \sum_{i=0}^{\beta-1} \binom{i+d-1}{i} = \frac{1}{(d-1)!} \sum_{i=0}^{\beta-1} (i+1) \dots (i+d-1).$$

We denote the vector of these coefficients by $\boldsymbol{\theta} \in \mathbb{R}^k$, and explicitly mention it in the notation for the polynomial $\pi(\mathbf{x}) = \pi(\mathbf{x}, \boldsymbol{\theta})$.

Example 14.1. Let $d = 2$ and $\beta = 3$. A polynomial of degree $\beta - 1 = 2$ has a general form

$$\pi(\mathbf{x}, \boldsymbol{\theta}) = \theta_0 + \theta_1 x^{(1)} + \theta_2 x^{(2)} + \theta_3 (x^{(1)})^2 + \theta_4 x^{(1)} x^{(2)} + \theta_5 (x^{(2)})^2$$

with the vector of unknown coefficients $\boldsymbol{\theta} = (\theta_0, \theta_1, \theta_2, \theta_3, \theta_4, \theta_5)'$. To verify the dimension of $\boldsymbol{\theta}$, compute

$$k = k(3,2) = \sum_{i=0}^{2} \binom{i+1}{i} = \binom{1}{0} + \binom{2}{1} + \binom{3}{2} = 1 + 2 + 3 = 6. \quad \square$$

For any $\mathbf{x}$ in the hypercube H_n centered at $\mathbf{x}_0$, we rescale this polynomial to get $\pi\big((\mathbf{x}-\mathbf{x}_0)/h_n,\, \boldsymbol{\theta}\big)$. Suppose there are N pairs of observations $(\mathbf{x}_i\,,\, y_i)$ such that the design points belong to H_n. Without loss of generality, we may assume that these are the first N observations, $\mathbf{x}_1, \dots, \mathbf{x}_N \in H_n$. The vector $\boldsymbol{\theta}$ can be estimated by the method of least squares. The estimator $\hat{\boldsymbol{\theta}}$ is the solution of the minimization problem (cf. (9.1)),

$$\sum_{i=1}^{N} \Big[\, y_i \;-\; \pi\big(\,(\mathbf{x}-\mathbf{x}_0)/h_n\,,\, \hat{\boldsymbol{\theta}}\big)\,\Big]^2 \;\to\; \min_{\hat{\boldsymbol{\theta}}}\,. \tag{14.2}$$

As in Section 9.1, we can define a system of normal equations (9.2) where $\mathbf{G}$ is the design matrix with dimensions $N \times k$. The columns of $\mathbf{G}$ are composed of the monomials in the polynomial $\pi\big(\,(\mathbf{x}-\mathbf{x}_0)/h_n,\, \boldsymbol{\theta}\big)$ evaluated at the N design points.

Let Assumption 9.2 hold, that is, we assume that the elements of the matrix $\big(\mathbf{G}'\mathbf{G}\big)^{-1}$ are bounded from above by $\gamma_0 N^{-1}$ with a constant γ_0 independent of n. Clearly, this assumption is a restriction on the design $\mathcal{X}$. The next proposition is a simplified version of Proposition 9.4.

Proposition 14.2. *Suppose Assumption 9.2 holds. Let $\hat{\theta}_0 = \pi(0,\, \hat{\boldsymbol{\theta}})$ be the estimate of the intercept, that is, the first component of the solution $\hat{\boldsymbol{\theta}}$ of (14.2). Then uniformly in $f \in \Theta(\beta, L, L_1)$, we have that*

$$\hat{\theta}_0 \;-\; f(\mathbf{x}_0) \;=\; b_0 \;+\; \mathcal{N}_0$$

where $|b_0| \le C_b h_n^{\beta}$, and $\mathcal{N}_0$ is a zero-mean normal random variable with the variance $\mathbb{V}ar_f\big[\mathcal{N}_0 \,|\, \mathcal{X}\big] \le C_v/N$. The positive constants C_b and C_v are independent of n.

Proof. From the definition of the Hölder class $\Theta(\beta, L, L_1)$, we obtain (cf. (9.3)),

$$y_i \;=\; p(\mathbf{x}_i, \mathbf{x}_0, f) \;+\; \rho(\mathbf{x}_i, \mathbf{x}_0, f) \;+\; \varepsilon_i\,,\;\; i = 1, \dots, n,$$

where the remainder term

$$|\rho(\mathbf{x}_i, \mathbf{x}_0, f)| \le \|\mathbf{x}_i \;-\; \mathbf{x}_0\|^{\beta} \le \big(\sqrt{d}h_n\big)^{\beta}.$$

Repeating the proofs of Lemmas 9.1 and 9.3, we find that the least-squares estimator $\hat{\boldsymbol{\theta}}$ actually estimates the vector of coefficients of the polynomial $p(\mathbf{x}, \mathbf{x}_0, f)$ in the above approximation. The deterministic error here that does not exceed $C_b h_n^{\beta}$, and the zero-mean normal stochastic term has the variance that is not larger than C_v/N. By definition, the zero-order term of the approximation polynomial θ_0 is equal to $f(\mathbf{x}_0)$. Hence, the estimate of the intercept

$$\hat{\theta}_0 \;=\; \theta_0 \;+\; b_0 \;+\; \mathcal{N}_0 \;=\; f(\mathbf{x}_0) \;+\; b_0 \;+\; \mathcal{N}_0$$

satisfies the claim of the proposition. □

Finally, we have arrived at the point where the influence of the higher dimension shows up. In Section 9.1, to obtain the minimax rate of convergence, we needed Assumption 9.5 which helped to control the stochastic term. This assumption required that the number N of the design points in the h_n-neighborhood of the given point x_0 is proportional to nh_n. Clearly, this assumption was meant to meet the needs of regular designs, deterministic or random. So, the question arises: How many design points can we anticipate in the regular cases in the d-dimensional H_n-neighborhood of $\mathbf{x}_0$? Simple geometric considerations show that at best we can rely on a number proportional to the volume of H_n.

Assumption 14.3. There exists a positive constant γ_1, independent of n, such that for all large enough n, the inequality $N \geq \gamma_1 n h_n^d$ holds. □

Now we are in the position to formulate the upper bound result.

Theorem 14.4. *Suppose that the design $\mathcal{X}$ satisfies the conditions of Assumptions 9.2 and 14.3 with $h_n = h_n^* = n^{-1/(2\beta+d)}$. Given $\mathcal{X}$, the quadratic risk of the local polynomial estimator $\hat{\theta}_0 = \pi(0, \hat{\boldsymbol{\theta}})$ described in Proposition 14.2 admits the upper bound*

$$\sup_{f \in \Theta(\beta, L, L_1)} \mathbb{E}_f\Big[\big(\pi(0, \hat{\boldsymbol{\theta}}) - f(\mathbf{x}_0)\big)^2 \,\Big|\, \mathcal{X}\Big] \leq r^* n^{-2\beta/(2\beta+d)}$$

where a positive constant r^* is independent of n.

Proof. Analogously to the proof of Theorem 9.6, from Proposition 14.2, the upper bound of the quadratic risk holds uniformly in $f \in \Theta(\beta, L, L_1)$,

$$\mathbb{E}_f\Big[\big(\pi(0, \hat{\boldsymbol{\theta}}) - f(\mathbf{x}_0)\big)^2 \,\Big|\, \mathcal{X}\Big] \leq C_b^2 (h_n)^{2\beta} + \frac{C_v}{N} \leq C_b^2 (h_n)^{2\beta} + \frac{C_v}{\gamma_1 n h_n^d}.$$

The balance equation in the d-dimensional case has the form $(h_n)^{2\beta} = 1/(nh_n^d)$. The optimal choice of the bandwidth is $h_n = h_n^* = n^{-1/(2\beta+d)}$, and the respective rate of convergence is $(h_n^*)^{\beta} = n^{-\beta/(2\beta+d)}$. □

14.2. Additive regression

The minimax rate of convergence $\psi_n = n^{-\beta/(2\beta+d)}$ in a d-dimensional Hölder regression rapidly slows down as d increases. One way to overcome this "curse of dimensionality" is to assume that the smoothness also grows with d. Indeed, if $\beta = d\beta_1$, then the exponent in the rate of convergence $\beta/(2\beta + d) = \beta_1/(2\beta_1 + 1)$ matches the one-dimensional case with the smoothness parameter β_1. However, this assumption is very restrictive.

Another approach is to impose some constraints on the structure of the regression model. Here we consider one example, a so-called *additive regression* model. To understand the role of a higher dimension, it suffices to look at the case $d = 2$ and a very basic regular design. Suppose that in the two-dimensional regression model, the design $\mathcal{X}$ is equidistant. Let $m = \sqrt{n}$ be an integer. The design points, responses, and random errors are all labeled by two integer indices i and j, where $i, j = 1, \dots, m$. We denote the design points by

$$\mathbf{x}_{ij} = (i/m,\, j/m). \tag{14.3}$$

Thus, the regression relation takes the form

$$y_{ij} \;=\; f(i/m,\, j/m) \,+\, \varepsilon_{ij} \tag{14.4}$$

where ε_{ij} are independent $\mathcal{N}(0, \sigma^2)$ random variables.

In the *additive regression model*, the regression function is the sum of two functions, both of which depend on one variable,

$$f(\mathbf{x}) \;=\; f(x^{(1)},\, x^{(2)}) \;=\; f_1(x^{(1)}) \,+\, f_2(x^{(2)}) \tag{14.5}$$

where f_1 and f_2 are the Hölder functions of a single variable, $f_1,\, f_2 \in \Theta(\beta, L, L_1)$.

This definition of the additive model is not complete, since we can always add a constant to one term and subtract it from the other one. To make the terms identifiable, we impose the following conditions:

$$\int_0^1 f_1(t)\, dt \;=\; \int_0^1 f_2(t)\, dt \;=\; 0. \tag{14.6}$$

Let $\mathbf{x}_0$ be a fixed point strictly inside the unit square. Without loss of generality, we will assume that this point coincides with one of the design knots, $\mathbf{x}_0 = (i_0/m,\, j_0/m)$. Clearly, we could treat the model of observations (14.4) as a two-dimensional regression. The value of the regression function $f(\mathbf{x}_0)$ at $\mathbf{x}_0$ can be estimated with the rate $n^{-\beta/(2\beta+2)}$ suggested by Theorem 14.4 for $d = 2$. A legitimate question at this point is whether it is possible to estimate $f(\mathbf{x}_0)$ with a faster rate exploiting the specific structure of the model. In particular, is it possible to attain the one-dimensional rate of convergence $n^{-\beta/(2\beta+1)}$? As the following proposition shows, the answer to this question is affirmative.

Proposition 14.5. *In the additive regression model* (14.4)-(14.6) *at any point* $\mathbf{x}_0 = (i_0/m,\, j_0/m)$, *there exists an estimator* $\hat{f}_n(\mathbf{x}_0)$ *such that*

$$\sup_{f_1, f_2 \in \Theta(\beta, L, L_1)} \mathbb{E}_f\Big[\Big(\hat{f}_n(\mathbf{x}_0) \,-\, f(\mathbf{x}_0)\Big)^2\Big] \le r^*\, n^{-2\beta/(2\beta+1)}$$

for all large enough n. Here a constant r^* is independent of n.

Proof. Select the bandwidth $h_n^* = n^{-1/(2\beta+1)}$ as if the model were one-dimensional. Consider the set of indices

$$I_n = I_n(i_0/m) = \big\{ i \, : \, |i/m - i_0/m| \leq h_n^* \big\}.$$

The number N of indices in the set I_n is equal to $N = |I_n| = 2\lfloor m h_n^* \rfloor + 1$. Note that $m h_n^* = \sqrt{n} n^{-1/(2\beta+1)} \to \infty$, and hence $N \sim 2\, m\, h_n^*$ as $n \to \infty$. To estimate f_1 at i_0/m, consider the means

$$\bar{y}_{i\cdot} = \frac{1}{m} \sum_{j=1}^{m} y_{ij} = \frac{1}{m} \sum_{j=1}^{m} \Big[f_1(i/m) + f_2(j/m) \Big] + \frac{1}{m} \sum_{j=1}^{m} \varepsilon_{ij}$$

$$= f_1(i/m) + \delta_n + \frac{1}{\sqrt{m}} \bar{\varepsilon}_i, \quad i \in I_n, \tag{14.7}$$

where the deterministic error

$$\delta_n = \frac{1}{m} \sum_{j=1}^{m} f_2(j/m)$$

is the Riemann sum for the integral $\int_0^1 f_2(t)\, dt = 0$, and the random variables

$$\bar{\varepsilon}_i = \frac{1}{\sqrt{m}} \sum_{j=1}^{m} \varepsilon_{ij} \sim \mathcal{N}(0, \sigma^2)$$

are independent for different $i \in I_n$.

Applying (14.6), we find that

$$|\delta_n| \leq L_0/m = L_0/\sqrt{n} = o\big((h_n^*)^\beta \big) \text{ as } n \to \infty$$

with a Lipschitz constant $L_0 = \max |f_2'|$ of any function $f_2 \in \Theta(\beta, L, L_1)$. Thus, we have a one-dimensional regression problem with N observations in the bin centered at i_0/m. Applying the one-dimensional local polynomial approximation (see Lemma 9.3) to the means (14.7), we can estimate f_1 at the point i_0/m with the deterministic error not exceeding

$$C_b (h_n^*)^\beta + |\delta_n| = C_b (h_n^*)^\beta \big(1 + o(1)\big).$$

The stochastic error of this estimator is normal with the zero expectation and the variance which is not larger than

$$\frac{C_v}{N} \Big(\frac{\sigma}{\sqrt{m}} \Big)^2 \sim \frac{C_v}{2 m h_n^*} \Big(\frac{\sigma}{\sqrt{m}} \Big)^2 = \frac{C_v \sigma^2}{2 n h_n^*}.$$

The constants C_b and C_v are independent of n. Here $(\sigma/\sqrt{m})^2 = \sigma^2/m = \sigma^2/\sqrt{n}$ represents the variance of the stochastic error of the means (14.7). So, the one-dimensional balance between the deterministic and stochastic errors holds, and the one-dimensional rate of convergence $(h_n^*)^\beta$ is guaranteed.

Similarly, we can estimate f_2 at j_0/m with the same one-dimensional rate. $\square$

Remark 14.6. In this section, we considered the simplest version of the additive regression model. In more general settings, the design may be deterministic or random, and the dimension may be any positive integer. The effect, however, is still the same: the one-dimensional rate of convergence is attainable. Clearly, this rate is minimax since in any higher dimension, this rate cannot be improved for the subset of one-dimensional regression functions. $\square$

Remark 14.7. Traditionally the additive regression model includes a constant intercept f_0. That is, the regression function has the form

$$f(x^{(1)}, x^{(2)}) = f_0 + f_1(x^{(1)}) + f_2(x^{(2)}). \tag{14.8}$$

For simplicity, it was omitted in the model considered in this section. To estimate f_0, we could split the sample of observations into two sub-samples of sizes $n/2$ each, but this would destroy the regularity of the design. It is more convenient to consider the model with two independent repeated observations y_{ij} and $\tilde{y}_{ij}$ at the design knots $(i/m, j/m)$. Then we can use the second set of observations to estimate the intercept,

$$\hat{f}_0 = \frac{1}{n} \sum_{i,j=1}^{m} \tilde{y}_{ij}. \tag{14.9}$$

Now we can replace the observations y_{ij} in (14.4) by $y_{ij} - \hat{f}_0$, and use these shifted observations to estimate f_1 and f_2 as done above. Then the statement of Proposition 14.5 would stay unchanged (see Exercise 14.93). $\square$

14.3. Single-Index Model

14.3.1. Definition. The additive regression model of the previous section provides an example of a specific structure of the nonparametric regression function. In this section we give another example, known as a *single-index model.* This name unites a variety of models. We present here a version that is less technical.

Consider a two-dimensional regression model with the equidistant design in the unit square $[0, 1]^2$. It is convenient to study the model with two independent repeated observations at every design knot $\mathbf{x}_{ij} = (i/m, j/m)$,

$$y_{ij} = f(i/m, j/m) + \varepsilon_{ij} \text{ and } \tilde{y}_{ij} = f(i/m, j/m) + \tilde{\varepsilon}_{ij} \tag{14.10}$$

where $1 \leq i, j \leq m$, and $m = \sqrt{n}$ is assumed to be an integer (cf. Remark 14.7). The random variables ε_{ij} and $\tilde{\varepsilon}_{ij}$ are independent $\mathcal{N}(0, \sigma^2)$.

The structure of the regression function f is imposed by the assumption that there exist a Hölder function $g = g(t)$ and an angle α such that

$$f(i/m, j/m) = g\big((i/m)\cos\alpha + (j/m)\sin\alpha \big), \quad 1 \le i, j \le m. \tag{14.11}$$

We will suppose that $0 \le \alpha \le \pi/4$. The restrictions on the function g are more elaborate and are discussed below.

Let $\beta \ge 2$ be an integer, and let g_* be a positive number. Assume that $\Theta(\beta, L, L_1)$ is the class of Hölder functions $g = g(t)$ in the interval $0 \le t \le \sqrt{2}$, and let

$$\Theta(\beta, L, L_1, g_*) = \Theta(\beta, L, L_1) \cap \{ g'(t) \ge g_* \}$$

be a sub-class of functions the first derivative of which exceeds g_*.

Introduce a class of functions $\mathcal{H} = \mathcal{H}(\beta, L, L_1, g_*)$ in the unit square $0 \le x^{(1)}, x^{(2)} \le 1$ by

$$\mathcal{H} = \Big\{ f = f(x^{(1)}, x^{(2)}) = g\big(x^{(1)}\cos\alpha + x^{(2)}\sin\alpha\big),$$

$$0 \le \alpha \le \pi/4, \;\; g \in \Theta(\beta, L, L_1, g_*) \Big\}.$$

This class is well defined because the variable $t = x^{(1)}\cos\alpha + x^{(2)}\sin\alpha$ belongs to the interval $0 \le t \le \sqrt{2}$. The functions in $\mathcal{H}$, if rotated at a proper angle, depend on a single variable, and are monotonically increasing in the corresponding direction. The point

$$t_{ij} = (i/m)\cos\alpha + (j/m)\sin\alpha \tag{14.12}$$

is the projection (show!) of $\mathbf{x}_{ij} = (i/m, j/m)$ onto the straight line passing through the origin at the angle α (see Figure 11). If we knew α, we could compute the projections t_{ij}, and the problem would become one-dimensional.

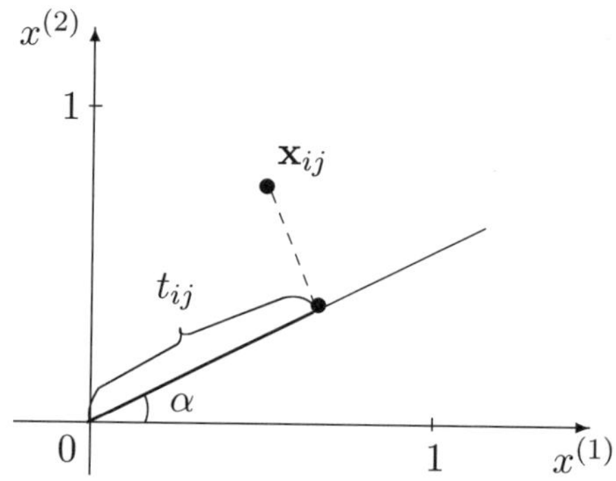

Figure 11. Projection of the design knot $\mathbf{x}_{ij}$ on the line passing through the origin at the angle α.

Let $\mathbf{x_0} = (i_0/m,\, j_0/m)$ be a fixed point. Our objective is to estimate the value $f(\mathbf{x_0})$ of the regression function at this point. Clearly, we can look at the observations (14.10) as the observations of the two-dimensional regression. The results of Section 14.1 would guarantee the minimax rate of estimation $n^{-2\beta/(2\beta+2)}$. Can this rate be improved to the one-dimensional rate $n^{-2\beta/(2\beta+1)}$? The answer is positive, and the algorithm is simple. First, estimate α by $\hat{\alpha}_n$, and then plug $\hat{\alpha}_n$ into the projection formula (14.12) for the one-dimensional design points,

$$\hat{t}_{ij} \;=\; (i/m)\cos\hat{\alpha}_n + (j/m)\sin\hat{\alpha}_n.$$

The two-sample model of observations (14.10) is convenient, because the first sample is used to estimate α, while the second one serves to estimate the regression function itself. We could work with one sample of size n, and split it into two independent sub-samples, but this would result in less regular designs.

14.3.2. Estimation of Angle. To estimate α, note that at any point $(x^{(1)}, x^{(2)}) \in [0,1]^2$, the partial derivatives of the regression function are proportional to $\cos\alpha$ and $\sin\alpha$, respectively,

$$\frac{\partial f(x^{(1)},\, x^{(2)})}{\partial x^{(1)}} \;=\; g'(x^{(1)}\cos\alpha + x^{(2)}\sin\alpha)\cos\alpha \tag{14.13}$$

and

$$\frac{\partial f(x^{(1)},\, x^{(2)})}{\partial x^{(2)}} \;=\; g'(x^{(1)}\cos\alpha + x^{(2)}\sin\alpha)\sin\alpha. \tag{14.14}$$

If we integrate the left-hand sides of (14.13) and (14.14) over the square $[0,1]^2$, we obtain the integral functionals that we may try to estimate. Unfortunately, these are functionals of partial derivatives of f, not of f itself. However, we can turn these functionals into the functionals of f if we integrate by parts.

Choose any function $\varphi \;=\; \varphi(x^{(1)},\, x^{(2)}),\;\; (x^{(1)},\, x^{(2)}) \in [0,1]^2$. Assume that φ is non-negative and very smooth, for example, infinitely differentiable. Assume also that it is equal to zero identically on the boundary of the unit square,

$$\varphi(x^{(1)},\, x^{(2)}) = 0 \text{ for } (x^{(1)},\, x^{(2)}) \in \partial[0,1]^2. \tag{14.15}$$

Multiplying the left-hand sides of (14.13) and (14.14) by φ and integrating by parts over $[0,1]^2$, we obtain an integral functional of f,

$$\Phi_l(f) \;=\; \int_0^1\!\!\int_0^1 \varphi(x^{(1)},\, x^{(2)})\frac{\partial f(x^{(1)},\, x^{(2)})}{\partial x^{(l)}}\, dx^{(1)}\, dx^{(2)} \tag{14.16}$$

$$=\; \int_0^1\!\!\int_0^1 w_l(x^{(1)},\, x^{(2)}) f(x^{(1)},\, x^{(2)})\, dx^{(1)}\, dx^{(2)} \tag{14.17}$$

where w_l are the weight functions

$$w_l(x^{(1)}, x^{(2)}) = -\frac{\partial \varphi(x^{(1)}, x^{(2)})}{\partial x^{(l)}}, \quad l = 1 \text{ or } 2.$$

The outside-of-the-integral term in (14.17) vanishes due to the boundary condition (14.15).

Thus, (14.13) and (14.14) along with (14.17) yield the equations

$$\Phi_1 = \Phi_1(f) = \Phi_0 \cos\alpha \;\text{ and }\; \Phi_2 = \Phi_2(f) = \Phi_0 \sin\alpha \tag{14.18}$$

with

$$\Phi_0 = \Phi_0(f) = \int_0^1 \int_0^1 \varphi(x^{(1)}, x^{(2)}) g'(x^{(1)} \cos\alpha + x^{(2)} \sin\alpha)\, dx^{(1)}\, dx^{(2)}.$$

Under our assumptions, uniformly in $f \in \mathcal{H}$, the values of $\Phi_0(f)$ are separated away from zero by some strictly positive constant,

$$\Phi_0(f) \,\geq\, \Phi_* \,>\, 0. \tag{14.19}$$

Now, given the values of the functionals Φ_1 and Φ_2, we can restore the angle α from the equation $\alpha = \arctan\big(\Phi_2/\Phi_1\big)$. Define the estimators of these functionals by

$$\hat{\Phi}_n^{(l)} \;=\; \frac{1}{n} \sum_{i,j=1}^{m} w_l(i/m,\, j/m)\, y_{ij}\,, \quad l = 1 \text{ or } 2. \tag{14.20}$$

Then we can estimate the angle α by

$$\hat{\alpha}_n \;=\; \arctan\Big(\,\hat{\Phi}_n^{(2)} / \hat{\Phi}_n^{(1)}\,\Big).$$

Note that the ratio $\hat{\Phi}_n^{(2)}/\hat{\Phi}_n^{(1)}$ can be however large, positive or negative. Thus, the range of $\hat{\alpha}_n$ runs from $-\pi/2$ to $\pi/2$, whereas the range of the true α is $[0, \pi/4]$. Next, we want to show that the values of $\hat{\alpha}_n$ outside of the interval $[0, \pi/4]$ are possible only due to the large deviations, and the probability of this event is negligible if n is large. As the following proposition shows, the estimator $\hat{\alpha}_n$ is $\sqrt{n}$-consistent with rapidly decreasing probabilities of large deviations. The proof of this proposition is postponed to the next section.

Proposition 14.8. *There exist positive constants a_0, c_0, and c_1, independent of f and n, such that for any x, $c_0 \leq x \leq c_1\sqrt{n}$, the following inequality holds:*

$$\mathbb{P}_f\big(\big|\hat{\alpha}_n - \alpha\big| > x/\sqrt{n}\,\big) \;\leq\; 4\, \exp\{-a_0 x^2\}.$$

14.3.3. Estimation of Regression Function. We use the second sample $\tilde{y}_{ij}$ of the observations in (14.10) to estimate the regression function $f(\mathbf{x}_0)$ at the given knot $\mathbf{x}_0 = (i_0/m,\, j_0/m)$. Recall that t_{ij}, as introduced in (14.12), is the projection of $(i/m,\, j/m)$ onto the line determined by the true angle α. Denote by $\hat{t}_{ij}$ the projection of $(i/m,\, j/m)$ onto the line determined by the estimated angle $\hat{\alpha}_n$, and let $\hat{u}_{ij}$ be the projection in the orthogonal direction given by the angle $\hat{\alpha}_n + \pi/2$, that is,

$$\hat{t}_{ij} = (i/m)\cos\hat{\alpha}_n + (j/m)\sin\hat{\alpha}_n,$$

and

$$\hat{u}_{ij} = -(i/m)\sin\hat{\alpha}_n + (j/m)\cos\hat{\alpha}_n.$$

Let the respective projections of the fixed point $\mathbf{x}_0 = (i_0/m, j_0/m)$ be denoted by $\hat{t}_0$ and $\hat{u}_0$. Introduce $\mathcal{T}$, a rectangle in the new coordinates (see Figure 12),

$$\mathcal{T} = \left\{ (t,\, u) \,:\, |t - \hat{t}_0| \le h_n^*,\ \ |u - \hat{u}_0| \le H \right\}$$

where $h_n^* = n^{-1/(2\beta+1)}$ and H is a constant independent of n and so small that $\mathcal{T} \subset [0,1]^2$.

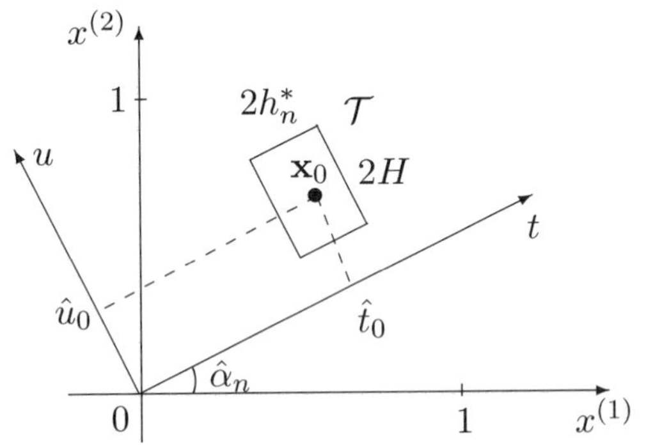

Figure 12. Rectangle $\mathcal{T}$ in the coordinate system rotated by the angle $\hat{\alpha}_n$.

Proposition 14.9. *For any design knot $\mathbf{x}_{ij} = (i/m,\, j/m) \in \mathcal{T}$, the observation $\tilde{y}_{ij}$ in (14.10) admits the representation*

$$\tilde{y}_{ij} = g(\hat{t}_{ij}) + \rho_{ij} + \tilde{\varepsilon}_{ij},\ \ 1 \le i, j \le m,$$

with the remainder term ρ_{ij} being independent of the random variable $\tilde{\varepsilon}_{ij}$, and satisfying the inequality

$$\max_{1\le i,j\le m} |\rho_{ij}| \le 2L_0|\hat{\alpha}_n - \alpha|$$

where $L_0 = \max|g'|$ is the Lipschitz constant of any $g \in \Theta(\beta, L, L_1, g_*)$.

Proof. Put $\rho_{ij} = g(t_{ij}) - g(\hat{t}_{ij})$. By definition, ρ_{ij} depends only on the first sub-sample y_{ij} of observations in (14.10), and hence is independent of $\tilde{\varepsilon}_{ij}$. We have $\tilde{y}_{ij} = g(\hat{t}_{ij}) + \rho_{ij} + \tilde{\varepsilon}_{ij}$. For any knot $(i/m,\, j/m)$, we obtain

$$|\rho_{ij}| = |g(t_{ij}) - g(\hat{t}_{ij})|$$
$$= \big|\, g\big(\,(i/m)\cos\alpha + (j/m)\sin\alpha\,\big) \;-\; g\big(\,(i/m)\cos\hat{\alpha}_n + (j/m)\sin\hat{\alpha}_n\,\big)\,\big|$$
$$\leq L_0\big(\,(i/m)\,\big|\cos\hat{\alpha}_n - \cos\alpha\,\big| \,+\, (j/m)\big|\,\sin\hat{\alpha}_n - \sin\alpha\,\big|\,\big)$$
$$\leq L_0\big(\,i/m \,+\, j/m\,\big)\,\big|\,\hat{\alpha}_n \,-\, \alpha\,\big| \leq 2\,L_0\,\big|\,\hat{\alpha}_n \,-\, \alpha\,\big|. \qquad \square$$

Further, let $\hat{f}(\mathbf{x}_0)$ be the local polynomial approximation obtained from the observations $\tilde{y}_{ij}$ at the design points $\hat{t}_{ij}$ where $(i/m,\, j/m) \in \mathcal{T}$. It means that $\hat{f}(\mathbf{x}_0) = \hat{\theta}_0$ where $\hat{\theta}_0$ is the least-squares estimator of the intercept. It can be obtained as a partial solution to the minimization problem

$$\sum_{(i/m,j/m)\in\mathcal{T}} \left(\tilde{y}_{ij} - \left[\hat{\theta}_0 + \hat{\theta}_1\frac{\hat{t}_{ij} - \hat{t}_0}{h_n^*} + \cdots + \hat{\theta}_{\beta-1}\Big(\frac{\hat{t}_{ij} - \hat{t}_0}{h_n^*}\Big)^{\beta-1}\right]\right)^2 \to \min_{\hat{\theta}_0,\dots,\hat{\theta}_{\beta-1}} .$$

To analyze this minimization problem, we have to verify the Assumptions 9.2 and 9.5 on the system of normal equations associated with it. Denote by $N(\mathcal{T})$ the number of design knots in $\mathcal{T}$, and let $\mathbf{G}'\mathbf{G}$ be the matrix of the system of normal equations with the elements

$$\big(\,\mathbf{G}'\mathbf{G}\,\big)_{k,\,k} \;=\; \sum_{(i/m,\,j/m)\,\in\,\mathcal{T}} \Big(\frac{\hat{t}_{ij} - \hat{t}_0}{h_n^*}\Big)^{k+l}, \quad k, l = 0, \dots, \beta - 1.$$

Note that the number of design knots $N(\mathcal{T})$, and the elements of the matrix $\mathbf{G}'\mathbf{G}$ are random because they depend on the estimator $\hat{\alpha}_n$.

Lemma 14.10. *(i) Uniformly in $\hat{\alpha}_n$, the number of design knots $N(\mathcal{T})$ satisfies*

$$\lim_{n\to\infty} \frac{N(\mathcal{T})}{4Hnh_n^*} = 1.$$

(ii) The normalized elements of the matrix $\mathbf{G}'\mathbf{G}$ have the limits

$$\lim_{n\to\infty} \frac{1}{N(\mathcal{T})}\,\big(\,\mathbf{G}'\mathbf{G}\,\big)_{k,\,l} \;=\; \frac{1 - (-1)^{k+l+1}}{2(k + l + 1)},$$

and the limiting matrix is invertible.

Proof. See the next section. $\square$

Theorem 14.11. *The estimator $\hat{f}(\mathbf{x}_0) = \hat{\theta}_0$ has the one-dimensional rate of convergence on $\mathcal{H}$,*

$$\sup_{f\in\mathcal{H}} n^{2\beta/(2\beta+1)}\,\mathbb{E}_f\Big[\,\big(\,\hat{f}(\mathbf{x}_0) \,-\, f(\mathbf{x}_0)\,\big)^2\,\Big] \;\leq\; r^*$$

for all sufficiently large n with a constant r^* independent of n.

Proof. Proposition 14.9 and Lemma 14.10 allow us to apply the expansion similar to Proposition 9.4. In the case under consideration, this expansion takes the form

$$\hat{\theta}_0 = f(\mathbf{x}_0) + b_0 + \mathcal{N}_0. \tag{14.21}$$

Conditionally on the first sub-sample of observations in (14.10), the bias term b_0 admits the bound

$$|\, b_0 \,| \leq C_b (h_n^*)^\beta + C_a \max_{1 \leq i, j \leq m} |\rho_{ij}| \leq C_b (h_n^*)^\beta + 2 L_0 C_a |\hat{\alpha}_n - \alpha|$$

where the constants C_a and C_b are independent of n. The stochastic term $\mathcal{N}_0$ on the right-hand side of (14.21) is a zero-mean normal random variable with the conditional variance that does not exceed $C_v / N(\mathcal{T}) \leq C_v/(2Hnh_n^*)$. Hence, uniformly in $f \in \mathcal{H}$, we have that

$$\mathbb{E}_f\Big[\big(\hat{f}(\mathbf{x}_0) - f(\mathbf{x}_0)\big)^2\Big] = \mathbb{E}_f\Big[\big(b_0 + \mathcal{N}_0\big)^2\Big]$$

$$\leq 2\,\mathbb{E}_f\big[b_0^2\big] + 2\,\mathbb{E}_f\big[\mathcal{N}_0^2\big]$$

$$\leq 2\mathbb{E}_f\Big[\big(C_b(h_n^*)^\beta + 2L_0 C_a |\hat{\alpha}_n - \alpha|\big)^2\Big] + 2\mathbb{E}_f\big[\mathcal{N}_0^2\big]$$

$$\leq 4\,C_b^2\,(h_n^*)^{2\beta} + 16\,L_0^2\,C_a^2\,\mathbb{E}_f\big[(\hat{\alpha}_n - \alpha)^2\big] + 2C_v/(2Hnh_n^*).$$

Note that with probability 1, $|\hat{\alpha}_n - \alpha| \leq \pi$. From Proposition 14.8, we can estimate the latter expectation by

$$\mathbb{E}_f\big[(\hat{\alpha}_n - \alpha)^2\big] = \int_0^\infty z^2\, d\mathbb{P}_f\big(|\hat{\alpha}_n - \alpha| \leq z\big)$$

$$\leq \big(c_0/\sqrt{n}\big)^2 + \int_{c_0/\sqrt{n}}^{c_1} z^2\, d\mathbb{P}_f\big(|\hat{\alpha}_n - \alpha| \leq z\big) + \int_{c_1}^{\pi} z^2\, d\mathbb{P}_f\big(|\hat{\alpha}_n - \alpha| \leq z\big)$$

$$\leq c_0^2/n + c_0^2/n + 2\int_{c_0/\sqrt{n}}^{c_1} z\, d\mathbb{P}_f\big(|\hat{\alpha}_n - \alpha| > z\big) + \pi^2\,\mathbb{P}_f\big(|\hat{\alpha}_n - \alpha| > c_1\big)$$

$$\leq 2c_0^2/n + 4\int_0^\infty \exp\{-a_0 n z^2\}\, d(z^2) + 4\,\pi^2\,\exp\{-a_0 n c_1^2\}$$

$$\leq 2c_0^2/n + 4/(a_0 n) + 4\pi^2 \exp\{-a_0 n c_1^2\} \leq C_1/n$$

for some positive constant C_1 and for all large enough n. Thus,

$$\mathbb{E}_f\Big[\big(\hat{f}(\mathbf{x}_0) - f(\mathbf{x}_0)\big)^2\Big]$$

$$\leq 4C_b^2 (h_n^*)^{2\beta} + C_1/n + 2C_v/(2Hnh_n^*) = O\big((h_n^*)^{2\beta}\big).$$

Here we used the facts that $(h_n^*)^{2\beta} = (nh_n^*)^{-1}$, and $C_1/n = o\big((h_n^*)^{2\beta}\big)$ as $n \to \infty$. □

14.4. Proofs of Technical Results

To prove Proposition 14.8, we need two lemmas.

Lemma 14.12. *For any n, the estimator $\hat{\Phi}_n^{(l)}$ given by (14.20) of the functionals $\Phi_l(f)$ defined in (14.16) admits the representation*

$$\hat{\Phi}_n^{(l)} = \Phi_l(f) + \rho_n^{(l)}(f)/\sqrt{n} + \eta_n^{(l)}/\sqrt{n}, \quad l = 1 \text{ or } 2,$$

where the deterministic remainder term is bounded by a constant $|\, \rho_n^{(l)}(f) \,| \leq C_\rho$, and the random variables $\eta_n^{(l)}$ are zero-mean normal with the variances bounded from above, $\mathbb{V}ar_f\big[\, \eta_n^{(l)} \,\big] \leq C_v$. The constants C_ρ and C_v are independent of n and f.

Proof. Recall that $m = \sqrt{n}$ is assumed to be an integer. Note that

$$\hat{\Phi}_n^{(l)} = \frac{1}{n} \sum_{i,j=1}^{m} w_l(i/m, j/m)\, f(i/m, j/m) + \frac{1}{n} \sum_{i,j=1}^{m} w_l(i/m, j/m)\, \varepsilon_{ij}$$

$$= \sum_{i,j=1}^{m} \int_{(i-1)/m}^{i/m} \int_{(j-1)/m}^{j/m} w_l(x_1, x_2)\, f(x_1, x_2)\, dx_2\, dx_1 + \frac{\rho_n^{(l)}}{\sqrt{n}} + \frac{\eta_n^{(l)}}{\sqrt{n}}$$

where

$$\rho_n^{(l)} = \sqrt{n} \sum_{i,j=1}^{m} \int_{(i-1)/m}^{i/m} \int_{(j-1)/m}^{j/m} \big[\, w_l(i/m, j/m)\, f(i/m, j/m) - w_l(x_1, x_2)\, f(x_1, x_2) \,\big]\, dx_2\, dx_1$$

and

$$\eta_n^{(l)} = \frac{1}{\sqrt{n}} \sum_{i,j=1}^{m} w_l(i/m, j/m)\, \varepsilon_{ij}.$$

The variance of the normal random variable $\eta_n^{(l)}$ is equal to

$$\mathbb{V}ar\big[\, \eta_n^{(l)} \,\big] = \frac{\sigma^2}{n} \sum_{i,j=1}^{m} w_l^2(i/m, j/m)$$

$$\underset{n\to\infty}{\longrightarrow} \sigma^2 \int_0^1 \int_0^1 w_l^2(x_1, x_2)\, dx_2\, dx_1 < C_v^2 < \infty.$$

The deterministic remainder term $\rho_n^{(l)}$ admits the upper bound

$$|\, \rho_n^{(l)} \,| \leq L_0 m \sum_{i,j=1}^{m} \int_{(i-1)/m}^{i/m} \int_{(j-1)/m}^{j/m} \big(\, |\, x_1 - i/m \,| + |\, x_2 - j/m \,|\, \big)\, dx_2\, dx_1$$

$$= L_0\, m \sum_{i,j=1}^{m} \frac{1}{m^3} = L_0$$

where $L_0 = \max |(w_l f)'|$ is the Lipschitz constant of the product $w_l f$. $\square$

Lemma 14.13. *Let Φ_*, $\hat{\Phi}_n^{(1)}$ and $\hat{\Phi}_n^{(2)}$ be as defined in (14.19) and (14.20). If y satisfies the conditions*

$$\max(C_\rho, C_v) \le y \le (4\sqrt{2})^{-1}\Phi_*\sqrt{n},$$

then for all sufficiently large n, and any $f \in \mathcal{H}$, the following inequality holds:

$$\mathbb{P}_f\Big(\big| \hat{\Phi}_n^{(2)}/\hat{\Phi}_n^{(1)} - \tan\alpha \big| \le \frac{12\,y}{\Phi_*\sqrt{n}} \Big) \ge 1 - 4\exp\Big\{ -\frac{y^2}{2C_v^2} \Big\}.$$

Proof. From Lemma 14.12, the random variable $\eta_n^{(l)}$, $l = 1$ or 2, is a zero-mean normal with the variance satisfying $\mathbb{V}ar_f\big[\eta_n^{(l)}\big] \le C_v$. Therefore, if $y \ge C_v$, then uniformly in $f \in \mathcal{H}$, we have

$$\mathbb{P}_f\big(|\eta_n^{(l)}| > y \big) \le 2\exp\big\{ -y^2/(2\,C_v^2) \big\},\ \ l = 1 \text{ or } 2.$$

Hence, with the probability at least $1 - 4\exp\big\{ -y^2/(2C_v^2) \big\}$, we can assume that $|\eta_n^{(l)}| \le y$ for both $l = 1$ and 2 simultaneously. Under these conditions, in view of (14.18) and Lemma 14.12, we obtain that

$$\big| \hat{\Phi}_n^{(2)}/\hat{\Phi}_n^{(1)} - \tan\alpha \big| = \left| \frac{\cos\alpha(\rho_n^{(2)} + \eta_n^{(2)})/\sqrt{n} - \sin\alpha(\rho_n^{(1)} + \eta_n^{(1)})/\sqrt{n}}{\cos\alpha\big(\Phi_0(f)\cos\alpha + (\rho_n^{(1)} + \eta_n^{(1)})/\sqrt{n}\big)} \right|$$

$$\le \left| \frac{\sqrt{2}(\cos\alpha + \sin\alpha)(C_\rho + y)}{\Phi_*\sqrt{n/2} - (C_\rho + y)} \right| \le \left| \frac{2(C_\rho + y)}{\Phi_*\sqrt{n/2} - (C_\rho + y)} \right|$$

where we used the facts that $\cos\alpha \ge 1/\sqrt{2}$ since $0 \le \alpha \le \pi/4$, and, at the last step, that $\sin\alpha + \cos\alpha \le \sqrt{2}$.
Further, by our assumption, $C_\rho \le y$ and $2y \le (1/2)\Phi_*\sqrt{n/2}$, therefore,

$$\big| \hat{\Phi}_n^{(2)}/\hat{\Phi}_n^{(1)} - \tan\alpha \big| \le \left| \frac{4y}{\Phi_*\sqrt{n/2} - 2y} \right| \le \frac{8y}{\Phi_*\sqrt{n/2}} \le \frac{12\,y}{\Phi_*\sqrt{n}}. \qquad \square$$

Proof of Proposition 14.8. In Lemma 14.13, put $y = (\Phi_*/12)\,x$. Then the restrictions on y in this lemma turn into the bounds for x, $c_0 \le x \le c_1\sqrt{n}$, where $c_0 = \max(C_\rho, C_v)(12/\Phi_*)$ and $c_1 = 2\sqrt{2}$. If we take $a_0 = (\Phi_*/12)^2/(2C_v^2)$, then

$$\mathbb{P}_f\Big(\big| \hat{\Phi}_n^{(2)}/\hat{\Phi}_n^{(1)} - \tan\alpha \big| > x/\sqrt{n} \Big) \le 4\exp\big\{ -a_0 x^2 \big\}.$$

Note that if $|\hat{\alpha}_n - \alpha| > x/\sqrt{n}$, then

$$|\tan\hat{\alpha}_n - \tan\alpha| = (\cos\alpha_*)^{-2}\,|\hat{\alpha}_n - \alpha| \ge |\hat{\alpha}_n - \alpha| > x/\sqrt{n}$$

where we applied the mean value theorem with some α_* between $\hat{\alpha}_n$ and α. Thus,

$$\mathbb{P}_f\Big(|\hat{\alpha}_n - \alpha| > x/\sqrt{n} \Big)$$

$$\leq \mathbb{P}_f\Big(\big| \hat{\Phi}_n^{(2)}/\hat{\Phi}_n^{(1)} - \tan\alpha \big| > x/\sqrt{n} \Big) \leq 4\exp\big\{ -a_0 x^2 \big\}. \qquad \square$$

Proof of Lemma 14.10. (i) For every design knot $(i/m, j/m)$, consider a square, which we call *pixel*, $[(i-1)/m, i/m] \times [(j-1)/m, j/m]$. Let $\mathcal{T}_*$ be the union of the pixels that lie strictly inside $\mathcal{T}$, and let $\mathcal{T}^*$ be the minimum union of the pixels that contain $\mathcal{T}$, that is, the union of pixels which intersections with $\mathcal{T}$ are non-empty. The diameter of each pixel is $\sqrt{2}/m = \sqrt{2/n}$, and its area is equal to $1/n$. That is why the number $N(\mathcal{T}_*)$ of the pixels in $\mathcal{T}_*$ is no less than $4n(H - \sqrt{2/n})(h_n^* - \sqrt{2/n})$, and the number $N(\mathcal{T}^*)$ of the pixels in $\mathcal{T}^*$ does not exceed $4n(H + \sqrt{2/n})(h_n^* + \sqrt{2/n})$. Since $1/\sqrt{n} = o(h_n^*)$, we find that

$$1 \leq \liminf_{n\to\infty} \frac{N(\mathcal{T}_*)}{4Hnh_n^*} \leq \limsup_{n\to\infty} \frac{N(\mathcal{T}^*)}{4Hnh_n^*} \leq 1.$$

Due to the inequalities

$$N(\mathcal{T}_*) \leq N(\mathcal{T}) \leq N(\mathcal{T}^*),$$

we can apply the squeezing theorem to conclude that the variable $N(\mathcal{T})$ also has the same limit,

$$\lim_{n\to\infty} \frac{N(\mathcal{T})}{4Hnh_n^*} = 1.$$

(ii) As $n \to \infty$, for any $k,\, l = 0, \dots, \beta - 1$, we have that

$$\frac{1}{N(\mathcal{T})}\big(\mathbf{G}'\mathbf{G}\big)_{k,l} \sim \frac{1}{4Hnh_n^*} \sum_{(i/m,\, j/m)\in\mathcal{T}} \Big(\frac{\hat{t}_{ij} - \hat{t}_0}{h_n^*} \Big)^{k+l}$$

$$\sim \frac{1}{4Hh_n^*} \int_{-H}^{H} \int_{\hat{t}_0 - h_n^*}^{\hat{t}_0 + h_n^*} \Big(\frac{t - \hat{t}_0}{h_n^*} \Big)^{k+l} dt\, du = \frac{1}{2} \int_{-1}^{1} z^{k+l}\, dz = \frac{1 - (-1)^{k+l+1}}{2(k+l+1)}.$$

The respective limiting matrix is invertible (see Exercise 9.66). $\square$

Exercises

Exercise 14.92. Prove that the number of monomials of degree i in d-variables is equal to $\binom{i+d-1}{i}$.

Exercise 14.93. Show that in the additive regression model with the intercept (14.8), the preliminary estimator (14.9) and the shifted observations $y_{ij} - \hat{f}_0$ allow us to prove the one-dimensional rate of convergence of the nonparametric components f_1 and f_2 as in Proposition 14.5.

Exercise 14.94. Let β_1 and β_2 be two positive integers, $\beta_1 \neq \beta_2$. A function in two variables $f(\mathbf{x})$, $\mathbf{x} = (x^{(1)}, x^{(2)})$, is said to belong to the *anisotropic Hölder class of functions* $\Theta(\beta_1, \beta_2) = \Theta(\beta_1, \beta_2, L, L_1)$, if f is bounded by L_1, and if for any point $\mathbf{x}_0 = (x_0^{(1)}, x_0^{(2)})$, there exists a polynomial $p(\mathbf{x}) = p(\mathbf{x}, \mathbf{x}_0, f)$ such that

$$|f(\mathbf{x}) - p(\mathbf{x}, \mathbf{x}_0, f)| \leq L\Big(|x^{(1)} - x_0^{(1)}|^{\beta_1} + |x^{(2)} - x_0^{(2)}|^{\beta_2} \Big)$$

where we assume that $\mathbf{x}$ and $\mathbf{x}_0$ belong to the unit square.

Suppose we want to estimate the value of the regression function f at a given design knot $(i_0/m,\, j_0/m)$ from the observations

$$y_{ij} = f(i/m,\, j/m) + \varepsilon_{ij}, \quad i,\, j = 1, \dots, m, \ m = \sqrt{n}.$$

Show that if the regression function belongs to the anisotropic class $\Theta(\beta_1, \beta_2)$, then there exists an estimator with the convergence rate $n^{-\tilde{\beta}/(2\tilde{\beta}+1)}$ where $\tilde{\beta}^{-1} = \beta_1^{-1} + \beta_2^{-1}$. Hint: Consider a local polynomial estimator in the bin with the sides h_1 and h_2 satisfying $h_1^{\beta_1} = h_2^{\beta_2}$. Show that the bias is $O(h_1^{\beta_1})$, and the variance is $O\big((nh_1h_2)^{-1} \big)$. Now use the balance equation to find the rate of convergence.

Chapter 15

Adaptive Estimation

In Chapters 8-11, we studied a variety of nonparametric regression estimation problems and found the minimax rates of convergence for different loss functions. These rates of convergence depend essentially on the parameter of smoothness β. This parameter determines the choice of the optimal bandwidth. An estimator which is minimax optimal for one smoothness does not preserve this property for another smoothness. The problem of *adaptation* consists of finding,if possible, an *adaptive estimator* which is independent of a particular β and is simultaneously minimax optimal over different nonparametric classes.

In this chapter, we will give examples of problems where the adaptive estimators exist in the sense that over each class of smoothness, the regression function can be estimated as if the smoothness parameter were known. We start, however, with a counterexample of an estimation problem in which the minimax rates are not attainable.

15.1. Adaptive Rate at a Point. Lower Bound

Consider regression observations on $[0, 1]$,

$$y_i = f(i/n) + \varepsilon_i, \quad i = 1, \dots, n,$$

where ε_i are standard normal random variables. Since the design is not the focus of our current interest, we work with the simplest equidistant design.

We assume that the smoothness parameter can take on only two values β_1 and β_2 such that $1 \leq \beta_1 < \beta_2$. Thus, we assume that the regression function f belongs to one of the two Hölder classes, either $\Theta(\beta_1) = \Theta(\beta_1, L, L_1)$ or $\Theta(\beta_2) = \Theta(\beta_2, L, L_1)$.

Let $x_0 = i_0/n$ be a given point in $(0, 1)$ which coincides with a design knot. We want to estimate $f(x_0)$ by a single estimator $\tilde{f}_n$ with the property that if $f \in \Theta(\beta_1)$, then the rate of convergence is $n^{-\beta_1/(2\beta_1+1)}$, while if $f \in \Theta(\beta_2)$, then the rate of convergence is $n^{-\beta_2/(2\beta_2+1)}$. Whether the true value of the smoothness parameter is β_1 or β_2 is unknown. The estimator $\tilde{f}_n$ may depend on both β_1 and β_2 but the knowledge of the true β cannot be assumed.

Formally, we introduce an *adaptive risk* of an estimator $\tilde{f}_n$ by

$$AR(\tilde{f}_n) = \max_{\beta \in \{\beta_1, \beta_2\}} \sup_{f \in \Theta(\beta)} \mathbb{E}_f\Big[n^{2\beta/(2\beta+1)} \big(\tilde{f}_n - f(x_0) \big)^2 \Big]. \tag{15.1}$$

The question we want to answer is whether there exists an estimator $\tilde{f}_n$ such that $AR(\tilde{f}_n) \leq r^*$ for some constant $r^* < \infty$ independent of n. The objective of this section is to demonstrate that such an estimator does not exist.

First, we define a class of estimators. For two given constants, $A > 0$ and a such that $\beta_1/(2\beta_1 + 1) < a \leq \beta_2/(2\beta_2 + 1)$, we introduce a class of estimators that are minimax optimal or sub-optimal over the regression functions of the higher smoothness,

$$\mathcal{F} = \mathcal{F}(A, a) = \Big\{ \tilde{f}_n : \sup_{f \in \Theta(\beta_2)} \mathbb{E}_f\Big[n^{2a} \big(\tilde{f}_n - f(x_0) \big)^2 \Big] \leq A \Big\}.$$

As the following proposition claims, the estimators that belong to $\mathcal{F}$ cannot attain the minimax rate of convergence on the lesser smoothness of regression functions.

Proposition 15.1. *There exists a constant $r_* = r_*(A, a)$ independent of n such that for any estimator $\tilde{f}_n \in \mathcal{F}(A, a)$, the following lower bound holds:*

$$\sup_{f \in \Theta(\beta_1)} \mathbb{E}_f\Big[\big(n/\ln n \big)^{2\beta_1/(2\beta_1+1)} \big(\tilde{f}_n - f(x_0) \big)^2 \Big] \geq r_* > 0. \tag{15.2}$$

Two important corollaries follow immediately from this result.

Corollary 15.2. *The adaptive risk $AR(\tilde{f}_n)$ in (15.1) is unbounded for any estimator $\tilde{f}_n$. Indeed, take $a = \beta_2/(2\beta_2 + 1)$ in the definition of the class $\mathcal{F}(A, a)$. Then we have that*

$$\sup_{f \in \Theta(\beta_2)} \mathbb{E}_f\Big[n^{2\beta_2/(2\beta_2+1)} \big(\tilde{f}_n - f(x_0) \big)^2 \Big] \leq A.$$

From Proposition 15.1, however, for all large n,

$$\sup_{f \in \Theta(\beta_1)} \mathbb{E}_f\Big[n^{2\beta_1/(2\beta_1+1)} \big(\tilde{f}_n - f(x_0) \big)^2 \Big] \geq r_* \big(\ln n \big)^{2\beta_1/(2\beta_1+1)}$$

with the right-hand side growing unboundedly as $n \to \infty$. Thus, the adaptive risk, being the maximum of the two supremums, is unbounded as well.

Corollary 15.3. *The contrapositive statement of Proposition 15.1 is valid. It can be formulated as follows. Assume that there exists an estimator $\tilde{f}_n$ that guarantees the minimax rate over the Hölder class of the lesser smoothness,*

$$\sup_{f\in\Theta(\beta_1)} \mathbb{E}_f\Big[n^{2\beta_1/(2\beta_1+1)}\big(\tilde{f}_n - f(x_0)\big)^2\Big] \le r^*$$

with a constant r^* independent of n. Then this estimator does not belong to $\mathcal{F}(A,a)$ for any a and A. As a consequence, from the definition of $\mathcal{F}(A,a)$ with $a = \beta_2/(2\beta_2+1)$, we find that

$$\sup_{f\in\Theta(\beta_2)} \mathbb{E}_f\Big[n^{2\beta_2/(2\beta_2+1)}\big(\tilde{f}_n - f(x_0)\big)^2\Big] \to \infty \text{ as } n\to\infty.$$

As Corollaries 15.2 and 15.3 explain, there is no adaptive estimator of a regression at a point. By this we mean that we cannot estimate a regression function at a point as if its smoothness were known.

Define a sequence

$$\psi_n = \psi_n(f) = \begin{cases} \big(n/(\ln n)\big)^{-\beta_1/(2\beta_1+1)} & \text{if } f\in\Theta(\beta_1),\\ n^{-\beta_2/(2\beta_2+1)} & \text{if } f\in\Theta(\beta_2).\end{cases} \tag{15.3}$$

The next question we ask about the adaptive estimation of $f(x_0)$ is whether it can be estimated with the rate $\psi_n(f)$. The answer to this question is positive. We leave it as an exercise (see Exercise 15.95).

The rest of this section is devoted to the proof of Proposition 15.1.

Proof of Proposition 15.1. Define two test functions $f_0(x) = 0$ and

$$f_1(x) = h_n^{\beta_1}\varphi\Big(\frac{x-x_0}{h_n}\Big) \text{ with } h_n = \Big(\frac{c\ln n}{n}\Big)^{1/(2\beta_1+1)}.$$

The choice of constant c will be made below. This definition is explained in detail in the proof of Theorem 9.16. In particular, $f_1\in\Theta(\beta_1,L,L_1)$ for some constants L and L_1.

Choose a constant a_0 such that $\beta_1/(2\beta_1+1) < a_0 < a$. Define a sequence

$$u_n = n^{2a_0}h_n^{2\beta_1} = n^{2a_0}\Big(\frac{c\ln n}{n}\Big)^{2\beta_1/(2\beta_1+1)}$$

$$= \big(c\ln n\big)^{2\beta_1/(2\beta_1+1)}\, n^{2[a_0-\beta_1/(2\beta_1+1)]} \ge n^{2[a_0-\beta_1/(2\beta_1+1)]} \tag{15.4}$$

for any fixed c and all large enough n, so that $u_n\to\infty$ at the power rate as $n\to\infty$. Take an arbitrarily small δ such that $\delta < \varphi^2(0)/4$. Note that if $\tilde{f}_n\in\mathcal{F}(A,a)$, then for all sufficiently large n, we have

$$u_n\mathbb{E}_{f_0}\big[h_n^{-2\beta_1}\tilde{f}_n^2\big] = u_n\mathbb{E}_{f_0}\big[u_n^{-1}n^{2a_0}\tilde{f}_n^2\big]$$

$$= n^{-2(a-a_0)}\mathbb{E}_{f_0}\big[n^{2a}\tilde{f}_n^2\big] \le n^{-2(a-a_0)}A < \delta.$$

Thus, $u_n \mathbb{E}_{f_0}\big[\, h_n^{-2\beta_1} \tilde{f}_n^2 \,\big] - \delta < 0$. Put $T_n = h_n^{-\beta_1} \tilde{f}_n$. We obtain

$$\sup_{f \in \Theta(\beta_1)} \mathbb{E}_f \Big[\, h_n^{-2\beta_1} \big(\tilde{f}_n - f(x_0) \big)^2 \Big] \geq \mathbb{E}_{f_1} \Big[\, h_n^{-2\beta_1} \big(\tilde{f}_n - f_1(x_0) \big)^2 \Big]$$

$$\geq \mathbb{E}_{f_1} \Big[\, h_n^{-2\beta_1} \big(\tilde{f}_n - f_1(x_0) \big)^2 \Big] + u_n \mathbb{E}_{f_0} \big[\, h_n^{-2\beta_1} \tilde{f}_n^2 \,\big] - \delta$$

$$= \mathbb{E}_{f_1} \big[\, (T_n - \varphi(0))^2 \,\big] + u_n \mathbb{E}_{f_0} \big[\, T_n^2 \,\big] - \delta.$$

Finally, we want to show that the right-hand side is separated away from zero by a positive constant independent of n. Introduce the likelihood ratio

$$\Lambda_n = \frac{d\mathbb{P}_{f_0}}{d\mathbb{P}_{f_1}} = \exp\Big\{ - \sum_{i=1}^{n} f_1(i/n)\xi_i - \frac{1}{2} \sum_{i=1}^{n} f_1^2(i/n) \Big\}$$

where $\xi_i = y_i - f_1(i/n)$, $i = 1, \dots, n$, are independent standard normal random variables with respect to $\mathbb{P}_{f_1}$-distribution. As in the proof of Theorem 9.16, we get

$$\sigma_n^2 = \sum_{i=1}^{n} f_1^2(i/n) = \|\varphi\|_2^2 \, n \, h_n^{2\beta_1+1} \big(1 + o_n(1)\big) = \|\varphi\|_2^2 \, (c \ln n) \big(1 + o_n(1)\big)$$

where $o_n(1) \to 0$ as $n \to \infty$. Introduce a standard normal random variable $Z_n = \sigma_n^{-1} \sum_{i=1}^{n} f_1(i/n)\, \xi_i$. Then the likelihood ratio takes the form

$$\Lambda_n = \exp\Big\{ - \sigma_n Z_n - \frac{1}{2} \|\varphi\|_2^2 \, (c \ln n) \big(1 + o_n(1)\big) \Big\}.$$

Note that if the random event $\{\, Z_n \leq 0 \,\}$ holds, then

$$\Lambda_n \geq \exp\Big\{ - \frac{1}{2} \|\varphi\|_2^2 \, (c \ln n) \big(1 + o_n(1)\big) \Big\} \geq n^{-c_1},$$

for all large n, where $c_1 = c \, \|\varphi\|_2^2$. From the definition of the likelihood ratio, we obtain the lower bound

$$\sup_{f \in \Theta(\beta_1)} \mathbb{E}_f \Big[\, h_n^{-2\beta_1} \big(\tilde{f}_n - f(x_0) \big)^2 \Big] \geq \mathbb{E}_{f_1} \Big[\, (T_n - \varphi(0))^2 + u_n \, \Lambda_n \, T_n^2 \Big] - \delta$$

$$\geq \mathbb{E}_{f_1} \Big[\, (T_n - \varphi(0))^2 + u_n \, n^{-c_1} \, T_n^2 \, \mathbb{I}(Z_n \leq 0) \Big] - \delta.$$

Now we choose c so small that $c_1 = c \, \|\varphi\|_2^2 < 2 \big(a_0 - \beta_1/(2\beta_1 + 1) \big)$. Then, by (15.4), $u_n \, n^{-c_1}$ increases and exceeds 1 if n is sufficiently large. Hence,

$$\sup_{f \in \Theta(\beta_1)} \mathbb{E}_f \Big[\, h_n^{-2\beta_1} \big(\tilde{f}_n - f(x_0) \big)^2 \Big] \geq \mathbb{E}_{f_1} \Big[\, (T_n - \varphi(0))^2 + T_n^2 \, \mathbb{I}(Z_n \leq 0) \Big] - \delta$$

$$\geq \mathbb{E}_{f_1} \Big[\, \mathbb{I}(Z_n \leq 0) \Big((T_n - \varphi(0))^2 + T_n^2 \Big) \Big] - \delta$$

$$\geq \mathbb{E}_{f_1} \Big[\, \mathbb{I}(Z_n \leq 0) \Big((-\varphi(0)/2)^2 + (\varphi(0)/2)^2 \Big) \Big] - \delta$$

$$\geq \frac{1}{2} \varphi^2(0) \, \mathbb{P}_{f_1} \big(Z_n \leq 0 \big) - \delta = \frac{1}{4} \varphi^2(0) - \delta = r_*$$

where r_* is strictly positive because, under our choice, $\delta < \varphi^2(0)/4$. $\square$

15.2. Adaptive Estimator in the Sup-Norm

In this section we present a result on adaptive estimation in the sup-norm. We will show that for the sup-norm losses, the adaptation is possible in the straightforward sense: the minimax rates are attainable as if the smoothness parameter were known.

First, we modify the definition (15.1) of the adaptive risk to reflect the sup-norm loss function,

$$AR_\infty(\tilde{f}_n) = \max_{\beta \in \{\beta_1, \beta_2\}} \sup_{f \in \Theta(\beta)} \Big(\frac{n}{\ln n} \Big)^{\beta/(2\beta+1)} \mathbb{E}_f \big[\, \| \tilde{f}_n - f \, \|_\infty \big]. \tag{15.5}$$

Since the log-factor is intrinsic to the sup-norm rates of convergence, there is no need to prove the lower bound. All we have to do is to define an adaptive estimator. As in the previous section, we proceed with the equidistant design and the standard normal errors in the regression model,

$$y_i = f(i/n) + \varepsilon_i, \ \ \varepsilon_i \stackrel{iid}{\sim} \mathcal{N}(0,1), \ \ i = 1, \dots, n.$$

Define f^*_{n,β_1} and f^*_{n,β_2} as the rate-optimal estimators in the sup-norm over the classes $\Theta(\beta_1)$ and $\Theta(\beta_2)$, respectively. Each estimator is based on the regressogram with the bandwidth $h^*_{n,\beta} = \big((\ln n)/n \big)^{1/(2\beta+1)}$, $\beta \in \{\beta_1, \beta_2\}$ (see Section 10.3.)

Now we introduce an adaptive estimator,

$$\tilde{f}_n = \begin{cases} f^*_{n,\beta_1} & \text{if } \| f^*_{n,\beta_1} - f^*_{n,\beta_2} \|_\infty \geq C \, \big(h^*_{n,\beta_1} \big)^{\beta_1}, \\ f^*_{n,\beta_2} & \text{otherwise,} \end{cases} \tag{15.6}$$

where C is a sufficiently large constant which depends only on β_1 and β_2. The final choice of C is made below.

Our starting point is the inequality (10.10). Since the notations of the current section are a little bit different, we rewrite this inequality in the form convenient for reference,

$$\| \, f^*_{n,\beta} - f \, \|_\infty \leq A_b \, (h^*_{n,\beta})^\beta + A_v \, \big(n \, h^*_{n,\beta} \big)^{-1/2} \, \mathcal{Z}^*_\beta, \ \ \beta \in \{\beta_1, \beta_2\}, \tag{15.7}$$

where A_b and A_v are constants independent of n. Using the notation defined in Section 10.3, we show that the distribution of $\mathcal{Z}^*_\beta$ has fast-decreasing tail probabilities,

$$\mathbb{P}\big(\, \mathcal{Z}^*_\beta \geq y \, \sqrt{2\beta^2 \ln n} \, \big) \leq \mathbb{P}\Big(\bigcup_{q=1}^{Q} \bigcup_{m=,0}^{\beta-1} |Z_{m,q}| \geq y\sqrt{2 \ln n} \Big)$$

$$\leq Q\beta \mathbb{P}\big(\, |Z| \geq y \, \sqrt{2 \ln n} \, \big)$$

where $Z \sim \mathcal{N}(0,1)$. Now, since $\mathbb{P}(|Z| \geq x) \leq \exp\{-x^2/2\}$ for any $x \geq 1$, we arrive at the upper bound

$$\mathbb{P}\big(\mathcal{Z}_\beta^* \geq y \sqrt{2\beta^2 \ln n} \big) \leq Q\beta n^{-y^2} \leq \beta n^{-(y^2-1)}. \tag{15.8}$$

Here we have used the fact that the number of bins $Q = 1/(2h_{n,\beta}^*) \leq n$ for all large enough n.

Theorem 15.4. *There exists a constant C in the definition (15.6) of the adaptive estimator $\tilde{f}_n$ such that the adaptive risk $AR_\infty(\tilde{f}_n)$ specified by (15.5) satisfies the upper bound*

$$AR_\infty(\tilde{f}_n) \leq r^* \tag{15.9}$$

with a constant r^* independent of n.

Proof. Denote the random event in the definition of the adaptive estimator $\tilde{f}_n$ by

$$\mathcal{C} = \Big\{ \| f_{n,\beta_1}^* - f_{n,\beta_2}^* \|_\infty \geq C \big(h_{n,\beta_1}^* \big)^{\beta_1} \Big\}.$$

If $f \in \Theta(\beta_1)$, then

$$\begin{aligned} &\big(n/(\ln n) \big)^{\beta_1/(2\beta_1+1)} \mathbb{E}_f \big[\| \tilde{f}_n - f \|_\infty \big] \\ &\leq \big(h_{n,\beta_1}^* \big)^{-\beta_1} \mathbb{E}_f \big[\| f_{n,\beta_1}^* - f \|_\infty \mathbb{I}(\mathcal{C}) \big] + \big(h_{n,\beta_1}^* \big)^{-\beta_1} \mathbb{E}_f \Big[\| f_{n,\beta_2}^* - f \|_\infty \mathbb{I}(\overline{\mathcal{C}}) \Big] \end{aligned} \tag{15.10}$$

where $\overline{\mathcal{C}}$ is the complementary random event to $\mathcal{C}$. The first term on the right-hand side of (15.10) is bounded from above uniformly in $f \in \Theta(\beta_1)$ because f_{n,β_1}^* is the minimax rate-optimal estimator over this class. If the complementary random event $\overline{\mathcal{C}}$ holds, then by the triangle inequality, the second term does not exceed

$$\big(h_{n,\beta_1}^* \big)^{-\beta_1} \Big[\mathbb{E}_f \big[\| f_{n,\beta_1}^* - f \|_\infty \big] + C \big(h_{n,\beta_1}^* \big)^{\beta_1} \Big] \tag{15.11}$$

which is also bounded from above by a constant.

Next, we turn to the case $f \in \Theta(\beta_2)$. As above,

$$\Big(n/(\ln n) \Big)^{\beta_2/(2\beta_2+1)} \mathbb{E}_f \big[\| \tilde{f}_n - f \|_\infty \big]$$

$$\leq \big(h_{n,\beta_2}^* \big)^{-\beta_2} \mathbb{E}_f \big[\| f_{n,\beta_2}^* - f \|_\infty \mathbb{I}(\overline{\mathcal{C}}) \big] + \big(h_{n,\beta_2}^* \big)^{-\beta_2} \mathbb{E}_f \Big[\| f_{n,\beta_1}^* - f \|_\infty \mathbb{I}(\mathcal{C}) \Big].$$

Once again, it suffices to study the case when the estimator does not match the true class of functions,

$$\begin{aligned} &\big(h_{n,\beta_2}^* \big)^{-\beta_2} \mathbb{E}_f \Big[\| f_{n,\beta_1}^* - f \|_\infty \mathbb{I}(\mathcal{C}) \Big] \\ &\leq v_n \Big[\big(h_{n,\beta_1}^* \big)^{-2\beta_1} \mathbb{E}_f \big[\| f_{n,\beta_1}^* - f \|_\infty^2 \big] \Big]^{1/2} \Big[\mathbb{P}_f \big(\mathcal{C} \big) \Big]^{1/2} \end{aligned} \tag{15.12}$$

where the Cauchy-Schwarz inequality was applied. The deterministic sequence v_n is defined by

$$v_n = \big(h^*_{n,\beta_2}\big)^{-\beta_2}\big(h^*_{n,\beta_1}\big)^{\beta_1} = \Big(\frac{n}{\ln n}\Big)^{\gamma}, \quad \gamma = \frac{\beta_2}{2\beta_2+1} - \frac{\beta_1}{2\beta_1+1} > 0.$$

The normalized expected value on the right-hand side of (15.12) is bounded from above uniformly over $f \in \Theta(\beta_2)$. Indeed, over a smoother class of functions $\Theta(\beta_2)$, a coarser estimator f^*_{n,β_1} preserves its slower rate of convergence. Formally, this bound does not follow from Theorem 10.6 because of the squared sup-norm which is not covered by this theorem. However, the proper extension is elementary if we use (15.7) directly (see Exercise 15.96.)

Hence, it remains to show that the probability $\mathbb{P}_f(\mathcal{C})$ in (15.12) vanishes fast enough to compensate the growth of v_n. From the definition of the random event $\mathcal{C}$ and the triangle inequality, we have

$$\mathcal{C} \subseteq \Big\{\, \|f^*_{n,\beta_1} - f\|_\infty \geq \frac{1}{2}\, C\,\big(h^*_{n,\beta_1}\big)^{\beta_1} \Big\} \cup \Big\{\, \|f^*_{n,\beta_2} - f\|_\infty \geq \frac{1}{2}\, C\,\big(h^*_{n,\beta_1}\big)^{\beta_1} \Big\}.$$

Now, note that the bias terms in (15.7) are relatively small,

$$A_b\,(h^*_{n,\beta_2})^{\beta_2} < A_b\,(h^*_{n,\beta_1})^{\beta_1} < \frac{1}{4}\, C\,(h^*_{n,\beta_1})^{\beta_1}$$

if the constant C in the definition of the adaptive estimator $\tilde{f}_n$ satisfies the condition $C > 4A_b$. Under this choice of C, the random event $\mathcal{C}$ may occur only due to the large deviations of the stochastic terms. It implies that $\mathcal{C} \subseteq \mathcal{A}_1 \cup \mathcal{A}_2$ with

$$\mathcal{A}_1 = \Big\{\, A_v\,\big(n\, h^*_{n,\beta_1}\big)^{-1/2}\, \mathcal{Z}^*_{\beta_1} \geq \frac{1}{4}\, C\,(h^*_{n,\beta_1})^{\beta_1} \Big\} = \Big\{\, \mathcal{Z}^*_{\beta_1} \geq \frac{C}{4\,A_v}\,\sqrt{\ln n}\,\Big\}$$

and

$$\mathcal{A}_2 = \Big\{\, A_v\,\big(n\, h^*_{n,\beta_2}\big)^{-1/2}\, \mathcal{Z}^*_{\beta_2} \geq \frac{1}{4}\, C\,(h^*_{n,\beta_1})^{\beta_1} \Big\}$$

$$\subseteq \Big\{\, A_v\,\big(n\, h^*_{n,\beta_1}\big)^{-1/2}\, \mathcal{Z}^*_{\beta_2} \geq \frac{1}{4}\, C\,(h^*_{n,\beta_1})^{\beta_1} \Big\} = \Big\{\, \mathcal{Z}^*_{\beta_2} \geq \frac{C}{4\,A_v}\,\sqrt{\ln n}\,\Big\}.$$

From the inequality (15.8), it follows that for a large C, the probabilities of the random events $\mathcal{A}_1$ and $\mathcal{A}_2$ decrease faster than any power of n. Choosing $C > 4\,A_v\,\sqrt{2\beta_2^2(1+2\gamma)}$, we can guarantee that the right-hand side of (15.12) vanishes as $n \to \infty$. □

Remark 15.5. The definition of the adaptive estimator $\tilde{f}_n$ and the proof of Proposition 15.4 contain a few ideas common to selection of adaptive estimators in different nonparametric models. First, we choose an estimator from minimax optimal estimators over each class of functions. Second, we focus on the performance of the chosen adaptive estimator over the alien class, provided the choice has been made incorrectly. Third, we use the fact that this performance is always controlled by the probabilities of large deviations that vanish faster than their normalization factors that are growing at a power rate. □

15.3. Adaptation in the Sequence Space

Another relatively less technical example of adaptation concerns the adaptive estimation problem in the sequence space. Recall that the *sequence space*, as defined in Section 10.5, is the n-dimensional space of the Fourier coefficients of regression function. We assume that each regression function $f(x)$, $0 \leq x \leq 1$, is observed at the equidistant design points $x = i/n$. This function is defined in terms of its Fourier coefficients c_k and the basis functions φ_k by the formula

$$f(i/n) = \sum_{k=0}^{n-1} c_k \varphi_k(i/n), \quad i = 1, \dots, n.$$

The transition from the original observations of the regression function f to the sequence space is explained in Lemma 10.16 (see formula (10.31)).

To ease the presentation, we will consider the following model of observations directly in the sequence space,

$$z_k = c_k + \xi_k/\sqrt{n} \text{ and } \tilde{z}_k = c_k + \tilde{\xi}_k/\sqrt{n}, \quad k = 0, \dots, n-1, \tag{15.13}$$

where ξ_k and $\tilde{\xi}_k$ are independent standard normal random variables. That is, we assume that each observation of the Fourier coefficient c_k is observed twice and the repeated observations are independent.

By Lemma 10.15, for any estimator $\hat{\mathbf{c}} = (\hat{c}_0, \dots, \hat{c}_{n-1})$ of the Fourier coefficients $\mathbf{c} = (c_0, \dots, c_{n-1})$, the quadratic risk $R_n(\hat{\mathbf{c}}, \mathbf{c})$ in the sequence space is equivalent to the quadratic risk of regression, that is,

$$R_n(\hat{\mathbf{c}}, \mathbf{c}) = \mathbb{E}_{\mathbf{c}}\Big[\sum_{k=0}^{n-1} (\hat{c}_k - c_k)^2 \Big] = \mathbb{E}_f\Big[\| \sum_{k=0}^{n-1} \hat{c}_k \varphi_k - f \|_{2,n}^2 \Big] \tag{15.14}$$

where $\mathbb{E}_{\mathbf{c}}$ refers to the expectation for the true Fourier coefficients $\mathbf{c} = (c_0, \dots, c_{n-1})$, and the discrete L_2-norm is defined as

$$\|f\|_{2,n}^2 = n^{-1} \sum_{i=1}^{n} f^2(i/n).$$

Next, we take two integers β_1 and β_2 such that $1 \leq \beta_1 < \beta_2$, and consider two sets in the sequence space

$$\Theta_{2,n}(\beta) = \Theta_{2,n}(\beta, L) = \Big\{ (c_0, \dots, c_{n-1}) : \sum_{k=0}^{n-1} c_k^2 \, k^{2\beta} \leq L \Big\}, \quad \beta \in \{ \beta_1, \beta_2 \}.$$

We associate $\Theta_{2,n}(\beta)$ with the smoothness parameter β because the decrease rate of Fourier coefficients controls the smoothness of the original regression function (cf. Lemma 10.13.)

As shown in Theorem 10.17, for a known β, uniformly in $\mathbf{c} \in \Theta_{2,n}(\beta)$, the risk $R_n(\hat{\mathbf{c}}, \mathbf{c}) = O(n^{-2\beta/(2\beta+1)})$ as $n \to \infty$. The rate-optimal estimator is the projection estimator which can be defined as

$$\hat{\mathbf{c}} = \big(z_0, \dots, z_M, 0, \dots, 0 \big)$$

where $M = M_\beta = n^{1/(2\beta+1)}$. In other words, $\hat{c}_k = z_k$ if $k = 0, \dots, M$, and $\hat{c}_k = 0$ for $k \geq M+1$.

Now, suppose that we do not know the true smoothness of the regression function, or, equivalently, suppose that the true Fourier coefficients may belong to either class, $\Theta_{2,n}(\beta_1)$ or $\Theta_{2,n}(\beta_2)$. Can we estimate these coefficients so that the optimal rate would be preserved over either class of smoothness? To make this statement more precise, we redefine the adaptive risk for sequence space. For any estimator $\tilde{\mathbf{c}} = (\tilde{c}_0, \dots, \tilde{c}_{n-1})$ introduce the adaptive quadratic risk by

$$AR(\tilde{\mathbf{c}}) = \max_{\beta \in \{\beta_1, \beta_2\}} \sup_{\mathbf{c} \in \Theta_{2,\, n}(\beta)} (M_\beta)^{2\beta}\, \mathbb{E}_{\mathbf{c}}\Big[\sum_{k=0}^{n-1} (\tilde{c}_k - c_k)^2 \Big] \tag{15.15}$$

where $M_\beta = n^{1/(2\beta+1)}$. The objective is to find an adaptive estimator $\tilde{\mathbf{c}}$ that keeps the risk $AR(\tilde{\mathbf{c}})$ bounded from above by a constant independent of n. To this end, introduce two estimators, each optimal over its own class,

$$\hat{\mathbf{c}}_\beta = \big(\hat{c}_{0,\,\beta}, \dots, \hat{c}_{n-1,\,\beta} \big) = \big(z_0, \dots, z_{M_\beta}, 0, \dots, 0 \big), \quad \beta \in \{\beta_1,\, \beta_2\}.$$

Further, define two statistics designed to mimic the quadratic risks of the respective estimators $\hat{\mathbf{c}}_\beta$,

$$R_\beta = \sum_{k=0}^{n-1} \big(\tilde{z}_k - \hat{c}_{k,\,\beta} \big)^2, \quad \beta \in \{\beta_1, \beta_2\}.$$

These statistics are based on the second set of the repeated observations $\tilde{z}_k$ in (15.13). From the definition of the quadratic risk (15.14), we have

$$\mathbb{E}_{\mathbf{c}}\big[R_\beta \big] = \mathbb{E}_{\mathbf{c}}\Big[\sum_{k=0}^{n-1} \big(c_k + \tilde{\xi}_k/\sqrt{n} - \hat{c}_{k,\,\beta} \big)^2 \Big] = R_n(\hat{\mathbf{c}}_\beta,\, \mathbf{c}) + 1.$$

Next, we give a natural definition of the adaptive estimator in our setting. The adaptive estimator is the estimator $\hat{\mathbf{c}}_\beta$ that minimizes the risk R_β, that is,

$$\tilde{\mathbf{c}} = \begin{cases} \hat{\mathbf{c}}_{\beta_1} & \text{if } R_{\beta_1} \leq R_{\beta_2}, \\ \hat{\mathbf{c}}_{\beta_2} & \text{if } R_{\beta_1} > R_{\beta_2}. \end{cases}$$

We give the proof of the following theorem at the end of the present section after we formulate some important auxiliary results.

Theorem 15.6. *There exists a constant r^* independent of n and such that the adaptive risk* (15.15) *is bounded from above,*

$$AR(\tilde{\mathbf{c}}) \leq r^*.$$

We have to emphasize that Remark 15.5 stays valid in this case as well. We have to understand the performance of the adaptive estimator if the correct selection fails. As we will show, this performance is governed by the large deviations probabilities of the stochastic terms. Before we prove the theorem, let us analyze the structure of the difference $\Delta R = R_{\beta_1} - R_{\beta_2}$ that controls the choice of the adaptive estimator. Put

$$\mathcal{M} = \big\{k : \ M_{\beta_2} + 1 \leq k \leq M_{\beta_1}\big\} \text{ and } \Delta M = M_{\beta_1} - M_{\beta_2} = M_{\beta_1}\big(1 + o_n(1)\big).$$

The following technical lemmas are proved in the next section.

Lemma 15.7. *The difference of the risk estimates satisfies the equation*

$$\Delta R = R_{\beta_1} - R_{\beta_2} = -S_n + M_{\beta_1}\big(1 + o_n(1)\big)/n + U_n^{(1)}/n - 2U_n^{(2)}/\sqrt{n}$$

with $S_n = S_n(\mathbf{c}) = \sum_{k\in\mathcal{M}} c_k^2$, and the random variables

$$U_n^{(1)} = \sum_{k\in\mathcal{M}} \big(\,\xi_k^2 - 1\,\big) \quad \text{and} \quad U_n^{(2)} = \sum_{k\in\mathcal{M}} z_k \tilde{\xi}_k.$$

The following random events help to control the adaptive risk

$$\mathcal{A}_1 = \big\{\,U_n^{(1)} \geq M_{\beta_1}\,\big\},\ \ \mathcal{A}_2 = \big\{U_n^{(2)} \leq -\sqrt{n}S_n/8\,\big\},$$

$$\mathcal{A}_3 = \big\{\,U_n^{(1)} \leq -M_{\beta_1}/4\,\big\}, \text{ and } \mathcal{A}_4 = \big\{\,U_n^{(2)} \geq M_{\beta_1}/(8\sqrt{n})\,\big\}.$$

Lemma 15.8. *(i) If the inequality $S_n > 4M_{\beta_1}/n$ holds, then*

$$\mathbb{P}_{\mathbf{c}}(\,\mathcal{A}_i\,) \leq \exp\big\{-\ A_i M_{\beta_1}\big\},\ \ i = 1 \text{ or } 2,$$

where A_1 and A_2 are positive constants independent of n.

(ii) If $S_n = o(M_{\beta_1}/n)$ as $n \to \infty$, then

$$\mathbb{P}_{\mathbf{c}}(\,\mathcal{A}_i\,) \leq \exp\big\{-\ A_i\, M_{\beta_1}\big\},\ \ i = 3 \text{ or } 4,$$

with some positive constants A_3 and A_4.

Proof of Theorem 15.6. As explained in the proof of Proposition 15.1, we have to understand what happens with the risk, if the adaptive estimator is chosen incorrectly, that is, if it does not coincide with the optimal estimator over the respective class. Let us start with the case $\mathbf{c} \in \Theta_{2,n}(\beta_1)$, while

$\tilde{\mathbf{c}} = \hat{\mathbf{c}}_{\beta_2}$. The contribution of this instance to the adaptive risk (15.15) is equal to

$$(M_{\beta_1})^{2\beta_1}\,\mathbb{E}_{\mathbf{c}}\Big[\,\mathbb{I}(\Delta R > 0)\sum_{k=0}^{n-1}\big(\,\hat{c}_{k,\,\beta_2} - c_k\,\big)^2\,\Big]$$

$$= (M_{\beta_1})^{2\beta_1}\,\mathbb{E}_{\mathbf{c}}\Big[\,\mathbb{I}(\Delta R > 0)\Big(\sum_{k=0}^{M_{\beta_2}}(z_k - c_k)^2 + \sum_{k=M_{\beta_2}+1}^{n-1} c_k^2\Big)\Big]$$

$$= (M_{\beta_1})^{2\beta_1}\,\mathbb{E}_{\mathbf{c}}\Big[\,\mathbb{I}(\Delta R > 0)\Big(\frac{1}{n}\sum_{k=0}^{M_{\beta_2}}\xi_k^2 + S_n + \sum_{k=M_{\beta_1}+1}^{n-1} c_k^2\Big)\Big]$$

where S_n is defined in Lemma 15.7. Note that

$$\mathbb{E}_{\mathbf{c}}\Big[\frac{1}{n}\sum_{k=0}^{M_{\beta_2}}\xi_k^2\Big] = \frac{M_{\beta_2}}{n} \ll \frac{M_{\beta_1}}{n} = \big(M_{\beta_1}\big)^{-2\beta_1},$$

and since $\mathbf{c} \in \Theta_{2,n}(\beta_1)$,

$$\sum_{k=M_{\beta_1}+1}^{n-1} c_k^2 \leq L(M_{\beta_1})^{-2\beta_1}.$$

Thus, even multiplied by $(M_{\beta_1})^{2\beta_1}$, the respective terms in the risk stay bounded as $n \to \infty$.

It remains to verify that the term $S_n\,(M_{\beta_1})^{2\beta_1}\,\mathbb{P}_{\mathbf{c}}(\Delta R > 0)$ also stays finite as $n \to \infty$. It suffices to study the case when $S_n > 4\,(M_{\beta_1})^{-2\beta_1} = 4\,M_{\beta_1}/n$, because otherwise this term would be bounded by 4. From Lemma 15.7,

$$\{\Delta R > 0\} = \big\{ - S_n + M_{\beta_1}\big(1 + o_n(1)\big)/n + U_n^{(1)}/n - 2\,U_n^{(2)}/\sqrt{n} > 0\,\big\}$$

can occur if at least one of the two random events $\mathcal{A}_1 = \big\{\,U_n^{(1)}/n \geq M_{\beta_1}/n\,\big\}$ or $\mathcal{A}_2 = \big\{ - 2\,U_n^{(2)}/\sqrt{n} \geq S_n/4\,\big\}$ occurs. Indeed, otherwise we would have the inequality

$$\Delta R < -(3/4)S_n + 2M_{\beta_1}\,(1 + o_n(1))/n < 0,$$

since by our assumption, $S_n > 4\,M_{\beta_1}/n$.

Lemma 15.8 part (i) claims that the probabilities of the random events $\mathcal{A}_1$ and $\mathcal{A}_2$ decrease faster than $\exp\big\{ - An^{1/(2\beta_1+1)}\big\}$ as $n \to \infty$, which implies that $S_n\,(M_{\beta_1})^{2\beta_1}\,\mathbb{P}_{\mathbf{c}}(\Delta R > 0)$ is finite.

The other case, when $\mathbf{c} \in \Theta_{2,\,n}(\beta_2)$ and $\tilde{\mathbf{c}} = \hat{\mathbf{c}}_{\beta_1}$, is treated in a similar fashion, though some calculations change. We write

$$(M_{\beta_2})^{2\beta_2}\,\mathbb{E}_{\mathbf{c}}\big[\,\mathbb{I}(\Delta R \leq 0)\sum_{k=0}^{n-1}(\hat{c}_{k,\,\beta_1} - c_k)^2\,\big]$$

$$(15.16) \qquad = (M_{\beta_2})^{2\beta_2}\,\mathbb{E}_{\mathbf{c}}\Big[\,\mathbb{I}(\Delta R \le 0)\Big(\frac{1}{n}\sum_{k=0}^{M_{\beta_1}} \xi_k^2 + \sum_{k=M_{\beta_1}+1}^{n-1} c_k^2\Big)\Big].$$

Since $\mathbf{c} \in \Theta_{2,n}(\beta_2)$,

$$(M_{\beta_2})^{2\beta_2} \sum_{k=M_{\beta_1}+1}^{n-1} c_k^2 \le L\,\big(M_{\beta_2}/M_{\beta_1}\big)^{2\beta_2} \to 0, \quad \text{as } n \to \infty.$$

It remains to verify that the first term in (15.16) is bounded. We obtain

$$(M_{\beta_2})^{2\beta_2}\,\mathbb{E}_{\mathbf{c}}\Big[\,\mathbb{I}(\Delta R \le 0)\,\Big(\frac{1}{n}\sum_{k=0}^{M_{\beta_1}} \xi_k^2\Big)\Big]$$

$$\le (M_{\beta_2})^{2\beta_2}\,\mathbb{E}_{\mathbf{c}}^{1/2}\Big[\Big(\frac{1}{n}\sum_{k=0}^{M_{\beta_1}} \xi_k^2\Big)^2\Big]\,\mathbb{P}_{\mathbf{c}}^{1/2}(\Delta R \le 0)$$

$$\le (M_{\beta_2})^{2\beta_2}\Big(\frac{2M_{\beta_1}}{n}\Big)\,\mathbb{P}_{\mathbf{c}}^{1/2}(\Delta R \le 0) \;=\; 2\,n^{\gamma}\,\mathbb{P}_{\mathbf{c}}^{1/2}(\Delta R \le 0).$$

Here we applied the Cauchy-Schwartz inequality, and the elementary calculations $\mathbb{E}_{\mathbf{c}}\big[\xi_k^4\big] = 3$, hence,

$$\mathbb{E}_{\mathbf{c}}\Big[\Big(\sum_{k=0}^{M_{\beta_1}} \xi_k^2\Big)^2\Big] = \sum_{k=0}^{M_{\beta_1}} \mathbb{E}_{\mathbf{c}}\big[\,\xi_k^4\,\big] + \sum_{\substack{k,\,l=0\\ k\ne l}}^{M_{\beta_1}} \mathbb{E}_{\mathbf{c}}\big[\,\xi_k^2\,\xi_l^2\,\big]$$

$$= 3\,M_{\beta_1} + M_{\beta_1}\,(M_{\beta_1} - 1) \le 4\,M_{\beta_1}^2.$$

The constant γ in the exponent above is equal to

$$\gamma = \frac{2\beta_2}{2\beta_2+1} + \frac{2}{2\beta_1+1} - 1 = \frac{2\beta_2}{2\beta_2+1} - \frac{2\beta_1}{2\beta_1+1} > 0.$$

If $\mathbf{c} \in \Theta_{2,n}(\beta_2)$, then $S_n \le L\,(M_{\beta_2})^{-2\beta_2} = L\,M_{\beta_2}/n = o\big(M_{\beta_1}/n\big)$ as $n \to \infty$. Note that the random event

$$\{\Delta R \le 0\} = \big\{ - S_n + M_{\beta_1}\,(1+o_n(1))/n + U_n^{(1)}/n - 2\,U_n^{(2)}/\sqrt{n} \le 0 \big\}$$

$$= \big\{ - U_n^{(1)}/n + 2\,U_n^{(2)}/\sqrt{n} \ge M_{\beta_1}\,(1+o_n(1))/n \big\}$$

occurs if either $\mathcal{A}_3 = \big\{ - U_n^{(1)} \ge M_{\beta_1}/4 \big\}$ or $\mathcal{A}_4 = \big\{ U_n^{(2)} \ge M_{\beta_1}/(8\sqrt{n}) \big\}$ occurs. Again, as Lemma 15.8 (ii) shows, the probabilities of these random events decrease faster than $n^{2\gamma}$, so that $n^{\gamma}\,\mathbb{P}_{\mathbf{c}}^{1/2}(\Delta R \le 0) \to 0$ as $n \to \infty$, and the statement of the theorem follows. □

15.4. Proofs of Lemmas

Proof of Lemma 15.7. By straightforward calculations, we obtain

$$\Delta R = R_{\beta_1} - R_{\beta_2} = \sum_{k \in \mathcal{M}} (\tilde{z}_k - z_k)^2 - \sum_{k \in \mathcal{M}} \tilde{z}_k^2$$

$$= \frac{1}{n} \sum_{k \in \mathcal{M}} \big(\tilde{\xi}_k - \xi_k\big)^2 - \Big[\sum_{k \in \mathcal{M}} c_k^2 + \frac{2}{\sqrt{n}} \sum_{k \in \mathcal{M}} c_k\, \tilde{\xi}_k + \frac{1}{n} \sum_{k \in \mathcal{M}} \tilde{\xi}_k^2 \Big]$$

$$= -\frac{2}{n} \sum_{k \in \mathcal{M}} \tilde{\xi}_k\, \xi_k + \frac{1}{n} \sum_{k \in \mathcal{M}} \xi_k^2 - \sum_{k \in \mathcal{M}} c_k^2 - \frac{2}{\sqrt{n}} \sum_{k \in \mathcal{M}} c_k\, \tilde{\xi}_k$$

$$= - \sum_{k \in \mathcal{M}} c_k^2 + \frac{1}{n} \Delta M + \frac{1}{n} \sum_{k \in \mathcal{M}} (\xi_k^2 - 1) - \frac{2}{\sqrt{n}} \sum_{k \in \mathcal{M}} \Big(c_k + \frac{\xi_k}{\sqrt{n}} \Big) \tilde{\xi}_k$$

where $\Delta M = M_{\beta_2} - M_{\beta_1} = M_{\beta_1}\big(1 + o_n(1)\big)$ is the number of elements in $\mathcal{M}$. So, the lemma follows. □

To prove Lemma 15.8 we need the following result.

Proposition 15.9. *The moment generating functions of the random variables* $U_n^{(1)}$ *and* $U_n^{(2)}$ *are respectively equal to*

$$G_1(t) = \mathbb{E}\Big[\exp\big\{t\, U_n^{(1)}\big\} \Big] = \exp\Big\{ - M_{\beta_1} \big(1 + o_n(1)\big) \big(t + (1/2) \ln(1-2t) \big) \Big\}$$

and

$$G_2(t) = \mathbb{E}\Big[\exp\big\{t\, U_n^{(2)}\big\} \Big] = \exp\Big\{ \frac{nt^2}{2(n-t^2)}\, S_n - \frac{\Delta M}{2} \ln(1 - t^2/n) \Big\}.$$

Proof. Note that $\mathbb{E}\big[\exp\{t\xi^2\}\big] = (1-2t)^{-1/2}$, $t \le 1/2$, where $\xi \sim \mathcal{N}(0,1)$. Therefore,

$$\mathbb{E}\big[\exp\{t\,(\xi^2 - 1)\} \big] = \exp\big\{ -t - (1/2) \ln(1-2t) \big\},$$

and the expression for $G_1(t)$ follows from independence of the random variables ξ_k^2.

Next, the moment generating function of $U_n^{(2)}$ can be expressed as

$$G_2(t) = \mathbb{E}\Big[\exp\big\{t\, U_n^{(2)}\big\} \Big] = \mathbb{E}\Big[\exp\Big\{t \sum_{k \in \mathcal{M}} \Big(c_k + \frac{\xi_k}{\sqrt{n}} \Big) \tilde{\xi}_k \Big\} \Big]$$

$$= \mathbb{E}\Big[\mathbb{E}\Big[\exp\Big\{t \sum_{k \in \mathcal{M}} \Big(c_k + \frac{\xi_k}{\sqrt{n}} \Big) \tilde{\xi}_k \Big\} \,\Big|\, \xi_k,\, k \in \mathcal{M} \Big] \Big]$$

$$= \prod_{k \in \mathcal{M}} \mathbb{E}\Big[\exp\big\{ (t^2/2n) \big(c_k \sqrt{n} + \xi_k \big)^2 \big\} \Big].$$

Now, for any real $a < 1$ and any b, we have the formula

$$\mathbb{E}\big[\exp\big\{ (a/2)\, (b + \xi)^2 \big\} \big] = (1-a)^{-1/2} \exp\big\{ ab^2 / (2(1-a)) \big\}.$$

Applying this formula with $a = t^2/n$ and $b = c_k\sqrt{n}$, we obtain

$$G_2(t) = \prod_{k\in\mathcal{M}} \exp\Big\{ \frac{n\,t^2}{2(n-t^2)}\,c_k^2 - \frac{1}{2}\ln(1-t^2/n)\Big\}$$

which completes the proof because $S_n = \sum_{k\in\mathcal{M}} c_k^2$. $\square$

Proof of Lemma 15.8. All the inequalities in this lemma follow from the exponential Chebyshev inequality (also known as Chernoff's inequality),

$$\mathbb{P}(U \ge x) \le G(t)\exp\{-t\,x\}$$

where $G(t) = \mathbb{E}\big[\exp\{t\,U\}\big]$ is the moment generation function of a random variable U.

It is essential that the moment generating functions of the random variables $U_n^{(1)}$ and $U_n^{(2)}$ in Proposition 15.9 are quadratic at the origin, $G_i(t) = O(t^2)$, $i = 1, 2$, as $t \to 0$. A choice of a sufficiently small t would guarantee the desired bounds. In the four stated inequalities, the choices of t differ.

We start with the random event $\mathcal{A}_1 = \big\{U_n^{(1)} \ge M_{\beta_1}\big\}$,

$$\mathbb{P}_{\mathbf{c}}(\mathcal{A}_1) \le G_1(t)\exp\big\{-t\,M_{\beta_1}\big\}$$
$$= \exp\big\{-M_{\beta_1}\big(1+o_n(1)\big)\big(t + (1/2)\ln(1-2t)\big) - t\,M_{\beta_1}\big\}.$$

We choose $t = 1/4$ and obtain

$$\mathbb{P}_{\mathbf{c}}(\mathcal{A}_1) \le \exp\Big\{-(1/2)(1-\ln 2)\,M_{\beta_1}\big(1+o_n(1)\big)\Big\} \le \exp\big\{-0.15\,M_{\beta_1}\big\}.$$

Similarly, if we apply Chernoff's inequality to the random variable $-U_n^{(2)}$ with $t = \sqrt{n}/10$, and use the fact that $\Delta M < M_{\beta_1} \le n\,S_n/4$, we get

$$\mathbb{P}_{\mathbf{c}}(\mathcal{A}_2) = \mathbb{P}_{\mathbf{c}}\big(-U_n^{(2)} \ge \sqrt{n}S_n/8\big)$$
$$\le \exp\Big\{\frac{n\,t^2}{2(n-t^2)}\,S_n - \frac{\Delta M}{2}\ln(1-t^2/n) - t\sqrt{n}\,S_n/8\Big\}$$
$$= \exp\Big\{\frac{nS_n}{198} - \frac{\Delta M}{2}\ln(99/100) - \frac{nS_n}{80}\Big\}$$
$$\le \exp\Big\{\frac{nS_n}{198} - \frac{nS_n}{8}\ln(99/100) - \frac{nS_n}{80}\Big\}$$
$$\le \exp\big\{-AnS_n\big\} \le \exp\big\{-4AM_{\beta_1}\big\}$$

where $A = -1/198 + (1/8)\ln(99/100) + 1/80 > 0$.

To prove the upper bound for the probability of $\mathcal{A}_3$, take $t = 1/8$. Then

$$\mathbb{P}_{\mathbf{c}}(\mathcal{A}_3) = \mathbb{P}_{\mathbf{c}}\big(-U_n^{(1)} \geq M_{\beta_1}/4 \big)$$

$$\leq \exp\Big\{ -M_{\beta_1}(1+o_n(1))\big(-t + (1/2)\ln(1+2t)\big) - t\,M_{\beta_1}/4 \Big\}$$

$$= \exp\big\{ -AM_{\beta_1}(1+o_n(1))\big\}$$

where $A = -1/8 + (1/2)\ln(5/4) + 1/32 > 0$.

Finally, if $nS_n = o(M_{\beta_1})$, then

$$G_2(t) = \exp\big\{ -(1/2)M_{\beta_1}(1+o_n(1))\ln(1-t^2/n)\big\}.$$

Put $t = \sqrt{n}/8$. Then

$$\mathbb{P}_{\mathbf{c}}(\mathcal{A}_4) = \mathbb{P}_{\mathbf{c}}\big(U_n^{(2)} \geq M_{\beta_1}/(8\sqrt{n})\big)$$

$$\leq \exp\Big\{ -(1/2)M_{\beta_1}(1+o_n(1))\ln(1-t^2/n) - tM_{\beta_1}/(8\sqrt{n})\Big\}$$

$$= \exp\big\{ -AM_{\beta_1}(1+o_n(1))\big\}$$

where $A = (1/2)\ln(63/64) + 1/64 > 0$. □

Exercises

Exercise 15.95. Let $\psi_n = \psi_n(f)$ be the rate defined by (15.3). Show that there exists an estimator $\tilde{f}_n$ and a constant r^* independent of n such that

$$\max_{\beta \in \{\beta_1, \beta_2\}} \sup_{f \in \Theta(\beta)} \mathbb{E}_f\Big[\Big| \frac{\tilde{f}_n - f(x_0)}{\psi_n(f)} \Big| \Big] \leq r^*.$$

Exercise 15.96. Use (15.7) to prove that the expectation in (15.12) is bounded, that is, show that uniformly in $f \in \Theta(\beta_2)$, the following inequality holds:

$$(h^*_{n,\beta_1})^{-2\beta_1}\mathbb{E}_f\big[\|f^*_{n,\beta_1} - f\|_\infty^2 \big] \leq r^*$$

where a constant r^* is independent of n.

Chapter 16

Testing of Nonparametric Hypotheses

16.1. Basic Definitions

16.1.1. Parametric Case. First of all, we introduce the notion of *parametric hypotheses testing.* Suppose that in a classical statistical model with observations $X_1, \dots, X_n$ that obey a probability density $p(x, \theta)$, $\theta \in \Theta \subseteq \mathbb{R}$, we have to choose between two values of the parameter θ. That is, we want to decide whether $\theta = \theta_0$ or θ_1, where θ_0 and θ_1 are known. For simplicity we assume that $\theta_0 = 0$ and $\theta_1 \neq 0$. Our primary hypothesis, called the *null hypothesis*, is written as

$$\mathbf{H}_0 \, : \, \theta = 0,$$

while the *simple alternative hypothesis* has the form

$$\mathbf{H}_1 \, : \, \theta = \theta_1.$$

In testing the null hypothesis against the alternative, we do not estimate the parameter θ. A substitution for an estimator is a *decision rule* $\Delta_n = \Delta_n(X_1, \dots, X_n)$ that takes on only two values, for example, 0 or 1. The case $\Delta_n = 0$ is interpreted as *acceptance of the null hypothesis*, whereas the case $\Delta_n = 1$ means *rejection of the null hypothesis* in favor of the alternative.

The appropriate substitution for the risk function is the error probability. Actually, there are probabilities of two types of errors. *Type I error* is committed when a true null hypothesis is rejected, whereas acceptance of a false null results in *type II error*. The respective probabilities are denoted by $\mathbb{P}_0\big(\Delta_n = 1\big)$ and $\mathbb{P}_{\theta_1}\big(\Delta_n = 0\big)$.

The classical optimization problem in hypotheses testing consists of finding a decision rule that minimizes the type II error, provided the type I error does not exceed a given positive number α,

$$\mathbb{P}_{\theta_1}\big(\Delta_n = 0\big) \to \inf_{\Delta_n} \text{ subject to } \mathbb{P}_0\big(\Delta_n = 1\big) \leq \alpha.$$

If n is large, then a reasonable anticipation is that α can be chosen small, that is, $\alpha \to 0$ as $n \to \infty$. This criterion of optimality is popular because of its elegant solution suggested by the fundamental Neyman-Pearson lemma (see Exercise 16.97).

A more sophisticated problem is to test the null hypothesis against a *composite alternative*,

$$\mathbf{H}_1 : \theta \in \Lambda_n$$

where Λ_n is a known set of the parameter values that does not include the origin, that is, $0 \notin \Lambda_n$. In our asymptotic studies, different criteria for finding the decision rule are possible. One reasonable criterion that we choose is minimization of the sum of the type I error probability and the supremum over $\theta \in \Lambda_n$ of the type II error probability,

$$r_n(\Delta_n) = \mathbb{P}_0\big(\Delta_n = 1\big) + \sup_{\theta \in \Lambda_n} \mathbb{P}_\theta\big(\Delta_n = 0\big) \to \inf_{\Delta_n}.$$

The key question in asymptotic studies is: How distant should Λ_n be from the origin, so that it is still possible to separate $\mathbf{H}_0$ from the alternative $\mathbf{H}_1$ with a high probability? By *separation between hypotheses* we mean that there exists a decision rule Δ_n^* such that the sum of the error probabilities $r_n(\Delta_n^*)$ is vanishing, $\lim_{n\to\infty} r_n(\Delta_n^*) = 0$.

16.1.2. Nonparametric Case. Our objective here is to extend the parametric hypotheses testing to the nonparametric setup. We replace the parameter θ by a regression function f from a Hölder class $\Theta(\beta) = \Theta(\beta, L, L_1)$, and consider the model of observations,

$$y_i = f(i/n) + \varepsilon_i \text{ where } \varepsilon_i \overset{iid}{\sim} \mathcal{N}(0, \sigma^2).$$

Suppose that we want to test $\mathbf{H}_0 : f = 0$ against the composite alternative $\mathbf{H}_1 : f \in \Lambda_n$ where the set of regression functions Λ_n is specified.

A general approach to the nonparametric hypotheses testing is as follows. Assume that a norm $\|f\|$ of the regression function is chosen. Let ψ_n be a deterministic sequence, $\psi_n \to 0$ as $n \to \infty$, which plays the same role as the rate of convergence in estimation problems. Define the set of alternative hypotheses

$$\Lambda_n = \Lambda_n(\beta, C, \psi_n) = \big\{ f : f \in \Theta(\beta) \text{ and } \|f\| \geq C\,\psi_n \big\} \tag{16.1}$$

with a positive constant C. Denote the corresponding sum of the error probabilities by

$$r_n(\Delta_n, \beta, C, \psi_n) = \mathbb{P}_0\big(\Delta_n = 1\big) + \sup_{f \in \Lambda_n(\beta, C, \psi_n)} \mathbb{P}_f\big(\Delta_n = 0\big).$$

We call the sequence ψ_n a *minimax separation rate* if (i) for any small positive γ, there exist a constant C^* and a decision rule Δ_n^* such that

$$\limsup_{n\to\infty} r_n(\Delta_n^*, \beta, C^*, \psi_n) \leq \gamma, \tag{16.2}$$

and (ii) there exist positive constants C_* and r_*, independent of n and such that for any decision rule Δ_n,

$$\liminf_{n\to\infty} r_n(\Delta_n, \beta, C_*, \psi_n) \geq r_*. \tag{16.3}$$

The meaning of this definition is transparent. The regression functions with the norm satisfying $\|f\| \geq C^* \psi_n$ can be tested against the zero regression function with however small prescribed error probabilities. On the other hand, the reduction of the constant below C_* holds the sum of error probabilities above r_* for any sample size n.

16.2. Separation Rate in the Sup-Norm

In general, estimation of regression function and testing of hypotheses (in the same norm) are two different problems. The minimax rate of convergence is not necessarily equal to the minimax separation rate. We will demonstrate this fact in the next section. For some norms, however, they are the same. In particular, it happens in the sup-norm.

The following result is not difficult to prove because all the preliminary work is already done in Section 12.1.

Theorem 16.1. *Assume that the norm in the definition of Λ_n is the sup-norm,*

$$\Lambda_n = \Lambda_n(\beta, C, \psi_n) = \big\{ f \,:\, f \in \Theta(\beta) \text{ and } \|f\|_\infty \geq C\,\psi_n \big\}.$$

Then the minimax separation rate coincides with the minimax rate of convergence in the sup-norm $\psi_n = \big((\ln n)/n\big)^{\beta/(2\beta+1)}$.

Proof. First, we prove the existence of the decision rule Δ_n^* with the claimed separation rate such that (16.2) holds. Let f_n^* be the regressogram with the rate-optimal bandwidth $h_n^* = \big((\ln n)/n\big)^{1/(2\beta+1)}$. Our starting point is the inequalities (15.7) and (15.8). For any $C > A_b + 2\beta A_v$, uniformly over $f \in \Theta(\beta)$, these inequalities yield

$$\mathbb{P}_f\big(\|f_n^* - f\|_\infty \geq C\,\psi_n\big) \leq \mathbb{P}_f\Big(A_b(h_n^*)^\beta + A_v(nh_n^*)^{-1/2}\,\mathcal{Z}_\beta^* \geq C\,(h_n^*)^\beta\Big)$$

$$(16.4) \quad = \mathbb{P}_f\big(A_b + A_v\, \mathcal{Z}_\beta^* / \sqrt{\ln n} \ge C \big) \le \mathbb{P}_f\big(\mathcal{Z}_\beta^* \ge 2\beta\sqrt{\ln n} \big) \le \beta\, n^{-1}$$

where we have applied (15.8) with $y = \sqrt{2}$. Put $C^* = 2\,C$, and define the set of alternatives by

$$\Lambda_n(\beta, C^*, \psi_n) = \big\{ f : f \in \Theta(\beta) \text{ and } \|f\|_\infty \ge C^*\,\psi_n \big\}.$$

Introduce a rate-optimal decision rule

$$\Delta_n^* = \begin{cases} 0, & \text{if } \|f_n^*\|_\infty < \frac{1}{2}\,C^*\,\psi_n, \\ 1, & \text{otherwise.} \end{cases}$$

Then, from (16.4), we obtain that as $n \to \infty$,

$$\mathbb{P}_0\big(\Delta_n^* = 1 \big) = \mathbb{P}_0\big(\|f_n^*\|_\infty \ge \frac{1}{2}\,C^*\,\psi_n \big) = \mathbb{P}_0\big(\|f_n^*\|_\infty \ge C\,\psi_n \big) \to 0.$$

Next, for any $f \in \Lambda_n(\beta, C^*, \psi_n)$, by the triangle inequality, as $n \to \infty$,

$$\mathbb{P}_f\big(\Delta_n^* = 0 \big) = \mathbb{P}_f\big(\|f_n^*\|_\infty < \frac{1}{2}\,C^*\,\psi_n \big) \le \mathbb{P}_f\big(\|f_n^* - f\|_\infty \ge C\,\psi_n \big) \to 0.$$

Hence (16.2) is fulfilled for any $\gamma > 0$.

The proof of the lower bound (16.3) is similar to that in Lemma 12.2. We repeat the construction of the Q test functions f_q, $q = 1, \dots, Q$, in (12.3) based on a common "bump" function φ. For any test Δ_n, introduce the random event $\mathcal{D} = \{ \Delta_n = 1 \}$. Then for any $\delta > 0$,

$$\mathbb{P}_0\big(\mathcal{D}\big) + \max_{1 \le q \le Q} \mathbb{P}_{f_q}\big(\overline{\mathcal{D}}\big) \ge \mathbb{P}_0\big(\mathcal{D}\big) + \mathbb{E}_0\big[\mathbb{I}\big(\overline{\mathcal{D}}\big)\, \xi_n \big]$$

$$\ge \mathbb{P}_0\big(\mathcal{D}\big) + (1-\delta)\mathbb{P}_0\big(\overline{\mathcal{D}} \cap \{\xi_n > 1-\delta\}\big) \ge (1-\delta)\mathbb{P}_0\big(\xi_n > 1 - \delta \big)$$

where

$$\xi_n = \frac{1}{Q} \sum_{q=1}^{Q} \exp\big\{ \ln\big(d\mathbb{P}_{f_q} / d\mathbb{P}_0 \big) \big\}.$$

As shown in Lemma 12.2, the random variable ξ_n converges to 1 as $n \to \infty$. Hence,

$$\liminf_{n \to \infty} \Big[\mathbb{P}_0\big(\Delta_n = 1 \big) + \max_{1 \le q \le Q} \mathbb{P}_{f_q}\big(\Delta_n = 0 \big) \Big] \ge 1 - \delta.$$

Note that if $C_* < \|\varphi\|_\infty$, then all the test functions f_q, $q = 1, \dots, Q$, belong to the set of alternatives

$$\Lambda_n(\beta, C_*, \psi_n) = \big\{ f : f \in \Theta(\beta) \text{ and } \|f\| \ge C_*\,\psi_n \big\}.$$

Thus the lower bound (16.3) holds with $r_* = 1 - \delta$ however close to 1. $\square$

16.3. Sequence Space. Separation Rate in the L_2-Norm

Analyzing the proof of Theorem 16.1, we find the two remarkable properties of the hypotheses testing in the sup-norm. First, the separation rate ψ_n coincides with the minimax rate of estimation, and the minimax optimal decision rule if very simple: the null hypothesis is accepted if the sup-norm of the estimator is small enough. Second, the choice of a sufficiently large constant C^* in the definition of the alternative hypothesis $\Lambda_n(\beta, C^*, \psi_n)$ guarantees the upper bound for arbitrarily small error probability γ. Moreover, C^* does not depend on the value of γ. It happens because the sup-norm is a very special type of norm. The distribution of $\|f_n^*\|_\infty$ is degenerate. If C^* is large enough, then $\mathbb{P}_f\big(\|f_n^*\|_\infty \geq C^*\,\psi_n\big) \to 0$ as $n \to \infty$.

In this section, we turn to the quadratic norm. To ease the presentation, we consider the problem of hypotheses testing in the sequence space. All the necessary definitions are given in Section 10.5. The observations z_k in the sequence space are

$$z_k = c_k + \sigma\,\xi_k/\sqrt{n}, \quad k = 0, \dots, n-1,$$

where the c_k's are the Fourier coefficients of the regression function f, that is, $f = \sum_{k=0}^{n-1} c_k \varphi_k$. Here the φ_k's form an orthonormal basis in the discrete L_2-norm. The errors ξ_k are independent standard normal random variables, and $\sigma > 0$ represents the standard deviation of the observations in the original regression model.

We use $\mathbf{c} = \big(c_0, \dots, c_{n-1}\big)$ to denote the whole set of the Fourier coefficients. As in Section 15.3, it is convenient to work directly in the sequence space of the Fourier coefficients. For ease of reference, we repeat here the definition of the following class:

$$\Theta_{2,n}(\beta) = \Theta_{2,n}(\beta, L) = \Big\{ \big(c_0, \dots, c_{n-1}\big) : \sum_{k=0}^{n-1} c_k^2\, k^{2\beta} \leq L \Big\}.$$

We want to test the null hypothesis that all the Fourier coefficients are equal to zero versus the alternative that their L_2-norm $\|\mathbf{c}\|_2 = \big(\sum c_k^2\big)^{1/2}$ is larger than a constant that may depend on n. Our goal is to find the minimax separation rate ψ_n. Formally, we study the problem of testing $\mathbf{H}_0$: $\mathbf{c} = \mathbf{0}$ against the composite alternative

$$\mathbf{H}_1 \; : \mathbf{c} \in \Lambda_n = \Lambda_n(\beta, C, \psi_n)$$

where

$$\Lambda_n = \big\{ \mathbf{c} : \mathbf{c} \in \Theta_{2,n}(\beta) \text{ and } \|\mathbf{c}\|_2 \geq C\,\psi_n \big\}. \tag{16.5}$$

In Section 13.1, it was shown that the squared L_2-norm of regression in $[0,1]$ can be estimated with the minimax rate $1/\sqrt{n}$. This is true in the sequence space as well. The proof in the sequence space is especially simple. Indeed, the sum of the centered z_k^2's admits the representation

$$\sum_{k=0}^{n-1}\Big(z_k^2-\frac{\sigma^2}{n}\Big)=\|\mathbf{c}\|_2^2-\frac{2\sigma}{\sqrt{n}}\sum_{k=0}^{n-1}c_k\,\xi_k+\frac{\sigma^2}{n}\sum_{k=0}^{n-1}(\xi_k^2-1)$$

$$=\|\mathbf{c}\|_2^2-\frac{2\sigma}{\sqrt{n}}\mathcal{N}+\frac{\sigma^2}{\sqrt{n}}\mathcal{Y}_n \tag{16.6}$$

where $\mathcal{N}$ denotes the zero-mean normal random variable with the variance $\|\mathbf{c}\|_2^2$. The variable $\mathcal{Y}_n$ is a centered chi-squared random variable that is asymptotically normal,

$$\mathcal{Y}_n=\sum_{k=0}^{n-1}(\xi_k^2-1)/\sqrt{n}\to\mathcal{N}(0,2).$$

The convergence rate $1/\sqrt{n}$ in estimation of $\|\mathbf{c}\|_2^2$ follows immediately from (16.6).

Now we continue with testing the null hypothesis against the composite alternative (16.5). We will show that the separation rate of testing in the quadratic norm is equal to $\psi_n=n^{-2\beta/(4\beta+1)}$. Note that this separation rate is faster than the minimax estimation rate in the L_2-norm $n^{-\beta/(2\beta+1)}$. The proof of this fact is split between the upper and lower bounds in the theorems below.

We introduce the rate-optimal decision rule, proceeding similar to (16.6). We take $M_n=n^{2/(4\beta+1)}$, so that the separation rate $\psi_n=M_n^{-\beta}$, and estimate the norm of the Fourier coefficients by

$$\hat{S}_n=\sum_{k=0}^{M_n}\big(z_k^2-\sigma^2/n\big).$$

Consider a class of decision rules Δ_n that depends on a constant b,

$$\Delta_n=\Delta_n(\beta,b)=\begin{cases}0, & \text{if } \hat{S}_n<b\,\psi_n^2=b\,n^{-4\beta/(4\beta+1)},\\ 1, & \text{otherwise.}\end{cases} \tag{16.7}$$

The following theorem claims that by choosing properly the constants in the definitions of the set of alternatives and the decision rule, we can make the error probabilities less than any prescribed number in the sense of the upper bound (16.2).

Theorem 16.2. *For any small positive γ, there exist a constant $C=C^*=C^*(\gamma)$ in the definition (16.5) of the set of alternatives $\Lambda_n=\Lambda_n(\beta,C^*,\psi_n)$,* *and a constant $b = b(\gamma)$ in the definition (16.7) of decision rule $\Delta_n = \Delta_n^* = \Delta_n^*(\beta, b)$ such that*

$$\limsup_{n\to\infty} r_n(\Delta_n^*) \le \gamma$$

where

$$r_n(\Delta_n^*) = \mathbb{P}_0\big(\Delta_n^* = 1\big) + \sup_{\mathbf{c}\in\Lambda_n(\beta, C^*, \psi_n)} \mathbb{P}_{\mathbf{c}}\big(\Delta_n^* = 0\big).$$

Proof. It suffices to show that for all sufficiently large n, probabilities of type I and II errors are bounded from above by $\gamma/2$, that is, it suffices to show that

$$\limsup_{n\to\infty} \mathbb{P}_0\big(\Delta_n^* = 1\big) \le \gamma/2 \tag{16.8}$$

and

$$\limsup_{n\to\infty} \sup_{\mathbf{c}\in\Lambda_n(\beta, C^*, \psi_n)} \mathbb{P}_{\mathbf{c}}\big(\Delta_n^* = 0\big) \le \gamma/2. \tag{16.9}$$

Starting with the first inequality, we write

$$\mathbb{P}_0\big(\Delta_n^* = 1\big) = \mathbb{P}_0\big(\hat{S}_n \ge b\,\psi_n^2\big) = \mathbb{P}_0\Big(\sum_{k=0}^{M_n}\big(z_k^2 - \sigma^2/n\big) \ge b\,\psi_n^2\Big)$$

$$= \mathbb{P}_0\Big(\frac{\sigma^2}{n}\sum_{k=0}^{M_n}(\xi_k^2 - 1) > b\,\psi_n^2\Big) = \mathbb{P}_0\Big(\sigma^2 n^{-1}\sqrt{2(M_n+1)}\,\mathcal{Y}_n > b\,\psi_n^2\Big)$$

where $\mathcal{Y}_n = \sum_{k=0}^{M_n}(\xi_k^2 - 1)/\sqrt{2(M_n+1)}$ is asymptotically standard normal random variable. Under our choice of M_n, we have that as $n\to\infty$,

$$n^{-1}\sqrt{M_n+1} \sim n^{1/(4\beta+1)-1} = n^{-4\beta/(4\beta+1)} = \psi_n^2.$$

Consequently,

$$\mathbb{P}_0\big(\Delta_n^* = 1\big) = \mathbb{P}_0\Big(\sqrt{2}\,\sigma^2\,\mathcal{Y}_n \ge b(1 + o_n(1))\Big) \to 1 - \Phi\Big(\frac{b}{\sqrt{2}\,\sigma^2}\Big),$$

as $n\to\infty$, where Φ denotes the cumulative distribution function of a standard normal random variable. If we choose $b = \sqrt{2}\,\sigma^2\,q_{1-\gamma/2}$ with $q_{1-\gamma/2}$ standing for the $(1-\gamma/2)$-quantile of the standard normal distribution, then the inequality (16.8) follows.

To verify the inequality (16.9), note that

$$\mathbb{P}_{\mathbf{c}}\big(\Delta_n^* = 0\big) = \mathbb{P}_{\mathbf{c}}\big(\hat{S}_n \le b\,\psi_n^2\big) = \mathbb{P}_{\mathbf{c}}\Big(\sum_{k=0}^{M_n}\big(z_k^2 - \sigma^2/n\big) \le b\,\psi_n^2\Big)$$

$$= \mathbb{P}_{\mathbf{c}}\Big(\|\mathbf{c}\|_2^2 - \sum_{k=M_n+1}^{n-1} c_k^2 - \frac{2\sigma}{\sqrt{n}}\sum_{k=0}^{M_n} c_k\,\xi_k + \frac{\sigma^2}{n}\sum_{k=0}^{M_n}(\xi_k^2 - 1) \le b\,\psi_n^2\Big).$$

Observe that for any $\mathbf{c} \in \Lambda_n(\beta, C^*, \psi_n)$, the variance of the following normalized random sum vanishes as $n \to \infty$,

$$\mathbb{V}ar_{\mathbf{c}}\left[\frac{2\sigma}{\sqrt{n}\|\mathbf{c}\|_2^2}\sum_{k=0}^{M_n} c_k\xi_k\right] \leq \frac{4\sigma^2}{n\|\mathbf{c}\|_2^2} \leq \frac{4\sigma^2}{n(C^*\psi_n)^2} = \left(\frac{2\sigma}{C^*}\right)^2 n^{-1/(4\beta+1)} \to 0,$$

which implies that

$$\|\mathbf{c}\|_2^2 - \frac{2\sigma}{\sqrt{n}}\sum_{k=0}^{M_n} c_k\,\xi_k = \|\mathbf{c}\|_2^2\left(1 + o_n(1)\right) \text{ as } n \to \infty,$$

where $o_n(1) \to 0$ in $\mathbb{P}_{\mathbf{c}}$-probability. Thus,

$$\mathbb{P}_{\mathbf{c}}\left(\Delta_n^* = 0\right) = \mathbb{P}_{\mathbf{c}}\Big(\frac{\sigma^2}{n}\sum_{k=0}^{M_n}(\xi_k^2 - 1) \leq -\|\mathbf{c}\|_2^2(1+o_n(1)) + \sum_{k=M_n+1}^{n-1} c_k^2 + b\,\psi_n^2\Big).$$

Put $\mathcal{Y}_n = \sum_{k=0}^{M_n}(\xi_k^2 - 1)/\sqrt{2(M_n+1)}$. Note that

$$\sum_{k=M_n+1}^{n-1} c_k^2 < \sum_{k=M_n+1}^{n-1}\left(\frac{k}{M_n}\right)^{2\beta} c_k^2 \leq M_n^{-2\beta}\,L.$$

Therefore,

$$\mathbb{P}_{\mathbf{c}}\left(\Delta_n^* = 0\right) \leq \mathbb{P}_{\mathbf{c}}\Big(\frac{\sigma^2}{n}\sqrt{2(M_n+1)}\,\mathcal{Y}_n \leq -(C^*\psi_n)^2\left(1 + o_n(1)\right) + M_n^{-2\beta}L + b\psi_n^2\Big)$$

where $\mathcal{Y}_n$ is asymptotically standard normal. Note that here every term has the magnitude $\psi_n^2 = M_n^{-2\beta}$. If we cancel ψ_n^2, the latter probability becomes

$$\mathbb{P}_{\mathbf{c}}\Big(\sqrt{2}\,\sigma^2\,\mathcal{Y}_n \leq (-C^* + L + b)\left(1 + o_n(1)\right)\Big) \to \Phi\Big(\frac{-C^* + L + b}{\sqrt{2}\,\sigma^2}\Big),$$

as $n \to \infty$. Choose $C^* = 2b + L$ and recall that $b = \sqrt{2}\sigma^2 q_{1-\gamma/2}$. We obtain

$$\frac{-C^* + L + b}{\sqrt{2}\,\sigma^2} = \frac{-b}{\sqrt{2}\sigma^2} = -\,q_{1-\gamma/2} = q_{\gamma/2}$$

where $q_{\gamma/2}$ denotes the $\gamma/2$-quantile of Φ. Thus, the inequality (16.9) is valid. □

Remark 16.3. In the case of the sup-norm, we can find a single constant C^* to guarantee the upper bound for any γ. In the case of the L_2-norm, it is not possible. Every γ requires its own constants C^* and b. □

As the next theorem shows, the separation rate $\psi_n = n^{-2\beta/(4\beta+1)}$ cannot be improved. If the constant C in the definition (16.5) of the set of alternatives $\Lambda_n(\beta, C, \psi_n)$ is small, then there is no decision rule that would guarantee arbitrarily small error probabilities.

Theorem 16.4. *For any constant r_*, $0 < r_* < 1$, there exists $C = C_* > 0$ in the definition (16.5) of the set of alternatives Λ_n such that for any decision rule Δ_n, the sum of the error probabilities*

$$r_n(\Delta_n) = \mathbb{P}_0\big(\Delta_n = 1\big) + \sup_{\mathbf{c}\in\Lambda_n(\beta, C_*, \psi_n)} \mathbb{P}_{\mathbf{c}}\big(\Delta_n = 0\big)$$

satisfies the inequality $\liminf_{n\to\infty} r_n(\Delta_n) \geq r_*$.

Proof. Let $M_n = n^{2/(4\beta+1)} = \psi_n^{-1/\beta}$. Introduce a set of 2^{M_n} binary sequences

$$\Omega_n = \big\{\boldsymbol{\omega} = \big(\omega_1, \dots, \omega_{M_n}\big),\ \omega_k \in \{-1, 1\},\, k = 1, \dots, M_n\big\}.$$

Define a set of alternatives $\Lambda_n^{(0)}$ with the same number of elements 2^{M_n},

$$\Lambda_n^{(0)} = \Big\{\mathbf{c} = \mathbf{c}(\boldsymbol{\omega}) \,:\, c_k = C_*\,\psi_n\,\omega_k\,/\sqrt{M_n} \text{ if } k = 1, \dots, M_n,$$
$$\text{and } c_k = 0 \text{ otherwise, } \boldsymbol{\omega} \in \Omega_n\Big\}$$

where a positive constant C_* will be chosen later. Note that if C_* is small enough, $C_*^2 < (2\beta + 1)L$, then

$$\Lambda_n^{(0)} \subset \Lambda_n(\beta, C_*, \psi_n). \tag{16.10}$$

Indeed, if $\mathbf{c} \in \Lambda_n^{(0)}$, then

$$\sum_{k=0}^{n-1} c_k^2\, k^{2\beta} = \frac{(C_*\psi_n)^2}{M_n} \sum_{k=1}^{M_n} k^{2\beta} \sim \frac{(C_*\psi_n)^2}{M_n} \frac{M_n^{2\beta+1}}{2\beta + 1} = \frac{C_*^2}{2\beta+1} < L.$$

Thus, every $\mathbf{c} \in \Lambda_n^{(0)}$ belongs to $\Theta_{2,n}(\beta, L)$. Also, the following identity takes place:

$$\|\mathbf{c}\|_2 = \Big[\sum_{k=1}^{M_n} (C_*\psi_n\omega_k)^2 / M_n\Big]^{1/2} = C_*\psi_n,$$

which implies (16.10).

Next, we want to show that for any decision rule Δ_n, the following inequality holds:

$$\liminf_{n\to\infty} \Big[\mathbb{P}_0\big(\Delta_n = 1\big) + \max_{\boldsymbol{\omega}\in\Omega_n} \mathbb{P}_{\mathbf{c}(\boldsymbol{\omega})}\big(\Delta_n = 0\big)\Big] \geq r_*. \tag{16.11}$$

Put

$$\alpha_n^2 = \Big(\frac{C_*\psi_n}{\sigma}\Big)^2 \frac{n}{M_n} = \Big(\frac{C_*}{\sigma}\Big)^2 n^{-1/(4\beta+1)} \to 0, \text{ as } n \to \infty.$$

Further, we substitute the maximum by the mean value to obtain

$$r_n(\Delta_n) \geq \mathbb{P}_0\big(\Delta_n = 1\big) + \max_{\boldsymbol{\omega}\in\Omega_n} \mathbb{P}_{\mathbf{c}(\boldsymbol{\omega})}\big(\Delta_n = 0\big)$$

$$\geq \mathbb{P}_0\big(\Delta_n = 1\big) + 2^{-M_n} \sum_{\boldsymbol{\omega}\in\Omega_n} \mathbb{P}_{\mathbf{c}(\boldsymbol{\omega})}\big(\Delta_n = 0\big)$$

$$= \mathbb{E}_0\Big[\mathbb{I}\big(\Delta_n = 1\big) + \mathbb{I}\big(\Delta_n = 0\big) 2^{-M_n} \sum_{\boldsymbol{\omega}\in\Omega_n} \exp\big\{L_n(\boldsymbol{\omega})\big\}\Big]$$

$$= \mathbb{E}_0\Big[\mathbb{I}\big(\Delta_n = 1\big) + \mathbb{I}\big(\Delta_n = 0\big)\,\eta_n\Big]$$

where

$$L_n(\boldsymbol{\omega}) = \ln\frac{d\,\mathbb{P}_{\mathbf{c}(\boldsymbol{\omega})}}{d\,\mathbb{P}_0} \quad \text{and} \quad \eta_n = 2^{-M_n} \sum_{\boldsymbol{\omega}\in\Omega_n} \exp\big\{L_n(\boldsymbol{\omega})\big\}.$$

Now, the log-likelihood ratio

$$L_n(\boldsymbol{\omega}) = \frac{n}{\sigma^2}\sum_{k=1}^{M_n}\big(c_k\,z_k - c_k^2/2\big) = \sum_{k=1}^{M_n}\big(\alpha_n\,\omega_k\,\xi_k - \alpha_n^2/2\big).$$

Here we have used the fact that, under $\mathbb{P}_0$, $z_k = \sigma\,\xi_k/\sqrt{n}$. In addition, the identities $\omega_k^2 = 1$ and

$$\sqrt{n}\,c_k/\sigma = (C_*\psi_n/\sigma)\sqrt{n/M_n}\,\omega_k = \alpha_n\,\omega_k$$

were employed. The random variable η_n admits the representation, which will be derived below,

$$\eta_n = \exp\Big\{-\frac{1}{2}\,\alpha_n^2\,M_n\Big\}\prod_{k=1}^{M_n}\Big[\frac{1}{2}\,e^{\alpha_n\,\xi_k} + \frac{1}{2}\,e^{-\,\alpha_n\,\xi_k}\Big]. \tag{16.12}$$

Even though this expression is purely deterministic and can be shown algebraically, the easiest way to prove it is by looking at the ω_k's as independent random variables such that

$$\mathbb{P}^{(\boldsymbol{\omega})}\big(\omega_k = \pm 1\big) = 1/2.$$

Using this definition, the random variable η_n can be computed as the expected value, denoted by $\mathbb{E}^{(\boldsymbol{\omega})}$, with respect to the distribution $\mathbb{P}^{(\boldsymbol{\omega})}$,

$$\eta_n = \mathbb{E}^{(\boldsymbol{\omega})}\Big[\exp\big\{L_n(\boldsymbol{\omega})\big\}\Big] = \prod_{k=1}^{M_n}\mathbb{E}^{(\boldsymbol{\omega})}\Big[\exp\big\{\alpha_n\xi_k\omega_k - \frac{1}{2}\,\alpha_n^2\big\}\Big]$$

$$= \exp\big\{-\alpha_n^2 M_n/2\big\}\prod_{k=1}^{M_n}\mathbb{E}^{(\boldsymbol{\omega})}\Big[\exp\big\{\alpha_n\,\xi_k\,\omega_k\big\}\Big]$$

so that the representation (16.12) for η_n follows.

Recall that ξ_k are independent standard normal random variables with respect to the $\mathbb{P}_0$-distribution, hence, $\mathbb{E}_0[\,\eta_n\,] = 1$. To compute the second moment of η_n, we write

$$\mathbb{E}_0[\,\eta_n^2\,] = \exp\big\{-\alpha_n^2 M_n\big\}\Big(\mathbb{E}_0\Big[\frac{1}{4}e^{2\alpha_n\xi_1} + \frac{1}{2} + \frac{1}{4}e^{-2\alpha_n\xi_1}\Big]\Big)^{M_n}$$

$$= \exp\big\{-\alpha_n^2 M_n\big\}\Big(\frac{1}{2}e^{2\alpha_n^2}+\frac{1}{2}\Big)^{M_n} = \exp\big\{-\alpha_n^2 M_n\big\}\Big(1+\alpha_n^2+\alpha_n^4+o(\alpha_n^4)\Big)^{M_n}$$

$$= \exp\Big\{-\alpha_n^2 M_n + \big(\alpha_n^2+\alpha_n^4/2+o(\alpha_n^4)\big)M_n\Big\} = \exp\big\{\alpha_n^4 M_n/2+o(\alpha_n^4 M_n)\big\}.$$

From the definition of α_n, we have

$$\alpha_n^4\, M_n = \Big(\,(C_*/\sigma)^2\, n^{-1/(4\beta+1)}\,\Big)^2 M_n = (C_*/\sigma)^4.$$

Thus, as $n\to\infty$, we find that $o(\alpha_n^4 M_n)\to 0$ and $\mathbb{E}_0[\,\eta_n^2\,]\to\exp\big\{C_*^4/(2\sigma^4)\big\}$. Then the variance $\mathbb{V}ar_0[\,\eta_n\,]\sim\exp\big\{C_*^4/(2\sigma^4)\big\}-1$ for large n. For any $\delta>0$, by the Chebyshev inequality,

$$\liminf_{n\to\infty}\mathbb{P}_0\big(\,\eta_n\ge 1-\delta\,\big)\ge 1-\delta^{-2}\Big(\exp\big\{C_*^4/(2\sigma^4)\big\}-1\Big).$$

The right-hand side can be made arbitrarily close to 1 if we choose sufficiently small C_*. Finally, we obtain that

$$\liminf_{n\to\infty} r_n(\Delta_n)\ge\liminf_{n\to\infty}\mathbb{E}_0\Big[\,\mathbb{I}(\,\Delta_n=1\,)+\mathbb{I}(\,\Delta_n=0\,)\,\eta_n\,\Big]$$

$$\ge (1-\delta)\liminf_{n\to\infty}\mathbb{P}_0\big(\,\eta_n\ge 1-\delta\,\big)\ge(1-\delta)\Big[\,1-\delta^{-2}\big(\exp\big\{C_*^4/(2\sigma^4)\big\}-1\,\big)\Big].$$

By choosing a small positive δ and then a sufficiently small C_*, we can make the right-hand side larger than any $r_*<1$, which proves the lower bound (16.11). □

Exercises

Exercise 16.97. (Fundamental Neyman-Pearson Lemma) Assume that for a given $\alpha>0$, there exists a constant $c>0$ such that $\mathbb{P}_0(L_n\ge c)=\alpha$ where $L_n=\sum_{i=1}^n \ln\Big(\,p(X_i,\theta_1)/p(X_i,0)\,\Big)$. Put $\Delta_n^*=\mathbb{I}(L_n\ge c)$. Let Δ_n be a decision rule Δ_n which probability of type I error $\mathbb{P}_0(\Delta_n=1)\le\alpha$. Show that the probability of type II error of Δ_n is larger than type II error of Δ_n^*, that is, $\mathbb{P}_{\theta_1}(\Delta_n=0)\ge\mathbb{P}_{\theta_1}(\Delta_n^*=0)$.

Bibliography

[Bor99] A.A. Borovkov, *Mathematical statistics*, CRC, 1999.

[Efr99] S. Efromovich, *Nonparametric curve estimation: Methods, theory and applications*, Springer, 1999.

[Eub99] R. L. Eubank, *Nonparametric regression and spline smoothing*, 2nd ed., CRC, 1999.

[Har97] J. Hart, *Nonparametric smoothing and lack-of-fit tests*, Springer, 1997.

[HMSW04] W. Härdle, M. Müller, S. Sperlich, and A. Werwatz, *Nonparametric and semiparametric models*, Springer, 2004.

[IH81] I.A. Ibragimov and R.Z. Has'minski, *Statistical estimation. Asymptotic theory*, Springer, 1981.

[Mas07] P. Massart, *Concentration inequalities and model selection*, Springer, 2010.

[Pet75] V.V. Petrov, *Sums of independent random variables*, Berlin, New York: Springer-Verlag, 1975.

[Sim98] J.S. Simonoff, *Smoothing methods in statistics*, Springer, 1996.

[Tsy08] A.B. Tsybakov, *Introduction to nonparametric estimation*, Springer, 2009.

[Was04] L. Wasserman, *All of statistics: A concise course in statistical inference*, Springer, 2003.

[Was07] ———, *All of nonparametric statistics*, Springer, 2005.

Index of Notation

Index